Everlasting Flowers between the Pages

# Emergence of Natural History



VOLUME 9

The titles published in this series are listed at *brill.com/enh*

# Everlasting Flowers between the Pages

*The Making of Seventeenth-Century Florilegia*

*By*

Jessie Wei-Hsuan Chen

BRILL

LEIDEN | BOSTON

This publication is part of the PhDs in the Humanities project 'Everlasting Flowers between the Pages' (project number PGW.18.016) and is supported by the Open Access Books funds, both financed by the Dutch Research Council (NWO). Additionally, the project received the 2020 Stacy Lloyd III Fellowship for Bibliographic Study from the Oak Spring Garden Foundation.

Cover illustration: *Anemona* from *Jardin de rares et curieux fleurs* by Franciscus de Geest, second half of the seventeenth century, 00004982, Oak Spring Garden Foundation, Upperville, VA.

The Library of Congress Cataloging-in-Publication Data is available online at https://catalog.loc.gov
LC record available at https://lccn.loc.gov/2025029308

Typeface for the Latin, Greek, and Cyrillic scripts: "Brill". See and download: brill.com/brill-typeface.

ISSN 2452-3283
ISBN 978-90-04-73513-2 (hardback)
ISBN 978-90-04-73514-9 (e-book)
DOI 10.1163/9789004735149



This book is printed on acid-free paper and produced in a sustainable manner.

# Contents

# Acknowledgements

The making of this book is comparable to the making of a historical garden. Similar to many gardens in the seventeenth century, this work went through adversity. This book is a revised version of my dissertation, defended at Utrecht University on 15 December 2023. Not even halfway into the Ph.D. trajectory, the Covid-19 pandemic hit. This paper garden and its gardener battled against the coronavirus and its long-lasting effects. The gardener was heavily injured, many seedlings died, much of the garden plan was abandoned, and many things were destroyed that never recovered. Yet, a garden is ever-changing and ever-growing. Out of the shambles, something green sprouted, colourful flowers budded, and hopefully, some buds bloomed.

Much like historical sites, the recorded history of this book cannot document every single name that contributed to its construction. For everyone who helped me procure a few sources and secondary literature, and/or listened to me talk about my project at length, I thank you. For the recorded history of the paper garden, I would like to, first and foremost, thank my dear friends Tamar Hestrin Grader, Mariëlle Ekkelenkamp, Annemarie Kok, and Rhiannon Estwick. I would not have survived the several darkest periods of the past years without all of your friendship, companionship, and support. This paper garden and its gardener remained alive because of you.

I would also like to express my sincere gratitude to my Ph.D. supervisor Sven Dupré and co-supervisor Florike Egmond for helping me (re)design and strengthen the garden structures and fill in the colourful details in the parterres. This garden would not have reached a presentable state without your critical but constructive feedback. To my colleagues Marieke Hendriksen, Jenny Boulboullé, Thijs Hagendijk, Mariana Pinto, Jill Briggman, and Marleen Schans from the ARTECHNE project, thank you for kindly adopting me into the fold during my Ph.D. research. I cherish the fond memories of experimenting with the RRR methods, discussing various research topics, and having regular lunches with the group. To my supervisor and colleague Marlise Rijks from the BIO&IMAGO project, the year in Brussels gave me the needed resources and encouragement at the final stage to take this book to publication.

This work would not have existed without the funding through the scheme "PhDs in the Humanities" from the Dutch Research Council (NWO), which also supported the open access publication of this volume. During the application process, I received notable help from Andrea van Leerdam, Lieke Schrijvers, Charlotte Hoitsma, Clim Wijnands, Anna-Luna Post, and Oscar Gelderblom and the internal selection committee at Utrecht University. For the publication, the colourful figures in this book were graciously financed by the Vrije Universiteit Brussel. Many people from Brill Publishers and the series *Emergence of Natural History* helped with getting this book published in a timely manner. I would especially like to thank the (anonymous) reviewers, as well as Dominik Hünniger, Alessandra Giliberto, and Melissa Allieri.

The Oak Spring Garden Foundation was another major supporter of this volume. The 2020 Stacy Lloyd III Fellowship for Bibliographic Study provided me with the financial resources to acquire materials and equipment for establishing a home lab that prevented the research from reaching a complete standstill during the multiple lockdowns during the pandemic. The delayed residency in 2022 further inspired and strengthened many arguments in this volume. I especially thank Peter Crane, Danielle Eady, Tony Willis, Nancy Collins, Kimberley Fisher, Allissa Montgomery, Sarah Causey, and Max Smith, for fulfilling my many image requests, hosting me in the library and garden, and/or sharing their knowledge and expertise. The Dumbarton Oaks Research Library and Collection also offered me a fruitful residency with their inaugural Plant Humanities Summer Program in 2019. Yota Batsaki, Anatole Tchikine, Ashley Buchanan, Taylor Johnson, Courtney Randolph, and the cohorts of the program contributed to the early development of many ideas in this volume.

Similarly, numerous libraries and museums in Europe and North America generously granted me in-person access to study their precious flower books and other valuable works in their collections, which was critical in understanding the materiality of early modern images and books. I would like to thank the following institutions and people: Bijzondere Collecties van de Universiteitsbibliotheek Utrecht (Marco van Egmond), Rijksmuseum (Marja Stijkel), Bijzondere Collecties van de Universiteit van Amsterdam (Klaas van der Hoek), Collectie Six, Teylers Museum (former curator Esther van Gelder), SMK—Statens Museum for Kunst (Hanne Kolind Poulsen and Niels Borring), Det Kongelige Bibliotek, The British Museum, RHS Lindley Library (Charlotte Brooks), Victoria and Albert Museum, Natural History Museum in London (digital consultation through Victoria Pickering), Oxford University Herbaria (Stephen Harris), Kupferstichkabinett (Hanka Gerhold), Universitäts- und Landesbibliothek Sachsen-Anhalt, Biblioteka Jagiellońska (Izabela Korczyńska), Biblioteca Nazionale Centrale di Roma (Andrea Cappa), National Gallery of Art in Washington (Brooks H. Rich), and The New York Botanical Garden (Stephen Sinon). The digital collections made accessible through the Biodiversity Heritage Library and Internet Archive were also indispensable when in-person consultation was not possible.

More thanks go to the individuals who have taken the time to discuss my project, topics, images, objects, and sources with me. Many arguments and interpretations were the fruit of such interpersonal exchanges. In no particular order, individuals include Jaya Remond, Ruth Sargent Noyes, Sarah A. Bendall, Marjolijn Bol, Claudia Swan, Anne Goldgar, Elizabeth A. Honig, Cammy Brothers, Sachiko Kusukawa, José Ramón Marcaida, Stefan Hanß, Surekha Davies, Giulia Simonini, Julia Smith, Francesca Bewer, Jo Kirby Atkinson, Birgit Reissland, Abbie Vandivere, Jaap den Hollander, Ewald Weijers, Sara Charles, Tinde van Andel, Anastasia Stefanaki, Elaine Ayers, Anna Svensson, Catherine Powell-Warren, Gerda Huisman, Ada Migliazza, Peter Mason, Ann-Sophie Lehmann, and Eric Jorink.

Working on a Ph.D. and a first book project came with many challenges, both at work and in life. To V.E. Mandrij, Alice Zamboni, Rosa Wevers, Mireille Cornelis, Simone Veld, Carlotta Capurro, Jan van Daal, Henrike Scholten, and Christien Schrover, thank you for the stimulating conversations, and for brightening the gloomy days and sharing the rare happy moments with me throughout the process. Last but certainly not least, to my dearest parents Shu-Hui and Chi-Ming and my sister Wei-Ning. It is not always easy to be a continent away from the family, but your love and support go beyond borders. Thank you for being there for me from afar for all these years.

# Notes

### Flower Names

Flowers are generally given common English names, which are often the names of the genera in modern plant classification. Historical names, as written in the sources, are given in italics when discussing specific visual or textual examples of a flower. Where it is beneficial to know the precise genus or species of the flowers for discussions, modern botanical names are provided; they are italicised and marked with L.

Eg. Poppy anemone, *Amenone tenuifol. Byzantina flore polyphillo.*, *Anemone coronaria* L.

### Translations

English quotes are obtained from various sources. If an early modern translation is available, the historical spelling is transcribed without updating it to modern English. The sources of the translations are provided in related footnotes.

### Links

All links to web sources were checked regularly throughout the writing and editing of this book, last accessed on 21 July 2025.

## Core Examples

### Intaglio-Printed Florilegia

*Florilegium, ab Hadriano Collaert caelatum et a Philip. Galleo editum. illustriss. eccelentissimoque dno D. Ioanni Medici. omnis generis elegantiarum admiratori et patrono, Philip. Gallaeus DD.*
Adriaen Collaert
Antwerp: Philips Galle, *c.*1590
Addressed as: the *Florilegium*, Collaert's *Florilegium*, Collaert's booklet

*Le jardin du roy tres chrestien Henry IV, roy de France et de Navarre dedie a la royne.*
Pierre Vallet
Paris: Pierre Vallet, 1608
Addressed as: *Le jardin du roy* ..., Vallet's florilegium, Vallet's flower book

*Florilegium novum hoc est: variorum maximeque rariorum florum ac plantarum singularium una cum suis radicibus & cepis, eicones diligenter in aere sculptae & ad vivum-ut plurimum expressae.*
Johann Theodor de Bry
Oppenheim: Johann Theodor de Bry, 1612
Addressed as: *Florilegium novum*, De Bry's florilegium, De Bry's flower book

*Florilegium Emanuelis Sweerti septimonti Batavi Amsteledami commorantis, tractans de variis floribus, et aliis indicis plantis ad vivum delineatum in duabus partibus, et quatuor linguis concinnatum.*
Emanuel Sweerts
Frankfurt: Anthonium Kempner, 1612
Addressed as: *Florilegium Emanuelis Sweerti* ..., *Florilegium Sweerti*, Sweerts' florilegium, Sweerts'flower book

*Hortus Eystettensis sive diligens et accurata omnium plantarum, florum, stirpium ex variis orbis terrae partibus singulari studio collectarum, quae in celeberrimis viridiariis arcem episcopalem ibidem cingentibus hoc tempore conspiciuntur, delineatio et ad vivum repraesentatio.*
Basilius Besler
Nuremberg: s.n., 1613

Addressed as: *Hortus Eystettensis*, Besler's florilegium, Besler's flower book, Besler's volume

*Hortus floridus, in quo rariorum & minus vulgarium florum icones ad vivam veramq[ue] formam accuratissime delineatae et secundum quatuor anni tempora divisae exhibentur.*
Crispijn van de Passe II
Arnhem: Ioannem Ianssonium Bibliopolam, 1614
Addressed as: *Hortus floridus*, Van de Passe II's florilegium, Van de Passe II's flower book, Van de Passe II's volume

**Watercolour Florilegia**

Gottorfer Codex
Hans Simon Holtzbecker
1649–1659
Copenhagen, National Gallery of Denmark (Statens Museum for Kunst)
kksgb2947–kksgb2950
Addressed as: Gottorfer Codex, Holtzbecker's florilegium, Holtzbecker's flower book, Holtzbecker's volume

Nassau Florilegium
Johann Jakob Walther
*c.*1650–1670
London, Victoria and Albert Museum
9174 & 9175
Addressed as: Nassau Florilegium

Eberhard Anckelmann Florilegium
Hans Simon Holtzbecker
1660
London, The British Museum
1888,1211.1.1–74
Addressed as: Eberhard Anckelmann Florilegium

*Hortus amoenissimus, omnigenis floribus, plantis, stirpibus, etc. etc. etc. Tam exoticis, quam indigenis, ubique locorum, in hortis celebrimis maximorum Botanophilorum, summa vigilantia et selectu conquisitis, passim consistus, Vivisque coloribus, secundum nativam florum, plantarum, stirpiumque,*

*formam, accurate, ac admirabili labore adumbratus ab industrio et solertis ingenij juvini viro D. Francisco de Geest vibrandi filio, Leovardiano Frisio.*
Francíscus de Geest
1668
Rome, National Central Library of Rome (Biblioteca Nazionale Centrale di Roma)
MS Varia 291
Addressed as: *Hortus amoenissimus*, Rome florilegium, Rome flower book, Rome volume

*Jardin de rares et curieux fleurs faicts P. Francois de Geest de Leovarde en Frise.*
Franciscus de Geest
Second half of the seventeenth century
Upperville, VA, Oak Spring Garden Foundation
00004982
Addressed as: *Jardin de rares et curieux fleurs*, Oak Spring florilegium, Oak Spring flower book, Oak Spring volume

*Bloemenboek 'Plusieurs Especes de Fleurs dessinees d'apres le Naturel'*
Multiple contributors for Agneta Block
Second half of the seventeenth century
Amsterdam, Rijksmuseum
RP-T-1948-119
Addressed as: *Bloemenboek*, Block's florilegium, Block's flower book, Block's volume

# Figures

# Introduction

## 1 Picture Books of Colourful Flowers from the Seventeenth Century

Colourful flowers of lily, tulip, narcissus, columbine, anemone (Fig. 1), sunflower, and pot marigold are a few of the more than 250 depictions of flowering plant species and their varieties in an early modern manuscript. The frontispiece of this manuscript gives the French title *Jardin de rares et curieux fleurs* (Garden of rare and curious flowers) and the cover spine shows the Dutch term *Bloem Boeck* (Flower Book).[1] This book of flowers was painted by the Friesian draftsperson Franciscus de Geest (*c.*1635–1702/1712) in the second half of the seventeenth century.[2] The depicted flowers, as the French title suggests, were those that were 'rare' and 'curious' at the time of the manuscript's production. The flower depictions are painted with watercolour on parchment and bound in gilt leather, making the manuscript a luxurious material object, fitting the material status of the rare and curious flowers.[3] De Geest's manuscript is commonly categorised and defined as a *florilegium* (plural *florilegia*); the term *flower book* is also used, and this volume uses them interchangeably.

A florilegium is a collection or selection of flower images. Flower books are more similar to picture books than to treatises, with images instead of text being the main source of information. *Florilegium* is a Latin word, which literally means 'gathering flowers.' Throughout the Middle Ages, it was most often understood as a collection of excerpts and became somewhat equivalent to the Greek word *anthologia* (anthology).[4] Around 1590 in Antwerp, the engraver Adriaen Collaert (*c.*1560–1618) and publisher Philips Galle (1537–1612)

1 Franciscus de Geest, *Jardin de rares et curieux fleurs …*, Upperville, VA, Oak Spring Garden Foundation, 00004982. See description in Tongiorgi Tomasi, *An Oak Spring Flora*, 84–87.

2 On the biography of De Geest, see Huisman, 'Franciscus de Geest and His "Delightful Gardens",' 24–29.

3 It should be noted that the parchments are inconsistent in their qualities, with some sheets much lower. Additionally, there are many interesting problems in the materials De Geest used. The two manuscripts attributed to De Geest (including the *Hortus amoenissimus* that will be discussed throughout this book) present some unique decays not seen in the other collections I consulted during this research. Huisman questioned, in her essay, why a painter would have invested the time and cost to produce such manuscripts without a patron, but the material issues point towards that De Geest might have produced the manuscripts as some sort of passion project(s), patching lower quality or less expensive materials to make the manuscripts. See Huisman, 'Franciscus de Geest and His "Delightful Gardens",' 47–48.

4 Segal and Alen, *Dutch and Flemish Flower Pieces*, vol. 2, 999.

DOI:10.1163/9789004735149_002

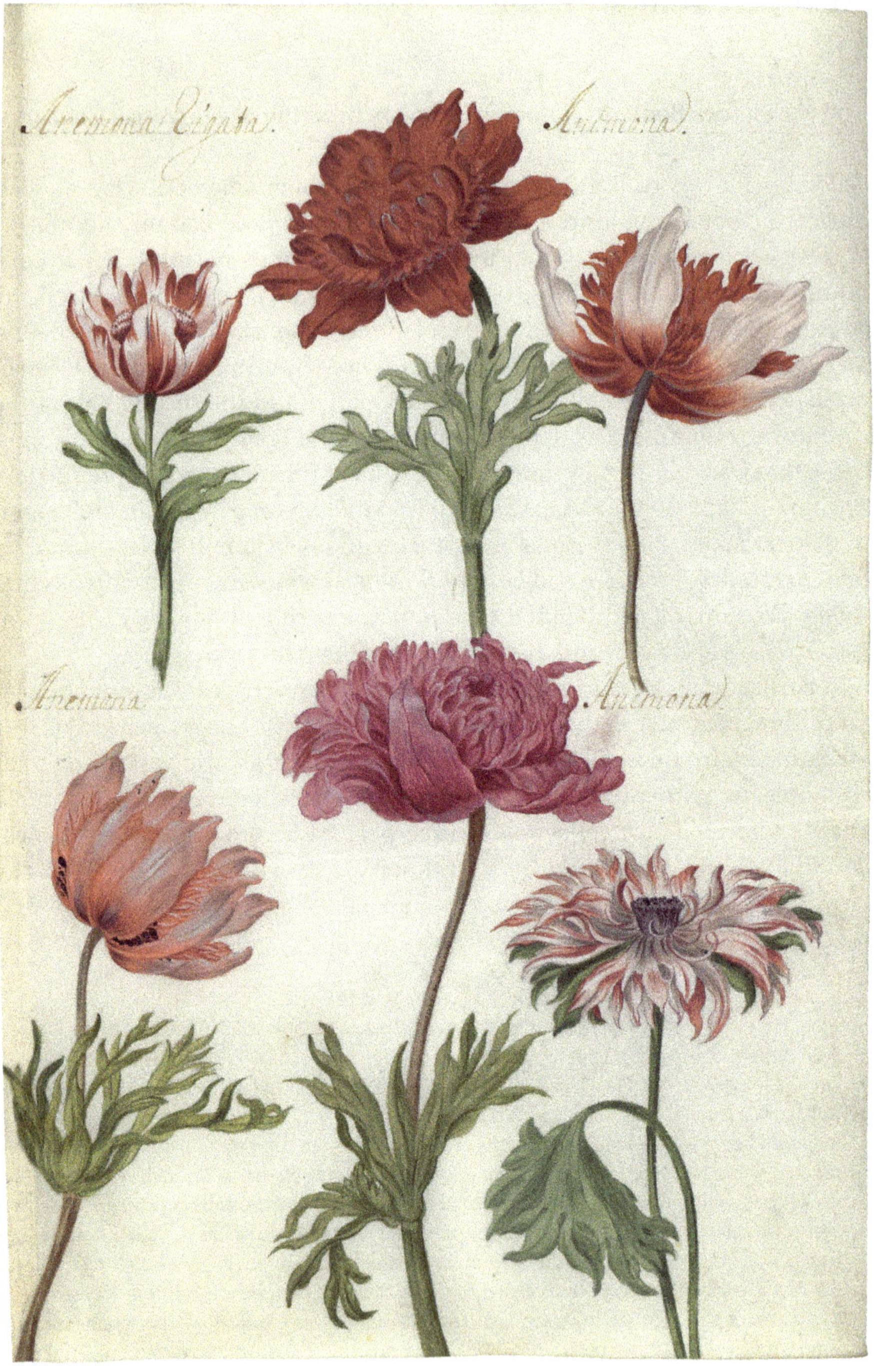

FIGURE 1 Poppy anemones from *Jardin de rares et curieux fleurs* (second half of the seventeenth century), Oak Spring Garden Foundation, Upperville, VA, 00004982

produced a small booklet titled *Florilegium*.[5] Collaert's booklet of quarto size has 23 engravings, with 21 of the plates each containing more than ten images of flowers.[6] This publication is thought to be the first to use the word *florilegium* in a title to refer to a pictorial collection of flowers.[7] Metal engraving and watercolour were the two major artistic techniques for flower book productions.[8] The primary readership and/or ownership of this genre was the group usually known as the *liefhebbers*.

What Collaert's *Florilegium* and De Geest's manuscript have in common is the predominant presence of the same kinds of flowers, including anemones, sunflowers (Fig. 2), and so on. What made these depicted flowers *rare* and, what is more, *curious* in the seventeenth century? The pictorial genre of the florilegium was highly connected to the garden culture and horticulture of the long seventeenth century in Europe. Florilegia usually feature flowering plants with prominent blooms and vibrant colours. These flowers were mostly cultivated in gardens, even though local wildflowers were not excluded. Many of the flowers and decorative plants were foreign to northern Europe and travelled from the Mediterranean, Asia Minor, and the Americas; other flowers were newly bred cultivars that became available in the long seventeenth century. While some, such as roses, have medicinal or utilitarian properties, many flowers in florilegia did not have clear nor important practical uses during the long seventeenth century and were mostly grown for decorative purposes. As constructing gardens and cultivating decorative plants required significant wealth and resources, rare plant collections—both living specimens in gardens and depicted ones in florilegia—were a form of self-representation through horticulture to show certain social status.[9]

Aside from having them as a symbol of status, what were the purposes of producing luscious books with only (coloured) pictures of flowers? In other words, *why* else did people make sumptuous picture books of rare and

5 Collaert, *Florilegium*. Bibliographic description numbered C39 in Bridson and White, *Plant, Animal & Anatomical Illustration in Art & Science*, 39.

6 Quarto is the size of the paper. The plates are smaller than the paper size and measured around 17.8 × 12.6cm. Surviving copies of printed florilegia of the same title often have the different numbers of plates. The plate numbers this book thus aim to give a general impression of how big each collection is.

7 Blunt and Stearn, *The Art of Botanical Illustration*, 13–14; Segal and Alen, *Dutch and Flemish Flower Pieces*, vol. 2, 1000.

8 A rare florilegium with woodcuts is Völler von Gellhausen, *Florilegium, das ist: Ein Blumen Buch*. A short introduction to this book can be found in Glardon, 'Review of Florilegium. Das Ist: Ein Blumen Buch,' 135–136.

9 Lauterbach, 'Commerce and Erudition,' 319–341.

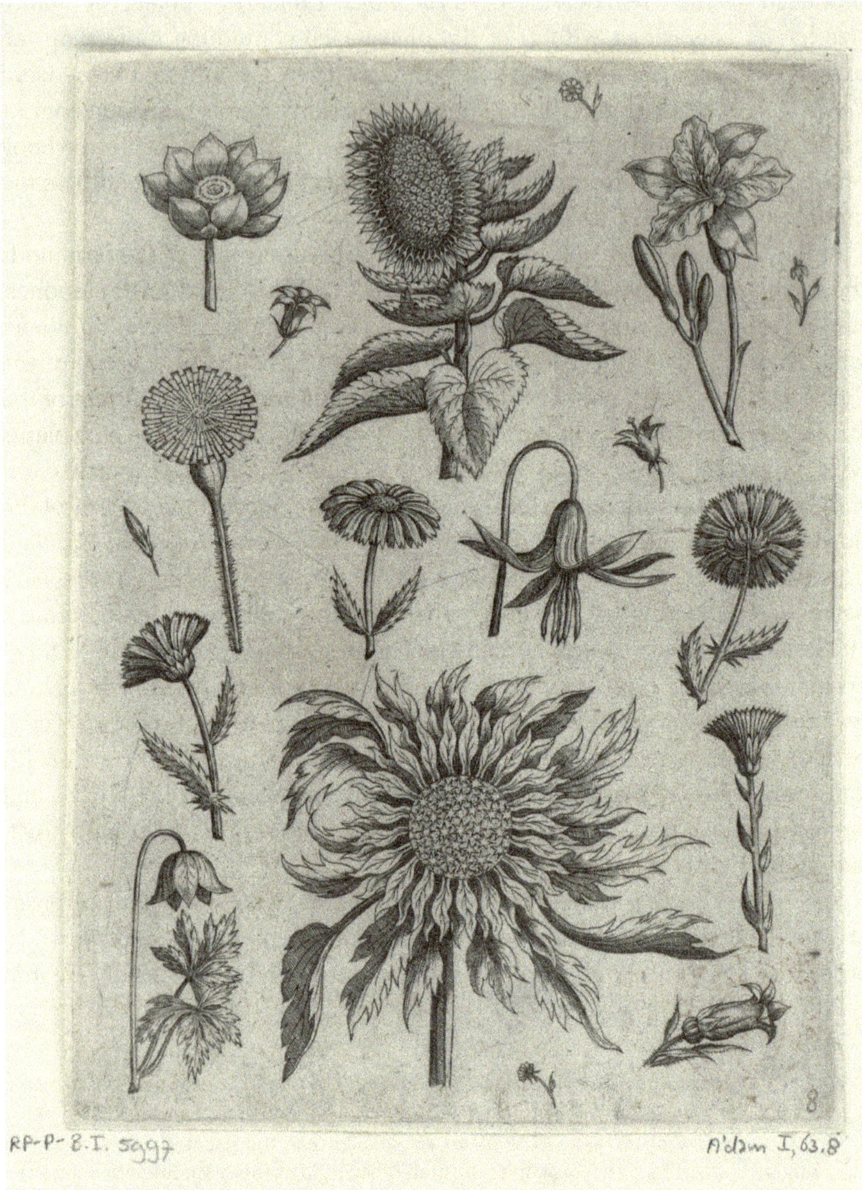


FIGURE 2 Sunflower plate from the *Florilegium* (*c.*1590), Rijksmuseum, Amsterdam, RP-P-BI-5997

curious flowers? More than being mere status symbols, this work argues that seventeenth-century florilegia are material objects of knowledge: specifically, objects that contain many forms of knowledge. It delves into what is captured and presented within the picture planes of the pages of flower books, as well as the behind-the-scenes making of these book pages. Taking into consideration various stages of the production, this volume also asks *how* these books were made: from the growing and cultivating of the flowers that are the modelled subjects of florilegia; to the compiling and presenting of visual and textual information about rare and curious flowers; to the making of the images of these floral rarities and curiosities. It unpacks the knowledge and expertise that the makers contributed throughout the production process, through an incorporation of hands-on and performative methods. By valuing makers' knowledge, this work asserts that knowing the historical making of flower books is an important step towards a better understanding of the genre of the florilegium.

The time period under examination is the long seventeenth century, which starts circa the middle of the sixteenth century to the beginning of the eighteenth century. This is the period when most of the flowers discussed in this book were considered rare and curious, even though these perceptions shifted throughout the long seventeenth century. This volume focuses on the works made in areas to the north of the Alps, with most examples from the Dutch Republic and Germany, some from France, and the floral culture extended to the Spanish Netherlands. While producing and collecting plant images was a common practice throughout Europe, the production of this particular genre was mostly a phenomenon of areas to the north of the Alps.

## 2 Towards a History of Knowledge

My aim for this volume is to reframe the understanding of the florilegium genre, drawing attention to several of the features and characteristics in flower books that are often overlooked or deemed unimportant by modern scholarship. As I will demonstrate, the relevance of these manuscripts is much more than a visual codification of certain rare species and varieties. Studying florilegia opens a window to a myriad of issues, such as strategies of representation, the fluid boundaries between text and image, and the production and transfer of knowledge, to name a few. To recontextualise the genre in the early modern period, I move *away* from the at-times limited dichotomy of art and science that has been the framework for past studies and push the direction *towards* the history of knowledge, which values different forms of knowledge that inform and collaborate with one another.[10]

Seventeenth-century florilegia are no stranger to the scholarship of natural history, with multiple fields contributing to the collective understanding of the genre from several perspectives. For example, the history of botanical books and illustrations is the most prolific sector for scholarship. Publications by Agnes Arber, Wilfred Blunt and William T. Stern, and Claus Nissen in the first half of the twentieth century have identified and defined the florilegium as a genre.[11] The authors described examples of many genres, including herbals and florilegia, to present the overall developments of botanical books and illustrations on an extended historical timeline. Gill Saunders further made an explicit connection in terms of aesthetic styles between florilegium images and flower images in (embroidery) pattern books from the early modern period.[12]

Expanding on the early modern visual culture, art historical studies have explored the florilegium within the several pictorial representations of flowers and plants. Publications by Beatrijs Brenninkmeijer-De Rooij, Eva de la Fuente Pedersen and Hanne Kolind Poulsen, and Sam Segal and Klara Alen have situated florilegium images within a plethora of early modern flower images. These works demonstrate a shared visual culture across medieval illuminations, early modern herbals, florilegia, drawings, prints, and oil paintings.[13] The majority of the scholarship on florilegia, however, is made up of scattered studies of individual collections and/or their makers. These works usually delve into questions specific to a particular printed title or manuscript, including identifying the depicted flowers and plants, or certain figures who contributed to the production process. A lot is known about the production history of Basilius Besler's (1561–1629) *Hortus Eystettensis* (1613) and Crispijn van de Passe II's (1594/1595–1670) *Hortus floridus* (1614) and the biographies of

10 The limitations of the art and science dichotomy and how this book responds to discourses in the history of knowledge are discussed in the latter half of this section. The main inspirations for pushing toward this direction include, Burke, *What is the History of Knowledge?*; Dupré and Somsen, 'The History of Knowledge and the Future of Knowledge,' 186–199; Östling, Larsson Heidenblad, and Nilsson Hammar, eds., *Forms of Knowledge*.

11 Arber, *Herbals: Their Origin and Evolution*. Blunt and Stearn, *The Art of Botanical Illustration*; Nissen, *Die botanische Buchillustration*.

12 Saunders, *Picturing Plants*.

13 Brenninkmeijer-De Rooij, *Roots of 17th-Century Flower Painting*; De la Fuente Pedersen and Kolind Poulsen, eds., *Flowers and World Views*; Segal and Alen, *Dutch and Flemish Flower Pieces*. The last title builds on Segal's work over the past decades, most notably *Flowers and Nature*.

the makers, as the two titles are the most studied printed florilegia.[14] Similarly, significant works have been done to detail the life and works of the German painter Hans Simon Holtzbecker (1610/20–1671) to establish him as a prolific master of watercolour florilegia.[15]

An issue with past studies is that they tend to either remain mostly on a survey or surface level or are research into detailed questions pertinent to a small number of famous collections and/or figures. This somewhat polarising and fragmented state of scholarship may have come from the underlying current for interpreting the genre as a bridge between art and science, even though this claim is not always explicit. Studying the genre through this lens has rarely been challenged. For example, the use of the word *botanical* itself in many pieces is highly skewed towards applying the perspective of the now-established discipline of botany and biology as a science.[16] Studies on historical herbals and the naturalists who wrote them sometimes acknowledge the florilegium genre, but often emphasise its lack of extensive textual information. Commentaries along the line that florilegium images are not botanically accurate and the genre has little to no scientific value, but they have high aesthetic qualities, are thus not difficult to find in these studies. On the other hand, art historical studies tend to venerate the biographies and artistic achievements of individual image makers. The rich research on Holtzbecker and Van de Passe II, for example, came from art historians deeming them masters of the genre, whose flowers were more skilfully drawn and/or painted than their peers' and thus deserving of more attention.[17] Even though the sentiment of bridging art and science may seem to be a positive one, it also means that flower books often do not belong to either (the history of) art or science, stagnating research questions into limited formats.

14 Keunecke, ed., *Hortus Eystettensis: zur Geschichte eines Gartens und eines Buches*; Barker, *Hortus Eystettensis: The Bishop's Garden and Besler's Magnificent Book*; Hofmann and Zöhl, *Hortus Eystettensis: Studien zur Entstehung des Kupferstichwerks und Zum Exemplar des Andrea Vendramin*. Savage, 'The Hortus Floridus of Crispijn Vande Pas the Younger,' 181–206; Veldman, *Crispijn de Passe and His Progeny*.

15 De Cuveland, *Der Gottorfer Codex von Hans Simon Holtzbecker*; Roth, ed., *Die Blumenbücher des Hans Simon Holtzbecker und Hamburgs Lustgärten*; Roth, ed., *Das Moller-Florilegium*; Poulsen, *The Green Florilegium*; Baumann, ed. *Der Gottorfer Codex*; Borring, Jensen, and Scocozza, 'The Seventeenth-Century Florilegium Gottorfer Codex,' 137–148.

16 This is similar to the pitfalls of applying the concept of Scientific Revolution to historical topics, but to a lesser degree, as discussed by Raj, 'Thinking Without the Scientific Revolution,' 445–458.

17 This is often emphasised through words and phrases such as an image maker being exceptional and/or a master of the genre.

In the past years, there have been significant discourses by historians to break the boundaries of art and science. As part of the collective efforts, a very new line (albeit still short) of interesting understandings of the genre emerged from recent scholarship in the 2020s during the writing of this volume. James Clifton reexamined Collaert's *Florilegium* within its religious context in early modern Antwerp and how people's religious practices and knowledge could lead to certain symbolic readings of the flowers.[18] Jaya Remond analysed the theory and practice of drawing plants in the early modern period to show how image makers interpreted and translated the shapes of nature into woodcuts, copperplates, and watercolours.[19] Turning to the reader's experience, Christoffer Basse Eriksen and Xinyi Wen compared the writings and notes of Henry Power and Margery Power and argued that many of Henry's natural philosophical writings were cross-pollinated by Margery's hand-colouring practice in her copy of Johann Theodor de Bry's (1561–1623) *Florilegium novum* (1612).[20] Although widely different in approaches and focuses, these pieces begin to consider the different kinds of traces left by the producers and/or consumers and what they mean within certain titles or copies. They show how works of natural history were diverse and not rigid in their productions and usages in the early modern period.

An important piece to move away from the art and science dichotomy that directly affects the understanding of seventeenth-century florilegia is a book chapter by Lorraine Daston for establishing a more period-oriented term of *epistemic images* instead of *scientific images*.[21] Daston defined epistemic images as those that were 'made with the intent not only of depicting the object of scientific inquiry but also of replacing it.'[22] Instead of visualising something intangible, epistemic images could stand in for the objects, granting access to the flora and fauna whose living counterparts were not available.[23]

18 Clifton, '"I Am Come into My Garden",' 37–66.

19 Remond, 'The Pictorial Idioms of Nature,' 273–296.

20 Eriksen and Wen, 'Colouring Flowers,' 21–43.

21 This is not to diminish the contribution of *scientific images*, as well as the idea of *art and science*, since they offered useful and important points of inquiry for the history of art and science to work together that led the field to its current state. Selected works that were useful in reflecting on the issues of scientific images, in the broadest sense, during this research include the essays in Ackerman et al., *The Natural Sciences and the Arts*; Prinz and Beyer, eds. *Die Kunst und das Studium der Natur vom 14. zum 16. Jahrhundert*; Lefèvre, Renn, and Schoepflin, eds. *The Power of Images in Early Modern Science*; Jorink and Ramakers, eds., 'Art and Science in the Early Modern Netherlands.'

22 Daston, 'Epistemic Images,' 18.

23 Christoph Lüthy and Alexis Smets gave a very different definition for epistemic images, which refers to 'any image that was made with the intention of expressing, demonstrating or illustrating a theory,' in 'Words, Lines, Diagrams, Images,' 398–439.

Daston showed how the Swedish naturalist Carl Linnaeus (1707–1778) and his contemporaries criticised florilegium images, most specifically the plates from the *Hortus Eystettensis*, for featuring too many garden cultivars that tried to pass as species. Yet, Linnaeus' own *Hortus Cliffortianus* is not too different from a florilegium, being commissioned by the wealthy George Clifford (1685–1760), whose garden was celebrated and contributed to many (rare) plant depictions in the lavishly illustrated treatise. Similarly, colours were perceived to be less important than the outlines of the plants to these naturalists, yet hand-coloured editions for the *Hortus Cliffortianus* and Linnaeus' other works were known.[24] Daston's piece offers greater nuances in reading epistemic images by recognising the discrepancies between what the naturalists claimed and did for their illustrations. However, assessing the florilegia by looking back to the seventeenth century still imposed onto them the perspective of eighteenth-century naturalists and what they thought to be a better epistemic program for depicting flowers and plants.

To the actual readership and ownership of the florilegium genre, the targeted audience of seventeenth-century *liefhebbers*, the unique forms of garden cultivars and colours *were* significant features to depict and represent through images.[25] *How* makers of florilegia delivered and communicated such features through the format of picture books may not have adhered to the epistemic values that related to the scientific inquiries of the period, but certainly required them to have knowledge and expertise. A broadened definition of epistemic images thus offers a more suitable starting point. In a recent journal issue edited by Alexander Marr and Christopher P. Heuer, the editors started with the definition 'image which seeks to produce, represent, or communicate knowledge.'[26] Marr and Heuer further asked what the role of production and reproduction methods, such as the materials and techniques used, play in understanding epistemic images. Applying such simple definitions and opening the lines of inquiry for epistemic images allows this shift towards the history of knowledge, especially towards the knowledge involved in the processes of making florilegia.

24 Daston, 'Epistemic Images,' 25.

25 Daston recognised that seventeenth-century epistemic program focused on 'the novel, the rare, and the anomalous,' but the referred group remained the naturalists who wrote about these particularities, Daston, 'Epistemic Images,' 27. *Liefhebbers*, however, were a different group from the naturalists, even though they shared some common grounds, literally and figuratively, in their pursuits of and interests in natural history.

26 Marr and Heuer, 'Introduction. The Uncertainty of Epistemic Images,' 253. See also an earlier reflection on the topic by Marr, 'Knowing Images,' 1000–1013.

A valuable aspect of the history of knowledge, as an approach instead of a discipline, is its open-endedness.[27] Recently, there has been a myriad of literature on the history of knowledge. In particular, Peter Burke argued, in his 2016 publication, that a plural form of *the histories of knowledges* is a more suitable mindset, as what kind(s) of knowledge is in a subject depends heavily on what the individuals and groups studying it considered knowledge to be.[28] For example, Burke recognised that,

> Even within a given culture, there are different kinds of knowledge: pure and applied, abstract and concrete, explicit and implicit, learned and popular, male and female, local and universal, knowing how to do something and knowing that something is the case.[29]

In 2019, with early modern examples, Sven Dupré and Geert Somsen demonstrated, in part, how early modern natural philosophers developed some of their epistemic values by learning from artisans and artisanal knowledge.[30] Different kinds of knowledge can investigate and question the very boundaries, or combinations of boundaries, that have been imposed onto the history of science and other fields. The divergent, and sometimes contrasting kinds of knowledge, are further celebrated in the edited volume by Johan Östling, David Larsson Heidenblad, and Anna Nilsson Hammar in 2020.[31] The editors did not see a problem with the history of knowledge lacking unified traditions and methodologies; instead, they emphasised how different forms of knowledge offer points of connections that could not have otherwise been made.

By moving towards the history of knowledge, at least towards the branch of thought that welcomes a diversity of knowledge(s) as outlined above, the richness of seventeenth-century florilegia started to present itself more prominently. As I will discuss through the six chapters of this volume, makers from the different stages of producing flower books were collectively knowledgeable.

---

27 There are some concerns about the history of knowledge as a discipline, as it is sometimes presented as a replacement or change of name for the history of science, which can create other types of problems, see Marchand, 'How Much Knowledge Is Worth Knowing?,' 126–149; Frumer, 'What Is and Isn't in a Name,' 150–166. On the other hand, treating it as a new discipline risks making it more limiting, going against why many have advocated for a move towards the history of knowledge in order to be more inclusive of different kinds of historical topics and research expertise.

28 Burke, *What is the History of Knowledge?*, [7].

29 Burke, *What is the History of Knowledge?*, [8].

30 Dupré and Somsen, 'The History of Knowledge and the Future of Knowledge,' 186–199.

31 Östling, Larsson Heidenblad, and Nilsson Hammar, eds., *Forms of Knowledge*.

They knew what flowers were rare and curious; they knew how to grow and cultivate the living rarities and curiosities; they knew the period's authoritative textual sources that described the flowers; they knew how to deliver and communicate the rare and curious features through a picture book format visually and textually. Seventeenth-century florilegia were not containers of lesser botanical knowledge than herbals and botanical treatises, but of different forms of knowledge that worked with one another through images of flowers.

Before elaborating on the makers as the group of actors for investigating the knowledge and expertise they attained for making florilegia, I need to highlight two points because they thread through the entire book, albeit often implicitly. First, this work concerns the 'materiality of knowledge.' Each chapter either deals with the knowledge of handling the tangible materials that were used in making the books of flowers or florilegia as objects that contain certain knowledge, sometimes both. The seminal 2006 volume, edited by Sachiko Kusukawa and Ian Maclean, examines the many ways in which knowledge was transmitted through words, images, and instruments in Europe.[32] The 2011 anthology by editors Sven Dupré and Christoph Lüthy and various authors zooms in on early modern Low Countries and further highlights the role of travelling material objects, including living flora, on the production and transmission of knowledge.[33] The 2011 exhibition and catalogue, curated and edited by Susan Dackermann, similarly shows how early modern Europeans used printed works for generating and pursuing different kinds of knowledge.[34] These works show that early modern knowledge production and transfer was a collaboration of text, images, and material objects. Text and images, moreover, have materiality as tangible broadsheets, woodcuts, engravings, etchings, and bound codices, among other formats.

Second, because this book enquires into the production and making processes, much of the knowledge that is researched is embodied and/or practical. Much connected to the materiality of knowledge, the works of Pamela H. Smith and her collaborators have further extended the research of knowledge production into active making. In the volumes edited by Smith and Benjamin Schmidt in 2007, and by Smith, Amy R.W. Meyers, and Harold J. Cook in 2014, the many authors explored the various kinds of artistic, craft, medical, and scientific practices by delving deeper into the *hows?* through an array of historical

32 Kusukawa and Maclean, eds., *Transmitting Knowledge.*
33 Dupré and Lüthy, eds., *Silent Messengers.*
34 Dackerman, ed., *Prints and the Pursuit of Knowledge in Early Modern Europe.*

evidence.[35] Smith's own works, including the most recent 2022 publication that updated the years of research since her 2004 book and the establishment of The Making and Knowing Project, have made clear how hands-on research is useful, if not critical, for better understanding the embodied and tacit knowledge in (historical) practices.[36] On a slightly different thread, the edited volume by Matteo Valleriani in 2017 highlights how the practical activities and steps (as documented in treatises, for example) are necessary for understanding the epistemic value of practical knowledge.[37] These works show that the *how-to* aspect of a practice is important for investigating embodied and practical knowledge.

## 3 Making Seventeenth-Century Florilegia

This book is interested in such activities carried out by the makers at various stages of the production process. More specifically, it focuses on the tasks and steps performed by the makers in the process of compiling a florilegium, making depictions of flowers, and growing rare flowers that potentially supplied live models for the depictions. For this reason, the word *knowledge* actually appears infrequently throughout the chapters, as this volume analyses different forms of knowledge through the *whats?* and *hows?* in the practices. Similarly, the word *makers* refers to the three groups of *florilegium compilers*, *image makers*, and *gardeners*.

I chose the words that aim to be as inclusive and neutral as possible, even though they may read as anachronistic at times. A compiler is a person who assembled the content of a florilegium; an image maker is a person who produced a picture of a flower; a gardener is a person who grew a flower in the garden. People who took on these roles did not come from one single trade or profession, and sometimes one person could take on multiple roles for a florilegium. For example, master printmakers and watercolour painters who were registered with guilds were both image makers for flower books. A learned person who had access to painting lessons and rare flowers, but was not a

35 Smith and Schmidt, eds., *Making Knowledge in Early Modern Europe*; Smith, Meyers, and Cook, eds., *Ways of Making and Knowing*.

36 Smith, *The Body of the Artisan*; Smith, *From Lived Experience to the Written Word*. 'Making and Knowing Project,' Columbia University, https://www.makingandknowing.org/; Bilak, Boulboullé, Klein, and Smith, 'The Making and Knowing Project,' 35–55.

37 Valleriani, ed., *The Structures of Practical Knowledge*.

professionally trained painter in a workshop, also made florilegia.[38] While flower books by professional and amateur image makers present different skill levels, the forms of knowledge they contain share similar patterns. Image makers thus better encompass these various people than printmakers, painters, and artists, with the last especially contributing to the problem of the art and science division. If known, the names and brief background information of relevant individuals will be given as encountered throughout the chapters, but their biographies are not the main interest of this book.

One additional group that was essential in the production of seventeenth-century flower books, but not part of this volume's investigation, is the early modern *liefhebbers*. The Dutch word *liefhebber* is often translated into *amateur* and/or *connoisseur* in modern English. The historical meaning of the term is complex and has been extensively discussed. What remains consistent in the secondary literature is that *liefhebbers* were a large part of the scholarly culture; many books in natural history were made possible by the funds, art and curiosity collections, and living flora and fauna collections of early modern *liefhebbers*.[39] Within the context of garden and horticulture, *liefhebbers* can be best understood as collectors of rare flowers and plants and/or garden enthusiasts.[40] In the Dutch version of Van de Passe II's *Hortus floridus*, the Dutch printmaker and florilegium compiler listed many *liefhebbers* and their gardens throughout the Dutch Republic. Van de Passe II was able to draw the many flowers in the florilegium thanks to these gardens that had living specimens. Manuscript flower books were often commissioned by *liefhebbers* to keep visual records of their living flowers, performing the function of an elaborate collection catalogue.[41]

The production of seventeenth-century florilegia was largely made possible by early modern *liefhebbers* supplying living rare and curious flowers to be

38 An interesting example of an amateur florilegium is an anonymous manuscript that contains more than 200 depictions of rare and curious flowers, now preserved at the Oak Spring Garden Foundation in Upperville, VA (not found in the online library catalogue). The way that the flowers are drawn and painted indicates that the person most likely did not receive (extensive) training on how to skilfully handle the materials and techniques. However, the flowers are still recognisable as the species and variations as stated in the captions. See descriptions in Tongiorgi Tomasi, *An Oak Spring Flora*, 78–80.

39 Selected works include Skogh, 'The Varied Role of the Amateur,' 489–498; Taylor, 'The Birth of the Amateur,' 499–522; Chapman and Weststeijn, 'Connoisseurship as Knowledge,' 7–41; Doherty, 'Painting Connoisseurship: *Liefhebbers* in the Studio,' 146–173.

40 See especially Goldgar, *Tulipmania*, 51–56.

41 Plant collectors were contributors to the making of herbals and botanical treatises as well, with their gardens often being the same sites for naturalists to study rare plants, see notably Egmond, *The World of Carolus Clusius*.

depicted and/or funds to make flower books, making the group both a major producer and consumer of the genre. *Liefhebbers* with the financial means for constructing gardens mostly came from middle and higher socio-economic levels of early modern societies. Most *liefhebbers* would have been learned, but not necessarily scholarly, and the languages for the textual components in florilegia cater to this readership. Prefaces and other kinds of text in florilegia are often multilingual. There is almost always a version in the scholarly language of Latin, and one of several vernacular languages of Dutch, French, and German. Women, albeit usually those from privileged social classes, were a big and important part of *liefhebbers* in garden culture and horticulture. Several early modern women were the most renowned cultivators of rare flowers and plants of their time, with their florilegia capturing the living collections on paper and parchment.[42] *Liefhebbers* who actively took on the tasks of compiling a florilegium, making flower depictions, and/or growing rare and curious flowers, also count as makers in this volume.[43]

Florilegia reflect the *liefhebbers*' understanding of the plant world, which is, in many ways, a highly wealthy and privileged perspective. *Liefhebbers* collected their rare and curious flowers and plants from all over the world through commercial, imperial, and colonial activities.[44] The plants have a rich global history, but *liefhebbers* and makers of florilegia encountered and experienced them primarily, if not exclusively, within the walls of European gardens. Aside from their places of origin, most of the flowers' native contexts and histories did not make it to the pages of florilegia. In other words, while some flowers would have been fairly traded, most of the harmful and ugly sides of exploiting lands, people, and natural resources when acquiring, transporting, and introducing

---

42 For sixteenth-century examples of women *liefhebbers*, see '11 Habsburg Women' in Egmond, *The World of Carolus Clusius*, 45–71. For seventeenth-century examples, see Sikkens-De Zwaan, 'Magdalena Poulle (1632–99),' 206–220; Vander Ploeg Fallon, 'Petronella de la Court and Agneta Block,' 95–108; Munroe, '"My Innocent Diversion of Gardening",' 111–123; Powell-Warren, *Gender and Self-Fashioning at the Intersection of Art and Science*.

43 Women have also been recognised as big contributors of natural historical images. There are many new studies that focus on individual image makers, but an older piece that provides useful information on a broader level is Tongiorgi Tomasi, '"La Femminil Pazienza",' 159–185.

44 There is vast literature on commercial, imperial, and colonial activities in the early modern world. A few titles that provided some food for thought during the writing of this book include Bleichmar, *Visible Empire*; Margócsy, *Commercial Visions*; Schmidt and Weststeijn, eds., *The Globalization of Netherlandish Art*.

many plants to Europe were left out.[45] As this volume focuses on the forms of knowledge displayed between the pages of florilegia, it does not address most of the larger issues at play in-depth. Nevertheless, it is important for me to acknowledge that florilegia were not made in a vacuum that was separated from the political, societal, and global problems in the early modern period.

Because the makers and the various kinds of activities in which they partook throughout the production process of seventeenth-century florilegia are the emphasis here, I employed hands-on and performative methods as a major aspect of research into historical practices for this book. The methods warrant a brief discussion here, as they are not the most commonly used methodologies for historical research across the board, even though they have been implemented more and more often. For this book, three kinds of makings were practised that can be broadly described as *historical reconstruction/reenactment*, *historical remaking*, and *historically inspired making*.[46] Each type of making offers an angle into the embodied and practical knowledge that will be discussed throughout this volume (Appendix 1).[47] These different angles cross-check and give nuance to the visual and textual information in seventeenth-century florilegia, leading to a more holistic understanding of the genre. Applying different forms of making during research was also integral to drawing connections and raising questions that were not apparent from closed studies of early modern flower books in libraries and museums. The making experiments focused on early modern watercolours for this book, but similar work for historical printmaking techniques was done in my previous research.[48]

For me, historical reconstruction/reenactment was instrumental in better grasping the materials and techniques used for early modern watercolour, which were much different from those used today. Reconstruction and reenactment is an active branch of performative methods and has led to many important insights into artistic, scientific, and technical processes from

45 For examples that are more specifically on plants, see Schiebinger and Swan, eds., *Colonial Botany*; Schiebinger, *Plants and Empire*.

46 There are many debates on what to call the different ways of making in historical studies. The terms selected here are those that best encapsulate the different kinds of making throughout my research process. For a variety of possible words and their uses, see 'Index of RRR Terminology' in Dupré et al, eds., *Reconstruction, Replication and Re-Enactment in the Humanities and Social Sciences*.

47 This appendix shows the blog post series Making Colours, which documents some portions of my hands-on research process and was initially posted on my personal website.

48 Chen, 'The Blocks, the Tools, and the Printed Plants'; 'Before Black There Was Red'; 'Remaking Sixteenth-Century Botanical Woodblocks.'.

the past.[49] Among historians of art and science, following the step-by-step instructions in historical recipes and treatises is the most commonly understood approach when the term is used. The aforementioned The Making and Knowing Project is such an example of historians reconstructing and reenacting artisanal techniques, such as lifecasting small flora and fauna.[50] Following this line, reconstructing recipes for making certain washes, pigments, paints, and so on, was fundamental for building the skillset and material knowledge that was comparable to what was required for early modern image makers to create (coloured) flower depictions. A challenge that arose, however, was that this form of reconstruction and reenactment is a heavily text-based practice, and the text reflects only one aspect of how makers worked.[51] Moreover, unlike historical recipes and scientific experiments that often have recorded steps and procedures, at least to a comprehensible degree, florilegium image makers left little traces of their working process in a textual form, especially the makers of the watercolour-painted manuscript ones.[52]

Historical remaking offered insight into how seventeenth-century florilegia image makers transformed painting materials into painted works by reproducing a small number of selected works. Dress historians and textile historians, as well as reenactors at living history museums, have long been making garments and/or other kinds of textiles by closely studying surviving objects and recreating them with historically oriented materials and techniques.[53] This

49 Selected works that were most impactful during this research include Staubermann, ed., *Reconstructions: Recreating Science and Technology of the Past*; Dupré et al., eds., *Reconstruction, Replication and Re-Enactment in the Humanities and Social Sciences*; Hendriksen, 'Rethinking Performative Methods in the History of Science,' 313–322. Additionally, some works on matters/materials and making processes were important in shaping the application of this method, notably Lehmann, 'How Materials Make Meaning,' 6–27; Ingold, *Making: Anthropology, Archaeology, Art and Architecture*; Anderson, Dunlop, and Smith, eds., *The Matter of Art*.

50 See 'Research Essays' on website 'Secrets of Craft and Nature in Renaissance France: A Digital Critical Edition and English Translation of BnF Ms. Fr. 640,' The Making and Knowing Project, https://edition640.makingandknowing.org/#/.

51 The reasons for writing how-to texts were not necessarily about recording the practices as faithful as possible. Rather, there were more motivations. For example, see DiMeo and Pennell, ed., *Reading and Writing Recipe Books, 1550–1800*; Leong, *Recipes and Everyday Knowledge*.

52 A semi-exception is the colouring instructions that are discussed in Chapter 4 and Chapter 6 on how to colour some intaglio-printed plates.

53 Many dress historians, however, use the word *reconstruction* for describing their method(s) of making. Here *remaking* is selected in order to distinguish it from the text-based reconstruction/reenactment, as discussed above. See especially Davidson and Hodson, 'Joining Forces: The Intersection of Two Replica Garments,' 204–210; Davidson, 'The Embodied Turn: Making and Remaking Dress as an Academic Practice,' 329–362; Bendall, *Shaping Femininity: Foundation Garments, the Body and Women in Early Modern England*; Bendall

object-based approach requires researchers to have a decent level of familiarity and skill with the materials and techniques, although mastering them is not a necessity. Remaking historical objects enables researchers to step into the shoes of the makers from the past, to better understand the specific ways of working of individual makers. The goal is not to identify the hands of a certain master for attributing a work to a name, but to know the various manners and possibilities of working with period materials and techniques that were available to most makers of the time.[54] With examples on both paper and parchment, remaking selected images from a few florilegia by a few makers illuminated the essential techniques, as discussed in Chapter 6, that allowed image makers to depict and visualise rare and curious flowers, despite individual differences in the handling of watercolour paints.

Historically inspired making has encouraged many inquiries into issues that were also present in the process of making and producing seventeenth-century florilegia, aside from materials and techniques. For example, the collaborative project Planting Pigments ColLAB explored making lake pigments using historical and experimental dyestuffs (Appendix 2). The instability of many plant-based colours visualised why florilegium image makers tended to use inorganic colours for durability.[55] Keeping cut flowers alive in order to draw and paint them for my personal creative project of *Hortus floridus Iesse*, a twenty-first-century reinterpretation of florilegia (Appendix 3), raised questions including how people historically preserved flowers, other than by painting them. Pressing the living models for the *Hortus floridus Iesse* led to observations of how image makers employed similar techniques to visually 'press' flowers, as discussed in Chapter 5. Working with current-day gardening techniques that have historical roots, Chapter 2 is entirely the fruit of my assisting the gardeners at the Oak Spring Garden Foundation to upkeep and maintain their formal garden and sowing seeds in the nursery. Historically inspired

---

and Dyer, eds., *Embodied Experiences of Making in Early Modern Europe*. Moreover, the public-facing works by dress historians were much helpful in offering insights into some of their research and remaking processes, see videos by Serena Dyer, https://www.youtube.com/@SerenaDyer, and the research project Refashioning the Renaissance, https://www.youtube.com/@refashioningtherenaissance9771 and https://refashioningdata.eu/home.

54 Reconstructing oil paintings is often used along the opposite line to investigate the painting techniques of certain masters for purposes of attribution.

55 Plant-based colours being impermanent has long been known, with advice against certain colours already common in the early modern period. The Planting Pigments ColLAB however, visualises how these colours may change as time passes. It also shows how different batches can produce different shades of colours, which is briefly discussed in Chapter 6.

making cannot and should not stand alone without historical reconstruction/reenactment and remaking, as it can easily lead to anachronistic projection if the researcher is not careful. However, there is great value in simply playing with (ahistorical) materials and techniques, as it offers a playground for free associations that are out of the bounds of disciplinary restrictions.

## 4 Overview of Core Examples and This Volume

The definition and basic information of the florilegium genre can, at times, be inconsistent and chaotic throughout the existing literature. The multiple mediums and functions can sometimes be confusing as well. This section thus serves to bring readers onto the same page for discussions in this volume. The core examples of this volume include six intaglio-printed florilegia and six watercolour-painted ones. Aside from Collaert's *Florilegium*, there are five major titles that form the basis of the genre for modern scholarship. All of them were first published between 1608 and 1614 and made with intaglio-printed plates of engravings and/or etchings.[56] In 1608, the publication of *Le jardin du roy …* by the French printmaker and embroidery designer Pierre Vallet (*c.*1575–1657) marks the first big florilegium in the line. The year 1612 saw two flower books, the *Florilegium novum* by the engraver and publisher De Bry, who was active in Oppenheim and Frankfurt, and the *Florilegium Emanuelis Sweerti …* by the Dutch printmaker and nurseryman Emanuel Sweerts (1552–1612). The monumental florilegium, both in its physical measurement and number of plates, of the *Hortus Eystettensis* by the German apothecary Besler of Nuremberg came out in 1613. The last in this line is the *Hortus floridus* by the Dutch printmaker Van de Passe II in 1614. These printed florilegia continued to be reissued throughout the first half of the seventeenth century, but most saw their last reprint before the 1680s. One exception is the *Neues Blumenbuch* by Maria Sibylla Merian (1647–1717) that was published in 1680, but the context of making the book was very different from that of the first wave.[57] Collaert's *Florilegium* and the five titles from the first quarter of the seventeenth century form one set of the discussed corpus for this book.

While the line is not hard drawn, there was a shift in the artistic techniques used for the genre throughout the seventeenth century, moving from

56 See Core Examples for a list of these titles.

57 Merian, *Neues Blumenbuch*. Surviving copies are scarce, accessible through facsimile, with a study by Bürger and Heilmeyer, *Maria Sibylla Merian: Neues Blumenbuch*. The most public accessible reproduction is Merian, *Neues Blumenbuch*, 4th Edition.

intaglio-printed to watercolour-painted collections. Surviving manuscript florilegia, such as the volume by De Geest, are mostly dated from the second half of the seventeenth century, but a few image collections from the late sixteenth century to the early seventeenth century already have traits of the genre.[58] It is difficult to pinpoint and estimate the number of manuscript florilegia that once existed, as many collections have been disassembled and dispersed, and/or with no public records for being in private hands.[59] It is not the goal of this volume to provide a comprehensive survey of the (partially) surviving collections. Rather, this book engages with a selection of manuscript flower books that are mostly intact as a collection for a better understanding of the genre as a whole. The collections that are the central points of analyses are the *Jardin de rares et curieux fleurs* and the *Hortus amoenissimus*, both made by De Geest; the Gottorfer Codex and the Eberhard Anckelmann Florilegium by the German painter Holtzbecker; the Nassau Florilegium by the Strasbourg painter Johann Jakob Walther (1650–1717); and the *Bloemenboek* for the Dutch rare plant collector Agneta Block (1629–1704).[60] These examples were all bound into codices in the late seventeenth century or early eighteenth century (and a couple disassembled again in the twentieth century due to conservation needs); there is evidence to suggest some collections might have existed as loose sheets stored in boxes during the seventeenth century.[61]

Early modern florilegia always had a primary role as picture books, which is the emphasis of this volume, even though they often had multiple functions. The first additional function was as a sales catalogue of rare and curious flowers and decorative plants, applying mostly to Sweerts' florilegium.[62] The second, and most significant function aside from being a picture book, was serving as a collection catalogue of rare flowers and decorative plants for a specific garden. Even though the genre is now called the *florilegium*, this word was not applied to every image collection that is now categorised as a flower book. Often, a collection would be titled with *hortus* (garden), such as the *Hortus Eystettensis*, as it was based on the garden of the Prince-Bishop Johann Konrad von Gemmingen

58 The most significant example is a collection called the Camerarius Florilegium that is also difficult to access, see descriptions in Wickert, *Das Camerarius-Florilegium*.

59 Printed facsimiles and reproductions have been a major means to access seventeenth-century florilegia. Recent digitisation of museum and library collections have made many previously unstudied flower books known to more people.

60 See Core Examples for a list of these titles.

61 Such as the Gottorfer Codex having sheets with damages of lost paints that indicate the sheets were dragged across each other, De la Fuente Pedersen and Kolind Poulsen, eds., *Flowers and World Views*, 104.

62 See Chapter 2 for an introduction to and a discussion on this florilegium.

of Eichstätt (1561–1612), who also commissioned Besler to produce the title.[63] There is no suggestion of a hard rule, but the tendency is that when the word *hortus* is used in the title, actual gardens and their rare flower/plant collections were involved, regardless of whether the flowers were observed and studied from one specific site or from a number of gardens within a region. Such a tight connection to historical garden sites further shows the intertwined nature of early modern horticulture and the genre of the florilegium.

A third additional function of seventeenth-century florilegia was working as a pattern book for other artistic practices. The line between florilegia and pattern books for different arts and crafts is often blurry. There are strong indications that suggest images from flower books could have been used as pictorial models for textile embroideries and musical instrument decorations.[64] Oil paintings of flower still lifes sometimes show reminiscences of florilegia images being copied onto canvas to assemble the painted bouquets.[65] Additionally, books specifically designed as pattern books sometimes feature plates of flowers that are stylistically similar to those in florilegia.[66] In many ways, Merian's *Neues Blumenbuch* is more in line with a pattern book by design that features rare and curious flower images, instead of a florilegium that was used as a pattern book. This volume does not delve into the practical function of florilegia working as pattern books, nor pattern books that include designs of rare and curious flowers. Such research depends on surviving objects that strongly show a direct use of florilegium images as patterns and with different methodologies, resulting in a much different book that is beyond the scope of this work. Such research also concerns the perspective and experience of the users instead of the makers of florilegia, with the latter the stated group of interest for this book.

Within this volume, I grouped the six chapters into three themes, but each chapter engages with stand-alone topics and discourses. Chapters 1 and 2 deal with the forms of knowledge in identifying and cultivating living flowers (including decorative plants) that were rare and curious by examining what are depicted in the images and discussed in the text of seventeenth-century flower

63 The term 'garden books' (Gartenbücher) was particularly used by Nissen, *Die botanische Buchillustration*, 66.

64 Images from herbals and natural history books of this time period have been identified to be used as models for different arts and crafts, but I have not seen a definitive case of this practice with florilegium images. See, for example, Vandervellen, *The Golden Age of Flemish Harpsichord Making*, 329–335; Learner, 'Embroidering the New Science,' 685–717.

65 Brenninkmeijer-De Rooij, *Roots of 17th-Century Flower Painting*, 62, 66; Van Eikema Hommes et al., 'Still-Life Sources,' 25–37.

66 Saunders, *Picturing Plants*, 54–55.

books. Chapters 3 and 4 analyse the images and text in intaglio-printed florilegia and how they work together to deliver and communicate different kinds of visual and verbal information about rare and curious flowers. Chapters 5 and 6 examine the visual strategies and material practices employed by image makers of florilegia, with an emphasis on watercolour-painted examples, to depict and present features that were critical for a flower to be rare and curious in the long seventeenth century.

Chapter 1 discusses what made flowers rare and curious in the seventeenth century. It unpacks the typical scope of the florilegium as a genre, or what went into a flower book, by looking into which flowers repeatedly appear between the pages of surviving printed and painted examples. The chapter then establishes the knowledge a florilegium compiler would have needed in order to identify and select floral rarities and curiosities by discussing the characteristics that made the flowers intriguing to *liefhebbers* in the long seventeenth century. Chapter 2 delves into how rare and curious flowers were grown and cultivated in early modern gardens. The chapter explores the nuts and bolts of seventeenth-century gardening techniques, such as soil prepping, (trans)planting, and the year-round care of flowers to bring them from seed to bloom. Through care and patience, this chapter underscores the practical knowledge of the gardeners who brought the floral rarities and curiosities to life.

Chapter 3 examines how printed florilegia extracted and highlighted visual and verbal knowledge of rare and curious flowers from sixteenth-century herbals. Through analysing how florilegium compilers referenced and excerpted images and text from herbals and botanical treatises, this chapter demonstrates how flower books actively engaged with authoritative plant studies of the time, as well as developing a different system of ordering flowers according to the compilers' careful observations of seasonality. Chapter 4 inspects how printed florilegia used text to supplement information that was difficult or could not be visualised. It first tracks the changes in the textual components in the reprints and different editions of printed florilegia, moving from providing flower descriptions to giving colouring instructions. By looking into how selected florilegium images work with flower descriptions and colouring instructions, this chapter shows how compilers verbally presented holistic and/or colour pictures of flowers through different kinds of writings that were familiar to early modern *liefhebbers*.

Chapter 5 explores how watercolour-painted florilegia used images to preserve rare and curious flowers. Turning to the genre of the herbarium, or collections of dried plants, this chapter investigates what can and cannot be preserved by pressing living specimens. Through the comparable artistic manipulations and visual techniques in early modern herbaria and florilegia, this chapter argues for an additional function of florilegia working as herbaria to preserve the rare and curious characteristics of flowers. Chapter 6 focuses on the material practice of making intaglio-printed and (hand-) coloured flower images. It discusses how the supports of paper and parchment allowed for the making of life-size images of flowers, as well as how the early modern materials and techniques of watercolour supported image makers in representing the array of colours among flowers. Through making life-sized and colourful images, this chapter illuminates the material and artisanal knowledge that enabled image makers to produce lifelike depictions of rare and curious flowers.

CHAPTER 1

# Rare and Curious Flowers in the Seventeenth Century

## 1 Introduction

The Flemish engraver Adriaen Collaert's (*c.*1560–1618) booklet of printed flowers titled *Florilegium* (*c.*1590), is often attributed as the first publication that applies the word *florilegium* for referring to a pictorial gathering of flowers.[1] More significant than *when* the word was first used, however, is that Collaert's *Florilegium* is not simply a gathering of any flowers. Rather, it is a selection of both a diversity and variety of flowers with certain characteristics that made them rare and curious in the seventeenth century. As the early modern systems of plant order are different from the Linnaean taxonomy that has become the standard today, in this volume, the word *diversity* is used to emphasise the presence of different flower genera or species, such as the inclusion of tulip, iris, narcissus and so on within a collection. The word *variety*, on the other hand, refers to the differences within the same flower genus or species, such as the presence of hyacinths in several colours. Collaert's *Florilegium* is a small collection with only 21 plates of flowers, but a combination of such flower diversity and variety is already visible.[2] Not every single flower on the 21 plates has identifiable features of real flowers. For example, several small buds and blooms on the sunflower plate (Fig. 2) are recognisably flower-like as motifs, but too generic for specific identification. The identifiable ones include a diversity of roses, lilies, crown imperials, sunflowers, thistles, ranunculi, irises, carnations, poppy anemones, and more. Additionally, there are usually multiple varieties of the listed flowers, such as sunflowers of different head sizes and poppy anemones with more petal counts than in their native form (called double flowers).

This chapter reconsiders the fundamental question of what a florilegium is by delving into what are typically included in seventeenth-century flower

1 Collaert, *Florilegium*. Strictly speaking, Collaert's series is more similar to a series or suit of prints instead of a codex. This book treats it as a codex to emphasise the collection as one unit, as surviving copies are often bound into thin books.

2 There are 24 plates in this title, including a frontispiece, an emblem page, and a flower still life in addition to the plates of flowers.

 DOI:10.1163/9789004735149_003

books. Using Collaert's booklet as the starting point, this chapter first demonstrates how florilegia balance a diversity and variety of rare and curious flowers in their scope. It then identifies what made flowers rare and curious in the sixteenth and seventeenth centuries. The goal here is not to reach a conclusive definition of the genre or a definitive parameter of the content in a florilegium. Historical image collections present a wide spectrum of scopes of flowers and/or plants depending on the interests of each collector. The boundaries between a florilegium and other kinds of book genres are often fluid and blurry as a result. As mentioned in the general introduction, the word *florilegium* itself was used by some individuals to title their pictorial collections, but the word *hortus* (garden) was equally, if not more often, applied to name collections with comparable scopes. A hard definition can thus be reductive and limiting. Instead, this chapter aims to illuminate the major set of characteristics that consistently appear in seventeenth-century florilegia.

## 2 Balancing a Diversity and Variety of Flowers

Collections of flower images for various purposes predate the publication of Collaert's *Florilegium*, but these pictorial gatherings do not have the same focus on showing both a diversity and variety of flowers. For example, the *Book of Flower Studies*, attributed to the Master of Claude de France and dated around 1510 and 1515, is a manuscript with 39 painted flowers (including a few plants) on parchments.[3] There remain many fundamental questions about this manuscript, including the primary function of the collection. It is not clear whether the book was made as a model or pattern book for the master (and his workshop) or was part of a larger collection of plant images. There is a high possibility that the volume may have been a pattern book, as many books of hours (or medieval prayer books) attributed to the master have pages with similar flower depictions as the main subject or as marginal decorations.[4] A spread with the alphabet written in pen and ink, with half of the letters on the back of a painted apothecary rose, also recalls a common feature of medieval model books for scribes.[5] The primary function could have affected the

3 Master of Claude de France, *Book of Flower Studies*, New York, The Metropolitan Museum of Art, 2019.197, https://www.metmuseum.org/art/collection/search/823979.

4 See the Books of Hours discussed in Wieck, *Miracles in Miniature*, 43–47. The Met attributes it as a model/pattern book in Weiss and Hollein, 'Report from the President and the Director,' 6.

5 Master of Claude de France, *Book of Flower Studies*, fol. 35v–36. An example of a scribal pattern book that displays many sets of the alphabet in a similar manner, but more elaborate in

viewership of the book. From the production perspective, however, the intentional gathering of flowers to form an image collection is similar to what is displayed in Collaert's *Florilegium*, especially when Collaert's booklet also has a layer of functioning as a pattern book. Several of the depicted flowers in the *Book of Flower Studies* are comparatively more stylised than naturalistic, but they are overall recognisable (Fig. 3). Many of the flowers are the same as the ones that appear in Collaert's booklet, and subsequently to florilegia of the seventeenth century, such as lily, iris, and carnation. The manuscript also contains fruit and crops like strawberry, hazelnut, and grapevine, but such plants also occasionally appear in florilegia, as will be discussed. Unlike the *Florilegium*, however, Master of Claude de France's collection does not have variations of the depicted flowers, as each of the 39 watercolours features a different species of flower or plant. As its given descriptive title suggests, the *Book of Flower Studies* is thus not a florilegium.

Conversely, image collections with only a variety but no diversity of flowers are often categorised as other genres. One major example is the *tulpenboek* (tulip book) from the second quarter of the seventeenth century.[6] The *Tulipmania* refers to a period in the 1630s in which a small group of individuals in the Dutch Republic participated in hyping up the tulip trade. While accounts reporting the event were often exaggerated, tulips indeed attracted more attention than other flowers in the sixteenth and seventeenth centuries.[7] It is the only flower that had image collections composed solely of that flower before the eighteenth century.[8] Tulip books were mostly produced during the *tulipmania,* and many flower books have a high percentage of tulip variations in the collections. A disassembled spread from a volume by Jacob Marrel (1613/1614–1681) (Fig. 4), one of the most prolific figures in this genre, shows the

font designs, is the manuscript by Gregorius Bock, *Scribal Pattern Book*, New Haven, Beinecke Rare Book and Manuscript Library, Beinecke MS 439, https://collections.library.yale.edu/catalog/10450154.

6 For examples, see *Tulpenboek (van Judith Leyster)*, Haarlem, Frans Hals Museum; *Tulpenboek*, Amsterdam, Six Collection; *Verzameling van een meenigte tulipaanen, naar het leven geteekend met hunne naamen, en swaarte der bollen, zoo als die publicq verkogt zijn, te Haarlem in den jaare A. 1637, door P. Cos, bloemist te Haarlem*, Wageningen, Wageningen University & Research—Library, Special Collections, R393F07, https://wur.on.worldcat.org/oclc/67675066.

7 See the seminal study on this historical event by Goldgar, *Tulipmania*.

8 Crazes for flowers such as hyacinth and orchid took place in the eighteenth and nineteenth centuries. The image collections made after the varieties of these flowers, however, tend to be more of a personal passion, rather than a (small) genre with many participants like the tulip books. See, for instance, Cribb and Tibbs, *A Very Victorian Passion: The Orchid Paintings of John Day, 1863 to 1888.*

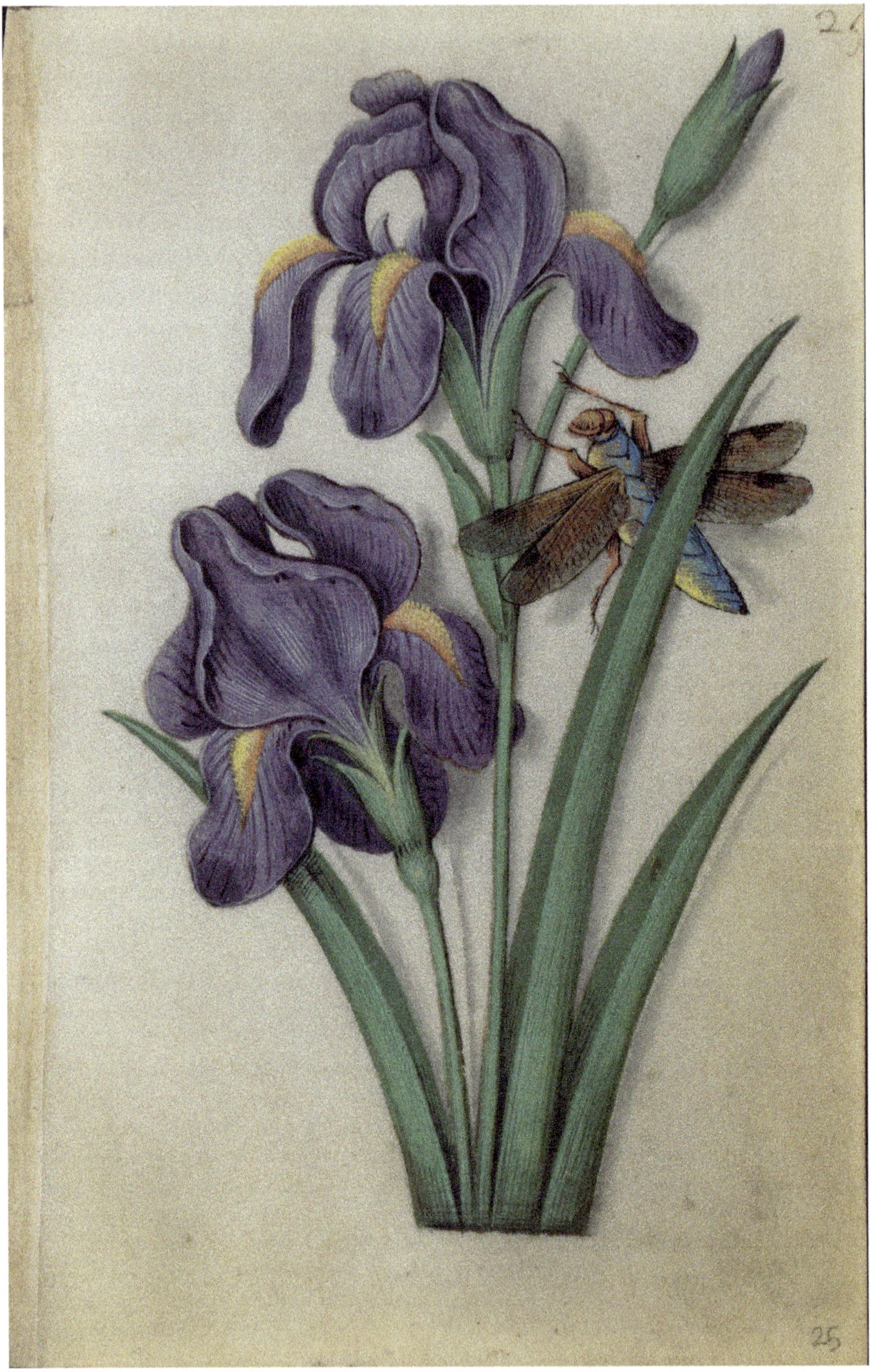

FIGURE 3 Irises from the *Book of Flower Studies* (*c.*1510–1515), The Metropolitan Museum of Art, New York, 2019.197

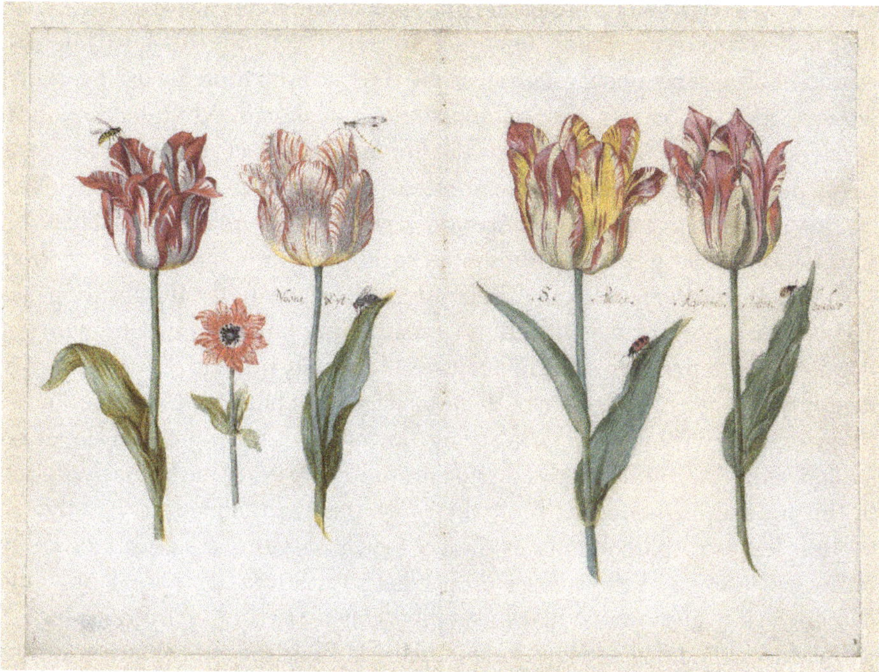

FIGURE 4 Variegated tulips and a poppy anemone from a tulip book by Jacob Marrel (*c.*1640), Rijksmuseum, Amsterdam, RP-T-1967-86

typical content of a *tulpenboek*.[9] Like the spread, a tulip book usually includes a wide variety of flamed tulips, each with different configurations of colours and stripes, but lacks flower diversity. There are occasional depictions of other flowers, such as the poppy anemone, between the two striped tulips in red and white (Fig. 4). However, such additions (including the insects and shells) are usually embellishments to make the pages more dynamic, as there is a clear visual hierarchy on the pages that draws most attention to the tulips and there are no systematic appearances for other flowers. Sometimes, the pricing of a particular variation is written near the tulip.[10] This monetary emphasis is rarely present in florilegia, even when a flower book could have the double

9 Jacob Marrel, 'Sheet from a Tulip Book,' *c.*1640, watercolour on parchment, 34 × 44.9cm, Rijkesmuseum, Amsterdam, http://hdl.handle.net/10934/RM0001.COLLECT.56132. There is another bound tulip book by Jacob Marrel in the same collection, see *Album met voorstellingen van tulpen*, Amsterdam, Rijksmuseum, RP-T-1950-266, http://hdl.handle.net/10934/RM0001.COLLECT.312980.

10 There are prices marked nearby many tulips in the tulip books in the Six Collection and the Wageningen University & Research—Library, Special Collections.

function of serving as a sales catalogue, such as the publication by the Dutch printmaker and nurseryman Emanuel Sweerts (1552–1612) that will be discussed in Chapter 2. The *tulpenboeken* are thus an even more focused genre of pictures responding to a specific and somewhat short-lived phenomenon in early modern horticulture, whereas the florilegium reflects the wider interests of the period in flowers and their varieties.

The inclusion of both a diversity and variety of flowers in an image collection is noticeable when examining florilegia of more considerable sizes, as Collaert's *Florilegium* is a small collection. *Size* here means the number of pages and depicted flowers in a collection instead of the physical measurements of the papers and parchments. The inclusion is usually not equal, with diversity regularly taking a higher share of the total pages, but both qualifiers are almost always clearly visible. The Oak Spring flower book of Franciscus de Geest (*c.*1635–1702/1712), for example, demonstrates this in exact terms. As described in the general introduction, there are more than 250 depictions painted on the 100 leaves of parchments in the volume, presenting a sizeable diversity of flowers.[11] Some of the species do not have variations, as there is only one depiction of a sunflower, a black calla lily (called *Serpentaria* in the volume), a hibiscus (called *Althea*), and so on. Others include several colour variations and/or double flowers, such as three double columbines in red-violet, red and white stripes, and blue-violet. De Geest's Oak Spring florilegium is substantial, but not immense, and the range of flower book sizes can vary drastically.

If Collaert's booklet is considered the smallest collection on the spectrum of florilegia produced in the long seventeenth century, the *Hortus Eystettensis* (1613) would be at the other end. The *Hortus Eystettensis* was produced by the Nuremberg apothecary and botanist Basilius Besler (1561–1629) for the Prince-Bishop of Eichstätt Johann Konrad von Gemmingen (1561–1612).[12] Unlike the books discussed thus far, this flower book has a clear function as a pictorial catalogue of a specific garden. Von Gemmingen commissioned Besler, who was in charge of the bishop's garden in Eichstätt, to put together a florilegium based on the flowers and plants on site. Besler focused on the rare and curious ones, as he eliminated some of the more common local species that were also growing in the garden.[13] The book took many years and teams of craftspeople

11 Franciscus de Geest, *Jardin de rares et curieux fleurs* ..., Upperville, VA, Oak Spring Garden Foundation, 00004982. See Introduction for more description on this collection.

12 Besler, *Hortus Eystettensis*. Bibliographic description numbered C50 in Bridson and White, *Plant, Animal & Anatomical Illustration in Art & Science*, 40.

13 Besler, *Hortus Eystettensis*, the third page (unnumbered) in the preliminary text. See also Littger and Dressendörfer, *Florilegium: The Book of Plants*, 18. This reproduction by Taschen is the most accessible and widely circulated secondary source, with clearly

to produce, and the bishop did not see its publication before his passing.[14] The *Hortus Eystettensis* is monumental both in its physical measurements and the number of plates, consisting of more than 350 plates printed on the largest size paper (around 57 × 46cm) available at the time.[15] The collection has more than 1,000 flower and plant depictions, covering all the previously mentioned flowers in this chapter, and many of their variations.

In addition to flowers with prominent and showy blooms, utilitarian plants such as sweet-smelling herbs and fruit, and other decorative plants make regular appearances in seventeenth-century flower books. For example, variations of lavender constitute multiple plates in Besler's volume.[16] Decorative plants, in particular, need clarification and emphasis here. Many plants that have now acquired common utilitarian functions, such as the tomato, the potato, or even the artichoke, were once grown as decorative plants because they were considerably few in number in the early modern period.[17] They were simply too rare to be used or consumed as common goods. The rarity of plants, which will be further discussed in the next section, and their decorative quality were closely connected in early modern gardens. It is this decorative quality that the genre of florilegium highlights, instead of whether a plant has prominent blooms.

As a genre, the florilegium has a predominant focus on flowers and decorative plants and does not attempt to collect images of all the plants under the sun. This specific focus becomes clear when examining in comparison an important collection from the second half of the sixteenth century. The *Libri picturati A16–30* is a vast collection from the southern part of the Low Countries that consists of more than 1,400 watercolours of plants and animals bound in fifteen volumes.[18] Many of the basic questions regarding the production and making of the collection remain difficult to answer, including the production

marked plate numbers and an index of plants; the rest of this book refers to the plate numbers as given in this reproduction.

14 Segal and Alen, *Dutch and Flemish Flower Pieces*, vol. 2, 1023. Besler was the figurehead and may have overstated his importance in the production, see Barker, *Hortus Eystettensis: The Bishop's Garden and Besler's Magnificent Book*, 14–18.

15 Barker, *Hortus Eystettensis: The Bishop's Garden and Besler's Magnificent Book*, 11.

16 Plate numbers 310, 311, 347 in Littger and Dressendörfer, *Florilegium: The Book of Plants.*

17 For the case of the tomato, see a recent publication that tracked down surviving records of where the plant appeared in the early modern Europe by Van Andel et al., 'Sixteenth-Century Tomatoes in Europe.'

18 On reproduced images, historical background, and bibliographic descriptions of the botanical volumes, see De Koning et al., eds., *Drawn after Nature: The Complete Botanical Watercolours of the 16th-Century Libri Picturati.* See also Wille, 'The Discovery of the Scientific Heritage of Karel van Sint Omaars (1533–1569),' 67–80; Egmond, 'Clusius, Cluyt, Saint Omer,' 11–67.

period and ownership. According to existing sources, the collection was most likely initiated and belonged to the Flemish nobleman Charles de Saint Omer (1533–1569), with the renowned early botanist Carolus Clusius (1526–1609) (who will become a familiar figure in subsequent chapters) acting as consultant. The *Libri picturati A16–30* has a noteworthy florilegium component. Volumes A21, A22, A25, and A30 feature many of the same flowers and their colour variations that commonly appear in florilegia. The full scope of the collection, however, far exceeds the florilegium and resembles the encyclopaedic characteristic of many sixteenth-century herbals. Crops, (fruit) trees, lichens, mushrooms, and so on, compose the majority of the collection. Flowers and decorative plants are part of the *Libri picturati A16–30*, but not its focal point. A seventeenth-century florilegium is thus a collection with a specific scope (but not size) depicting diverse and varied flowers with decorative qualities.

## 3 The Rare and the Curious

In addition to the decorative quality of a plant, the word *rarity* is also related to the concept of *curiosity* relevant to most of the flowers that appear in the genre of the florilegium. Recalling De Geest's Oak Spring volume, the flower book starts with a frontispiece showing a cartouche around the title 'garden of rare and curious flowers' written in French.[19] This phrase is perhaps most demonstrative of what the genre of the seventeenth-century florilegium is in a nutshell: a pictorial garden of rare and curious flowers. Between the two words, *rare/rarity* (and likewise *rare* in French, *rariorum* in Latin, etc.) is the more frequent choice in seventeenth-century florilegia, and *curious/curiosity* and its iterations across European languages appears sparingly in comparison.[20] Rarity and curiosity were not always interchangeable, but were closely related in the early modern period.[21] As Neil Kenny has untangled, *curiosity*, in particular, is a complex word and concept with a high level of nuances in the many variations of the word across several European languages, even though

19 *Jardin de rares et curieux fleurs …*

20 For example, the frontispiece of the *Hortus floridus* (1614), see citation in note 34, states '*In quo* rariorum *& minus vulgarium florum Icones ad vivam veramq[ue] formam accuratissime delineatae.*' The term *horti curiosorum* is used for the descriptions for some plates, including the snowdrop plate (Fig. 5) that will be discussed in the next sub-section.

21 The words *rares* and *curieux* can be considered as synonyms to a large degree in the early modern period, as disentangled by Kenny, *Curiosity in Early Modern Europe*, 113.

they mostly convey a sense of desire.[22] *Curious* and *curieux* are not simply directly translated, and related words such as *wonder* and *marvel* further muddle the water.[23]

What is useful for understanding the genre of the florilegium is not a clear distinction between all these different words in English and other European languages, but noting that they still belong to the same semantic field, despite having a spectrum of meanings. It is reasonable that many flowers were *curious* because they were *rare* in the period. Rare flowers were not easily acquired, studied, and/or described and thus their specimens, images, and knowledge of them were desirable. Under this light, seventeenth-century florilegia do not simply contain a diversity and variety of flowers. Rather, they depict a diversity and variety of *rare* and *curious* flowers. After all, flower gardens constructed for pleasure have been part of garden culture since antiquity, but it was only in the late sixteenth century that flower books emerged, and they flourished in the seventeenth century.[24]

What the florilegium as a genre depicts is objects from a specific intersection of multiple phenomena and events that were developing side by side and around the same time: the thriving culture of curiosity collecting and the growing number of plant diversity and varieties that was largely made possible by the trade, colonisation, and/or horticultural activities of the period.[25] Curiosity cabinets more often invoke images of a packed room with natural objects, paintings, scientific instruments, and more, but outdoor spaces such as gardens and menageries were part of the collecting culture as well. Sometimes, a garden was an extended space for displaying overflowing items. More often, items in gardens were curated collections within themselves to reflect or represent a theme or concept, such as the biblical paradise.[26]

In several cases, seventeenth-century *liefhebbers* who commissioned florilegia to be made after their rare flowers also collected other curiosities. A slightly later example is Frederick III, Duke of Holstein-Gottorp (1597–1659) and his Gottorf Castle. Between 1649 and 1659, the duke commissioned the

22 Kenny, *Curiosity in Early Modern Europe*, 14.

23 Kenny, *Curiosity in Early Modern Europe*, 14.

24 See the general history of horti- and floriculture on a longer historical timeline in Goody, *The Culture of Flowers*.

25 Gill Saunders laid out these connections briefly in her book, here I delve deeper for the analyses, see *Picturing Plants*, 44, 48.

26 On the indoor and outdoor connections of curiosity cabinets, see particularly Hunt, 'Curiosities to Adorn Cabinets and Gardens,' 193–203; De Jong, 'Nature and Art. The Leiden Hortus as "*Musaeum*",' 37–60; Egmond, 'Precious Nature,' 47–65; Tice, 'Collecting in the Garden,' 315–331.

German painter Hans Simon Holtzbecker (1610/20–1671) from Hamburg to produce the Gottorfer Codex, a large florilegium with more than 350 sheets of parchment containing flower depictions based on what was growing in his garden.[27] In addition to the rare flowers, the duke purchased the collection of natural objects, artifacts, and antiquities that belonged to the Dutch physician Bernardus Paludanus (1550–1633), who was renowned for his curiosity collection; the duke also commissioned to have a Copernican celestial globe made.[28] How these indoor and outdoor curiosity collections may have connected to one another remains an area of uncertainty that will benefit from future research.[29]

Early modern collecting culture existed in paper form as well. The renowned paper museum of Cassiano dal Pozzo (1588–1657) is such an example. The Roman patron and collector Cassiano, together with his younger brother Carlo Antonio dal Pozzo, assembled a vast collection of approximately 10,000 watercolours, drawings, and prints.[30] The depicted subjects include antiquities, architecture, zoology, botany, geology, social customs and ceremonies, costumes, portraits, topography, and military maps, among others.[31] Considering the curiosity-collecting culture in both object and paper formats, it is not surprising that rare flowers with certain characteristics and their images became curiosity items of *naturalia* for either vast (indoor) collections or as standalone collections in the long seventeenth century.

The study of early modern curiosity and its collecting culture is a thriving field. From case studies of specific collections of well-known figures, and the mobility of curiosity objects, to a particular group of items being collected, the large body of literature branches into many aspects of curiosity and related terms and cannot be fully covered here.[32] Two pieces of work, however, provide

27 Hans Simon Holtzbecker, Gottorfer Codex, Copenhagen, National Gallery of Denmark (Statens Museum for Kunst), kksgb2947–kksgb2950, https://open.smk.dk/en/artwork/image/KKSgb2947.

28 De la Fuente Pedersen and Kolind Poulsen, eds., *Flowers and World Views*, 133, 192. For a comprehensive inventory and description of the Gottorf Castle as a site and collection, see Spielmann and Drees, *Gottorf im Glanz des Barock*.

29 Many florilegium commissioners are said to have curiosity collections, but usually no more than that remark.

30 For an overview of the collection, see the exhibition catalogue *The Paper Museum of Cassiano dal Pozzo*.

31 For more on each subject within the collection, see the many titles in the book series 'The Paper Museum of Cassiano dal Pozzo' by Harvey Miller / Royal Collection Trust.

32 Selected examples include Impey and MacGregor, eds., *The Origins of Museums*; Findlen, *Possessing Nature*; Bredekamp, *The Lure of Antiquity and the Cult of the Machine*; Smith and Findlen, eds., *Merchants & Marvels*; Von Schlosser, *Art and Curiosity Cabinets of*

useful starting points for framing rare and curious flowers within the early modern curiosity and curiosity-collecting culture. Larraine Daston and Katherine Park have unpacked the changing definitions of wonder and other relevant terms by examining how medieval and early modern naturalists understood the natural world. In particular, the authors discussed the opposition of garden cultivars as wonders of nature and wonders of art.[33] Building on this emphasis on garden cultivars as both natural and artificial, the following sub-sections and the next chapter will show that many rare and curious flowers and plants were not *found* curiosities, but cultivated ones through human interventions in natural organisms. Additionally, the edited volume by R.J.W. Evans, Alexander Marr, and their contributors presented a wide range of case studies to show what curiosity means within specific topics and participating groups of people.[34] An important issue that was brought to attention was the 'host of actors, such as artisans and merchants' that did not belong to the group of social elites, but were indispensable in the curiosity collecting culture.[35]

The genre of the florilegium was made possible by such a wide range of actors from several levels of early modern societies. While it took resources and means to construct a garden and fill it with living rare flowers, which tended to be only possible for highly privileged groups within a society, actors from many fields engaged with compiling and producing florilegia. Compared with other book genres, flower books have a somewhat fluid division of labour in the production process. Some projects have a more conventional structure of commission, such as the *Hortus Eystettensis* with a patron (the bishop) asking a compiler (Besler) and image makers (the drafts-people and engravers) to produce the book. This is similar to many manuscript florilegia, such as the Gottorfer Codex with a clear patron (the duke) and an image maker (Holtzbecker). Other flower books were produced by actors without their own rare flower collections and who acquired depictions through other means. The *Hortus floridus* (1614), which will be further discussed in the next section, was largely the product of the Dutch printmaker Crispijn van de Passe II (1594/1595–1670), with contributions from his printmaker father and brothers.[36]

*the Late Renaissance*; Bass et al., *Conchophilia: Shells, Art, and Curiosity in Early Modern Europe*.

33 Daston and Park, *Wonders and the Order of Nature*, 261–265.

34 Evans and Marr, eds., *Curiosity and Wonder from the Renaissance to the Enlightenment*.

35 Evens and Marr, eds., *Curiosity and Wonder from the Renaissance to the Enlightenment*, 15–16.

36 Crispijn van de Passe II, *Hortus floridus*. Bibliographic description numbered C53 in Bridson and White, *Plant, Animal & Anatomical Illustration in Art & Science*, 41. For the production history of this title, see Veldman, *Crispijn de Passe and His Progeny*, 205–206;

Van de Passe II compiled the text and made the images largely by drawing flowers in the gardens he visited throughout the Low Countries. Similarly, De Geest's Oak Spring volume appears to be entirely his own brainchild, as no concrete documentation of patronage has yet been found for this work.[37] As a draftsperson not recorded to have been related to wealthy or powerful figures, De Geest most likely would not have had the means for a rare flower collection himself and drew and painted his flowers by visiting local gardens instead.[38] Because of this complex cast of personnel, this book uses the term *compiler* to refer to the primary person that put together the pictorial collections of rare and curious flowers under discussion, regardless of the individual's profession as a bishop, an apothecary, an image maker, or anything else.

What precisely, then, made the flowers in seventeenth-century florilegia rare and curious to the people of the period? What characteristics did compilers need to identify in a flower to add it to a florilegium? The second word in the simple definition of a florilegium, as given in the general introduction, 'a collection or a selection of flower images,' is worth emphasising for spotting the rare and the curious. To select, the compiler needed to know what was desired or what qualified to be part of a curiosity collection. Identification, in this regard, is not to identify the species or functions of the plants, as in sixteenth-century herbals, but to identify the rare and curious qualities in flowers and decorative plants. The rest of this chapter discusses the three major characteristics a flower that could have that made them rare and curious specifically to the compilers of florilegia, and by extension to the plant collectors or garden enthusiasts who grew these flowers in their gardens. Not every flower in a florilegium has one or all three of the characteristics, but a seventeenth-century flower book always includes flowers that have these characteristics in the collection.

### 3.1 *Growing Distance*

A simple reason why some flowers were rare and curious in the early modern period is because they were foreign to areas to the north of the Alps. As only a small number of specimens would have been travelling, they were few in number and thus rare and curious. Additionally, most introduced flowers and

Segal and Alen, *Dutch and Flemish Flower Pieces*, vol. 2, 1011–1015. Van de Passe II might have enlisted the help of his brother to engrave some of the plates, indicated by the monograms on some plates, if additional hands were brought onto the project; see Tongiorgi Tomasi, *An Oak Spring Flora*, 60.

37 Tongiorgi Tomasi, *An Oak Spring Flora*, 84–87.

38 Gerda C. Huisman proposed some possibilities of gardens that De Geest might have visited around Leeuwarden and Friesland in 'Franciscus de Geest and His "Delightful Gardens",' 39–41.

plants would have been growing in a rather different environment than their original habitats, wild and domesticated species alike, so it was not always easy to keep them alive. While many flowers and plants are highly adaptive and they eventually multiplied, or even became naturalised, after a certain period of time, there was usually an extended moment for them to remain novelties and curiosities. This does not mean that florilegia compilers were not interested in flowers that were native to or already naturalised in (northern) Europe. Common or local flowers made appearances in flower books as well. For example, there is another flower book, titled *Hortus amoenissimus* and now preserved at the National Central Library of Rome, that is also attributed to Franciscus de Geest.[39] The Rome and Oak Spring volumes have many of the same rare flower depictions with the same design and composition. Additionally, the Rome florilegium includes some local flowers and plants that were not garden varieties, as they were common in the wild.[40] Nonetheless, the overall percentage of rare and curious flowers from afar, described as *exotic* on the titlepage, still outweigh the endemic ones in the *Hortus amoenissimus*.

The word *exotic*, in the early modern period, indicated not only the foreign attributes of natural and artificial objects that were not native to (northern) Europe, but also delightful, wonderous, and curious attributes that were desirable for collecting.[41] Many flowers that were already familiar or even common to sixteenth- and seventeenth-century gardens in areas north of the Alps, such as the pot marigold or calendula, were native to southern Europe and the Mediterranean and travelled north throughout the Middle Ages.[42] This practice of introducing foreign plants to northern European gardens continued in the early modern period (and to today), and the distances over which plants travelled grew ever longer. Southern Europe and the Mediterranean remained an active source, and plants from Asia Minor and the Americas added significantly to the diversity and variety of flowers and plants in European gardens, both to the north and south of the Alps.[43] The flowers and plants

39 Franciscus de Geest, *Hortus amoenissimus*, Rome, National Central Library of Rome (Biblioteca Nazionale Centrale di Roma), MS Varia 291 http://digitale.bnc.roma.sbn.it/tecadigitale/manoscrittoantico/BNCR_Ms_VARIA_0291/BNCR_Ms_VARIA_0291/201. Huisman discussed some possible figures as patrons or commissioners, but no certainty was concluded, in 'Franciscus de Geest and His "Delightful Gardens",' 47–48.

40 The titlepage writes, '[…] *tam exocitis, quam indigenis, ubique locorum, in hortis celebrrimis maximorum Botarophilorum,* […]' Both exotic (*exocitis*) plants and endemic (*indigenis*) ones are included in the volume according to the titlepage. This is also verified through the modern plant identifications by Sam Segal for the facsimile in Menghini, ed., *Hortus Amoenissimus Omnigenis Floribus, Plantis, Stirpibus. …*

41 Schmidt, *Inventing Exoticism*, 325; see also Chapter 4 'Exotic Pleasures,' 227–324.

42 Segal and Alen, *Dutch and Flemish Flower Pieces*, vol. 1, 54.

43 Saunders, *Picturing Plants*, 44.

introduced to (northern) Europe in the sixteenth century, and some before the mid-seventeenth century, were most relevant to both living and pictorial collections of rare and curious flowers. While other plant-based goods from South and East Asia, including indigo from East India, were largely imported to Europe mostly through the VOC and highly important to many facets of life in the same period as florilegia production, they were valued for their utilitarian qualities.[44] Living Asian plants, often with multiple qualities of medicinal, utilitarian, and decorative, would only become important to European gardens and plant collections in later centuries.[45]

Van de Passe II's *Hortus floridus* often mentions the places of origin, as assumed or understood during the period, for the depicted flowers. The *Hortus floridus* is a quarto-size book in horizontal format with close to 100 plates (usually 97 plates) of flowers. This florilegium is a highly complex book for bibliographical description.[46] While mainly attributed to Van de Passe II, the second half of the collection, the *Altera pars* that is almost always bound with the first half of the *Hortus floridus*, was most likely produced by his father and for another project.[47] Aside from the alterations for reprints in 1616 and 1617, which is usual in all types of books, there are several editions in Latin, Dutch, French, and English.[48] The editions in different languages are not straightforward translations, as they often contain different types of information in the text. The 1614 Latin edition with descriptions of flowers is of focus here, but other editions will become important in later chapters. Even though Van de Passe II was writing from a very specific perspective of what were rare and curious to gardens in the Dutch Republic, this aspect of rareness and curiousness caused by a flower's foreignness would have been largely applicable to gardens in the Spanish Netherlands and Germany, and, to an extent, the northern part of France. This is indicated by the fact that most seventeenth-century florilegia compilers selected similar or the same flowers as in the *Hortus floridus*, despite not having explicit and textual information to outline their rareness and curiousness.

Some of the flowers Van de Passe II depicted and considered rare and curious were wildflowers in southern Europe. For example, the compiler included

---

44 See especially Nadri, *The Political Economy of Indigo in India, 1580–1930*.

45 Indian plants were highly interesting for naturalists of the seventeenth century, but they were not done by studying plants in European gardens, see eg. Friedrich, Brendecke, and Ehrenpreis, eds. *Transformations of Knowledge in Dutch Expansion*.

46 See the thorough detangling by Savage, 'The Hortus Floridus of Crispijn Vande Pas the Younger,' 181–206; Segal and Alen, *Dutch and Flemish Flower Pieces*, vol. 2, 1012–1015.

47 Veldman, *Crispijn de Passe and His Progeny*, 208–210.

48 Segal and Alen, *Dutch and Flemish Flower Pieces*, vol. 2, 1012–1015.

two varieties of asphodels (*Asphodilus; Asphodelus*), one is bigger and has white flowers (*maior albo flore*) and one has yellow flowers (*flore luteo*). In addition to the descriptions of the forms and colours of the flowers, he wrote, 'In French Provence, Italy and Spain it grows wild: in Germany and Belgium it occurs only when sown.'[49] This shows that the asphodel was only available in the Low Countries and Germany due to cultivation, and what was common (which does not mean that it was not admired) in warmer places in Europe was rare and curious to northern gardens. Several entries in the volume further tie the connection of rare flowers to curiosities more explicitly. For the plate of a purple clematis (*Clematis altera Purpureo flore*), Van de Passe II wrote, 'In Italy and Spain it grows freely by the sides of hedges and roads; but with us Dutch it is only found in the gardens of the curious.'[50] Similarly, the compiler included

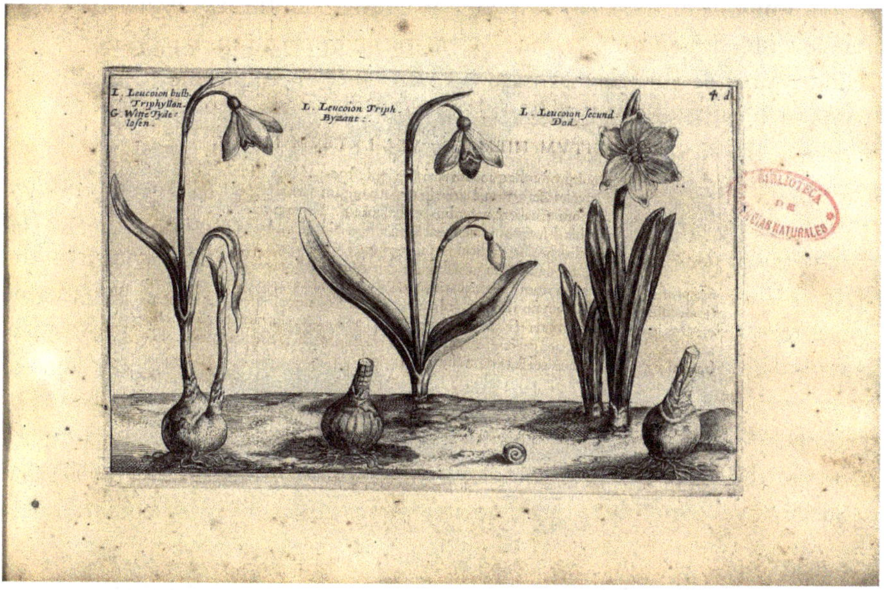

FIGURE 5 Snowdrop plate from the *Hortus floridus* (1614), Biblioteca Real Jardín Botánico, Madrid, Spain, A PAS, https://bibdigital.rjb.csic.es/idviewer/13646/207

49 '*In Gallie Provincia, Italia Hispania, sponte nascitur: apud Germ: & Belgas, non nisi satus provenit.*' in Van de Passe II, *Hortus floridus*, XXXVI (spring garden). The cited quotes for the *Hortus floridus* are all translated by Spencer Savage in *Garden of Flowers: All 104 Engravings from the Hortus Floridus of 1614*, 72.

50 '*In Italia & Hispania sponte passim juxta sepes & vias provenit; apud Belgas vero nostros, non nisi in hortis curiosorum invenitur.*' in Van de Passe II, *Hortus floridus*, [1] (autumn garden); Savage, *Garden of Flowers*, 148.

three kinds of snowdrops (*Galanthus* L.) in the florilegium (Fig. 5); the first one, the *Leucoion bulb. Triphyllon* as engraved on the plate, '[…] abounds in Italy; but is not to be found here except in the gardens of the curious.'[51]

Further away geographically, Asia Minor was a major source of rare and curious flowers for (northern) European gardens. The Ottoman Empire had an active garden culture and a love for flowers. The development of early modern European horticulture was largely enabled by travelling Europeans bringing garden flowers and plants home with them from the empire.[52] A dozen flowers and their variations that commonly made appearances in florilegia—crocus, autumn crocus (*Colchicum* L.), snowflake (*Leucojum* L.), dogtooth violet (*Erythronium* L.), star-of-Bethlehem (*Ornithogalum* L.), cyclamen, allium, hyacinth, lily, fritillary, ranunculus, and tulip—had already arrived in Europe before the 1560s from or through the Ottoman Empire. This introduction of a large number of flowers and plants to Europe continued until around the mid-seventeenth century.[53] The flowers' place of origin in Asia Minor is also regularly provided by Van de Passe II in many entries, usually indicated through words such as *Persicum, Constantinopoli,* or *Byzantinum*. The depiction in the middle of the snowdrop plate (Fig. 5) is a variation from the Ottoman Empire, as shown in the name *Leucoion Triph. Byzant*.[54] Aside from the genus *Tulipa* (L.) and the obvious craze for it, the many flowers in the genus *Fritillaria* (*L.*) form some of the early modern favourites. The crown imperial (written on the plate as the *Corona imperalis*), according to Van de Passe II, is a remarkable flower for its double tiers of flowers. The flower also travelled a long distance, as it was 'foreign to us, first sent by Clusius from Constantinople, but Persian in origin, and hence called also by him Persian lily, belongs to the genus of wild lilies.'[55] There was, however, an even rarer 'Persian lily,' as the compiler wrote for the entry of the *Lilium Persicum* (*Fritillaria persica* L.), 'This Persian or Susian lily is certainly to be set down among exotic and foreign plants, nor is

51 '*In Italia frequens est, hic nisi in hortis curiosorum minime invenitur.*' in Van de Passe II, *Hortus floridus*, 4 (winter garden); Savage, *Garden of Flowers*, 212.

52 An important piece on this subject is Atasoy, 'Links Between the Ottoman and the Western World on Floriculture and Gardening,' 61–79. See also Pouryan, 'Iranian World Plant Species in the European Network of Botanical Information Exchange in the Sixteenth Century,' 351–396.

53 Atasoy, 'Links Between the Ottoman and the Western World on Floriculture and Gardening,' 73–74.

54 Printed as *Leucoium bulbosum Byzantinum* in the description. Van de Passe sometimes gave a flower slightly different Latin names in the descriptions and on the plates.

55 '*Corona Imperialis, peregrina nobis planta, primum a Clusio Constantinopoli missa, sed origine Persica, quare & ab eodem lilium Persicum dicta, ex genere silvestrium liliorum;* […]' in Van de Passe II, *Hortus floridus*, XII (spring garden); Savage, *Garden of Flowers*, 24.

it very common but only found in the better-cultivated gardens.'[56] These examples show that, travelling a longer distance than southern Europe, flowers from Asia Minor added greatly to the diversity and variety in living and pictorial rare flower collections.

From the other direction, across the Atlantic Ocean, the Americas provided even more diversity and variety of rare and curious flowers, usually through the Spanish connection.[57] The new diversity and varieties from southern Europe and Asia Minor tended to be decorative flowers without major utilitarian functions. Comparatively, there is a higher percentage of utilitarian plants from the Americas grown as curiosities. For example, the *Hortus Eystettensis* included plates of the potato plant, variations of tomatoes, chilli peppers, tobacco plants, cacti, and so on.[58] Similarly, cacti occupied several leaves of the Gottorfer Codex. Van de Passe II also included the tobacco plant with a detailed morphological description to accompany the engraving of the tobacco flowers and indicated the place of origin as Peru through the name *Hyoscyamus Peruvianus* on the plate.[59] With many American plants, and to some extent those from Asia Minor, the stated place of origin does not always reflect reality, as there were often several degrees of transfers by the time a plant travelled from afar and settled in a European garden. Peru is the word of choice for Van de Passe II to refer to South American flowers and plants, but it does not mean that they came from the geographical region that is known as Peru today.

For example, the sunflower is called *Chrysanthemum Peruvianum* (Peruvian chrysanthemum), and two variations appear in the *Hortus floridus*. The compiler wrote that the flower was 'Discovered in Peru and in other countries of America, it is brought from thence.'[60] Many of Van de Passe II's flower descriptions are paraphrases of or references from the herbals of sixteenth-century physicians and botanists Rembert Dodoens (1517–1585) and Carolus Clusius,

56 *'Lilium hoc Persicum vel Susianum inter exoticas & peregrinas omnio plantas referendum, nec valde frequens est, sed tatum in cultioribus hortis invenitur;* [...]' in Van de Passe II, *Hortus floridus*, XXV (spring garden); Savage, *Garden of Flowers*, 50.

57 On travelling objects from the Americas, including plants and their images, that contributed to knowledge production since he early modern period, see works such as Bleichmar, *Visual Voyages: Images of Latin American Nature from Columbus to Darwin*; Thurner and Pimentel, eds., *New World Objects of Knowledge*.

58 Plates 345 (potato); 319, 320 (tomatoes); 324–331 (chili peppers); 312, 340, 341 (tobacco plants); 354, 359, 360 (cacti) in Littger and Dressendörfer, *Florilegium: The Book of Plants*.

59 The name is printed as the *Petum maius, siue latifolium* in Van de Passe II, *Hortus floridus*, 12 (autumn garden); Savage, *Garden of Flowers*, 170.

60 *'In Peru alijsque Americae Provintijs repertum sertur.'* in Van de Passe II, *Hortus floridus*, 2 (autumn garden); Savage, *Garden of Flowers*, 152.

and this will be further discussed in Chapter 3. Dodoens named the sunflower *Chrysanthemum Peruvianum*, and his description and depiction constitute one of the first published works on the flower in Europe. However, the Low Countries was one of the later places to grow sunflowers. The physician-botanist wrote that he saw a live one in the garden of a certain Jean de Brancion in Antwerp, but it was much smaller than the specimen in the royal garden in Spain. Dodoens' knowledge of the sunflower in Spain was not firsthand, but came from Christine Bertolf, known as Madame Hopperus, whose husband worked for the Spanish court. Madame Hopperus may have also given Brancion the very seed(s) that grew to be the sunflower observed by Dodoens.[61] The sunflower(s) in Antwerp most likely came from Spain instead of directly from the Americas, but whether the sunflowers growing in Spain originally came from Peru, Mexico, or other places in South America is less certain.[62] Regardless of the specific geographical locations from which the flowers and plants came, American plants became a major addition to the diverse and varied rare and curious plant collections in European gardens.

### 3.2 *Growing Petals*

The second characteristic that contributed to a flower's rareness and curiousness was an unusually large number of petals on its head, such as the double flower phenomenon. The definition of a double flower can be highly specific and complex. For this book, a double flower simply refers to a flower that has more petals or petal-like structures than the typical ones in their wild state.[63] The extra petals give a bloom a fuller appearance, basically doubling its usual volume. A double flower is caused by mutation, and the phenomenon can occur in the wild. More often, however, it is a cultivated effort, as double-flowered plants are sometimes sterile or produce only limited viable seeds.[64] Double flowers had been a known phenomenon since antiquity, with Pliny the Elder writing that double roses were common in the Roman Empire.[65] In the early modern period, the number of recorded kinds of double flowers increased

61 See more on this account and other similar plant exchanges in Egmond, *The World of Carolus Clusius*, 53–54.

62 Charles B. Heiser spent quite some of his career on studying the long historical migration of the sunflower, see 'The Sunflower among the North American Indians,' 432–448; 'The Origin and Development of the Cultivated Sunflower,' 161–167; 'The Domesticated Sunflower in Old Mexico?' 447–449.

63 Ferguson, *Double Flowers*, 33.

64 Ferguson, *Double Flowers*, 14–15.

65 Ferguson, *Double Flowers*, 13.

because, in theory, this phenomenon can occur in almost any flower.[66] Doubles of the flowers that were already common in northern European gardens and of the new species that travelled from afar (usually from the direction of Asia Minor) both regularly appear in florilegia, but with a higher percentage of double-flower varieties belonging to the former group. The doubles of the then-already common garden flowers were not necessarily low in quantity. For example, Van de Passe II included several kinds of roses, remarking that among all of the flowers, the roses are easily the most celebrated. One of the variations is the white rose with a full flower (*Rosa alba, pleno flore*), and 'The greater part of this variety of plants is abundant.'[67] Nonetheless, all the varieties in the *Hortus floridus* are double flowers, and there is no pictorial representation of a single rose in the collection.

The range of petal counts among the double flowers is considerably wide. Some of the doubles in florilegia have twice as many petals as the usual flowers, whereas others have many more times as many petals. A variation of the slender-leaved narcissus (*Narcissus tenuifol*) is described as 'with white somewhat double flower' (*flore albo, pleniore*). Van de Passe II wrote that this flower is only found in gardens and has '[…] six broad keeled white petals displayed in the form of a star, the centre of which, in place of a cup, six other smaller petals occupy.'[68] This variety merely doubles its petal count. In contrast, the compiler described a variation of the slender-leaved anemone (*Anemone tenuifol*), the 'Byzantine with many-petalled flower' (*Byzantina flore polyphillo*), as such:

> [T]he flower is remarkable with ten or more broader downy leaves round the circumference, on the outside pale-red but inside resplendent with a deeper colour; a larger and almost endless number of slender smaller petals, of a crimson or even pale purple colour, fill the centre of the flower.[69]

---

66 There is a double sunflower in oil painting by Bartolomeo Bimbi, but double sunflower was not encountered in florilegia during this research. See short discussion on the Bimbi double sunflower and other sunflowers in Italian collections in Tongiorgi Tomasi and Hirschauer, *The Flowering of Florence*, 99–101.

67 'Plurima haec varietate saecunda est.' in Van de Passe II, *Hortus floridus*, 12 (summer garden); Savage, *Garden of Flowers*, 126.

68 '[…] *sex latis & carinatis folijs albis stellari forma extantibus, cuius medium loco calicis alia sex minora folia itidem candida nonnihil flavescentis coloris hinc inde sparsim mostrantia, occupant;* […]' in Van de Passe II, *Hortus floridus*, XXI (spring garden); Savage, *Garden of Flowers*, 42.

69 '[…] *flos a decem pluribus, amplioribus ac lanuginosis in circuitu folijs, exterius suaverubentibus, at interius saturatiore colore splendentibus, conspicuous; cuius medium plura ac paene insinita tenuia foliola, sanguinei coloris vel etiam dilute purpurascentis, implent.*' in Van de Passe II, *Hortus floridus*, XIIII (spring garden); Savage, *Garden of Flowers*, 28.

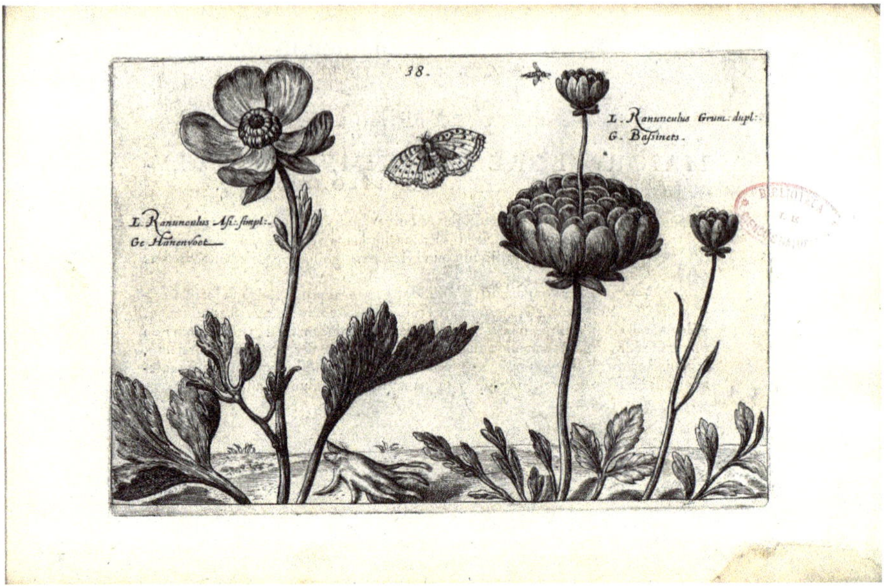


FIGURE 6 Ranunculus plate from the *Hortus floridus* (1614), Biblioteca Real Jardín Botánico, Madrid, Spain, A PAS, https://bibdigital.rjb.csic.es/idviewer/13646/85

For the two poppies 'with a manifold flower' (*Papaver multiplici flore*), the compiler described them as full of 'countless little petals split at their ends' on the flower heads.[70] From countable to countless, flowers with ever-growing petal numbers appear time and time again in seventeenth-century florilegia.

Within double flowers, an even rarer phenomenon that adds more petals to a flower is when one is proliferous, describing when a flower has a smaller bloom stalking out of the first bloom, usually from its centre (Fig. 6).[71] This phenomenon is seen in only a handful of examples.[72] In the *Hortus floridus*, the proliferous Asiatic ranunculus (*Ranunculus Asiaticus, flore pleno sanguineo geminato*) elicits one of the rare occasions when Van de Passe II's words are more illustrative than his images (Fig. 6). While the image shows a small

70 '[…] *infinitis minoribus coacervatis in medio foliolis repletum candidi vel subalbidi coloris.*' and '[…] *ex infinitis quoque foliolis in extremitate scissis plenum,* […]' in Van de Passe II, *Hortus floridus*, 15 (summer garden); Savage, *Garden of Flowers*, 132.

71 Published about a century after the most printed florilegia, a small book by a certain John Hill presented a study of proliferous flowers that could apply to those that were sought after in florilegia, see *Of the Origin and Production of Proliferous Flowers*.

72 For a short study on a proliferous carnation within the circle of Holtzbecker and his patrons, see De la Fuente Pedersen and Kolind Poulsen, eds., *Flowers and World Views*, 206–207.

double flower stalking out of a larger double ranunculus from a side view, the compiler described,

> The double flower of the second *Ranunculus* of the same colour, is so remarkable with its multiplicity of petals, that it is found far to exceed the hundredth number of petals, and what is surprising, puts forth from the centre of the flower another thin stalk which is fruitful with a similar but smaller floret.[73]

In addition to proliferous double ranunculi, the Gottorfer Codex has a couple of proliferous double pot marigolds (Fig. 7). The second flower from the left shows the same two-tier structure as Van de Passe II's ranunculus, with a smaller bloom growing out of the centre of the first one. The figure next to the proliferous one on the left, however, shows another kind of proliferous pot marigold. Instead of a single sub-flower from the centre, there are multiple small flowers peeking out from around the bottom of the first flower head.[74] There is a similar depiction in the *Hortus Eystettensis*, for which Besler praised the proliferous variety as one of the most elegant of its species.[75] As proliferous flowers are observed to only occur among double flowers, this made them highly sought-after rarities and curiosities.

### 3.3 *Growing Colours*

The third major characteristic that added to the rareness and curiousness of a seventeenth-century flower is its colour (s). More specifically, its unusual colour (s). While some flowers naturally have multiple colours in one species, the early modern period saw a growing number of new and rare colours becoming available, adding greatly to flower varieties. Newly introduced species from faraway places may have had different colours than the endemic ones, such as irises from Asia Minor having different hues and shades from those native to

73 '*Alterins Ranunculi flos plenus eiusdem coloris, adeo multiplicatis folijs conspicuus, ut centisimum foliorum numerum longe superasse repertus sit, & quod mirum, ex floris medio alium tenuem cauliculum profert simili quamvis minore flosculo faecundum.*' in Van de Passe II, *Hortus floridus*, XXXVIII (spring garden); Savage, *Garden of Flowers*, 76.

74 Proliferous pot marigolds or calendula, in particular, also strike as *monstrous* in appearance. This may have been another layer to have certain flowers in a curiosity collection, similar to what has been argued for citrus fruit, but there is no text in florilegia that explicitly connects mutated flowers to the concept of monster/monstrous. See Freedberg and Baldini, *Citrus Fruit*.

75 Besler, *Hortus Eystettensis*, quintus ordo. fol. 3 (summer garden); Littger and Dressendörfer, *Florilegium: The Book of Plants*, plate 206.

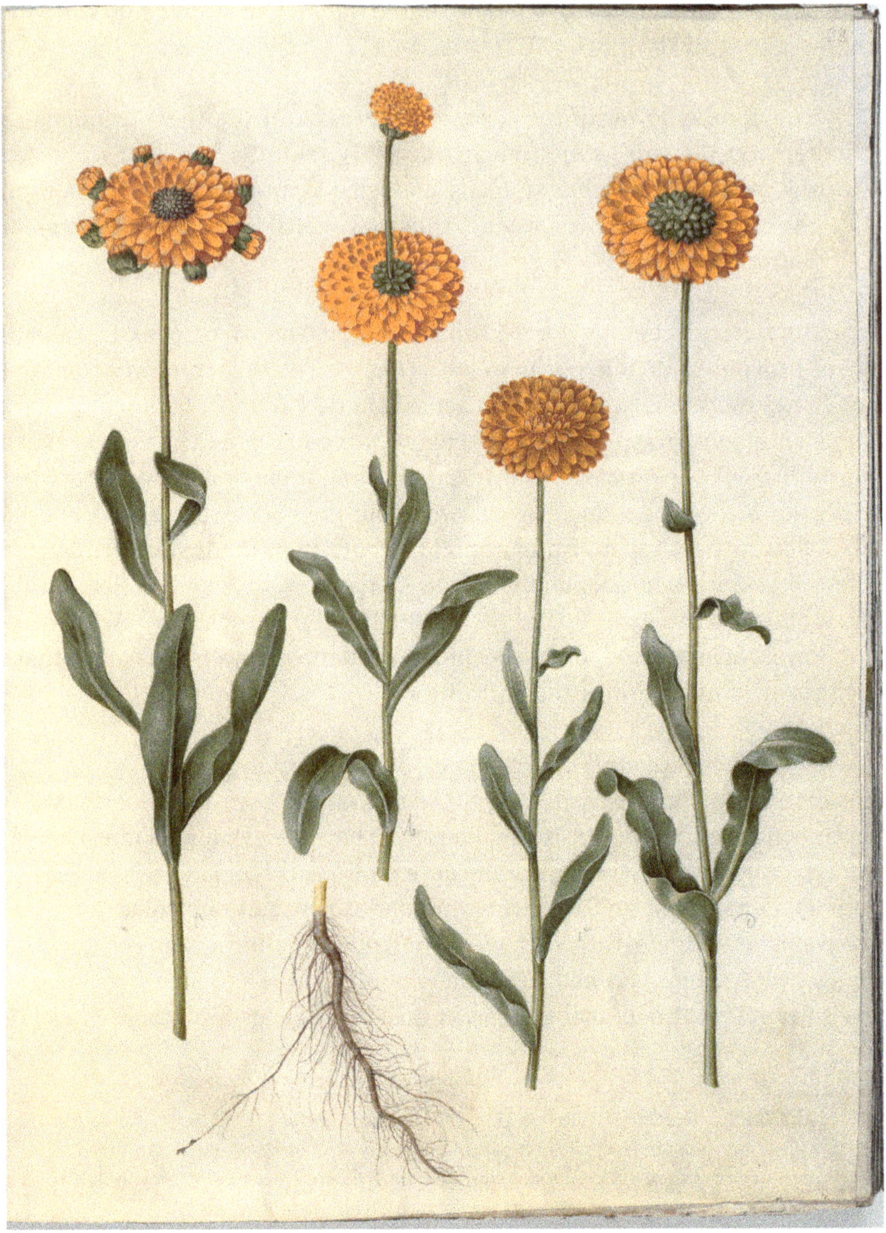

FIGURE 7 Pot marigolds from the Gottorfer Codex (1649–1659), SMK, National Gallery of Denmark, Copenhagen, KKSgb2950/28

the Low Countries. Moreover, many of the new colours were created through hybridisation in cultivation, but the reason behind the new colours was not always known.[76] For example, Van de Passe II presented two rose champions (*Lychnis coronaria*). One is a single rose champion 'with white flower' (*flore albo*) and another a double 'with a full red flower' (*flore rubro pleno*).[77] The two colours of red and white can be found in sixteenth-century herbals, but both the double red and the white appeared later in the century. Two herbals that will become particularly important in the next chapters show this trajectory. In 1568, Dodoens published a small herbal titled *Florum, et coronariarum odoratarumque nonnullarvm herbarum historia*, in which the aforementioned description and depiction of the sunflower first appeared.[78] This small herbal was published as the second instalment before all of Dodoens' small books were compiled, decades later, into his famous big herbal, the *Stirpium historiae pemptades sex* (1583) and its Dutch translation the *Cruydt-Boeck* (1618). The *Florum* describes only fragrant and decorative flowers, which was a departure from works of this period, and it is considered a blueprint for what would become the genre of the florilegium.[79] Dodoens recorded only red (*rubro*) for the flowers of the rose champion.[80]

More than ten years later in 1581, Dodoens' contemporary, the Flemish physician and botanist Matthias Lobelius (1538–1616), provided in another herbal, the *Kruydtboeck*, the same information of the red flower in the text and used the same woodcut of the rose champion as Dodoens for illustration. Additionally, two more woodcuts were added, one with the caption 'double red rose champion' (*Dobbel roode Christus-ooghen.*) and another with 'white and red single rose champion' (*Witte en roode enckel incarnaat Christus-ooghen.*).[81] Dodoens and Lobelius belonged to the same circle in-and-around Antwerp and studied flowers and plants in more-or-less the same gardens, and their herbals reflected the growing flower colours visible at these sites. Van de Passe II further selected to show only the two new varieties, but not the common single red rose champion.

76 Records of how tradespeople and practitioners did so usually appear later and became much more documented after the eighteenth century, see Kingsbury, *Hybrid: The History and Science of Plant Breeding*.

77 Van de Passe II, *Hortus floridus*, 16 (summer garden); Savage, *Garden of Flowers*, 134.

78 Dodoens, *Florum, et coronariarum odoratarumque nonnullarum herbarum historia*, 295–298.

79 Particularly argued by Wheelock Jr., *From Botany to* Bouquets, 25.

80 Dodoens, *Florum*, 42–45.

81 Lobelius, *Kruydtboeck*, 408–409.

Many flowers see this growth in colours, and the number of new varieties differs from flower to flower. While the rose champion grew only one more colour in sixteenth- and seventeenth-century records, a number of flowers were known for their wide range of colours during this period, including the anemone, carnation, and tulip. The number of colours grew both in terms of hues and shades, with florilegia paying ever more attention to subtle colour changes. The anemone is a significant example and will make recurring appearances throughout the rest of this book. The *Anemone* L. is a vast genus and there was distinction between cultivated and wild anemones already in antiquity. The major species that experienced this rapid increase in colours in the early modern period is now classified as the poppy anemone (*Anemone coronaria* L.).[82] The use of the word *anemone* in this volume, however, stands for all the flowers that were given this name in flower books, regardless of their classification today, with a high percentage referring to the poppy anemone.

The *Tulipa* (L.) is similarly a vast genus, but all the flowers that were called tulip are recognisably tulips despite their colour and structural differences. Flowers that were called *anemones* in florilegia, on the other hand, were sometimes given different names by other groups, such as the physician-botanists, and can have rather different appearances from one another. The Ottoman Empire and other places in Asia Minor introduced many anemone cultivars that had more petal counts and in different colours, such as the aforementioned red slender-leaved anemone of 'Byzantine with many-petalled flower' (*Byzantina flore polyphillo*), which includes all three characteristics of rareness and curiousness discussed here. The cultivation of anemones was equally active in Europe, as it is well-documented that many gardens were known for their specific cultivars. For example, the Italian Duke Francesco Caetani's (1594–1683) specialty was double anemones, and some colour variations in his garden are elaborately described as 'all scarlett and pale yellow' and 'the first flowers are white and purple, and last dove-coloured and peach.'[83]

Returning to herbals, Dodoens did not include an exhaustive description of the anemone in his 1568 *Florum*, but drastically expanded the section for the 1583 *Stirpium historiae pemptades sex*. In the *Florum*, the physician provided one woodcut of a cultivated anemone, describing the colour of the flower

82 An older study by Katharine Johnstone is useful here, as she also considered historical images for understanding anemone cultivars with some historical contexts in 'Some Cultivated Anemones and Their Histories,' 14–28.

83 Masson, 'Italian Flower Collectors' Gardens in Seventeenth-Century Italy,' 76–77. France, similarly, produced many single and double cultivars, including anemones with colours of fire, rosy pink, white, browns, and other colours, see Hyde, *Cultivated Power: Flowers, Culture, and Politics in the Reign of Louis XIV*, 67–69.

as a light red-purple (*diluto rubore purpureos*).[84] In contrast, the 1583 herbal included five woodcuts of anemone varieties.[85] The recorded colours now include several kinds of purples, such as the same light red-purple (*dilutiore purpura rubens*), blue-purple (*caerulei & cyaneo*), and dark violet (*violae nigrae purpureo*).[86] Moreover, 'the colour of the red poppy' (*Rhoeadis colore rubens*), white (*candidus*), and 'the yellow of ranunculus' (*Ranunculi floribus concolor*) are described as the colours for different anemone varieties as well.[87] Similar to the herbal descriptions, different shades and tones of purples and reds form the majority of anemone varieties across florilegia, with white and occasionally yellow in the mix. For example, the Gottorfer Codex shows more than 70 variations of single and double anemones (Fig. 8), displaying such an array of purples and reds, some whites, and one very light yellow. Van de Passe II presented fourteen anemone variations across six plates. The observation that there are many colours of anemones appears frequently in the text, with one stating that in the flower 'so great variety of colours is seen that it is very difficult to express in words.'[88] The fourteen anemones include six that have predominantly one solid colour across the whole flower that can more-or-less be described in one colour term; the rest consist of multiple colours in one flower that take a much longer sentence to describe. The six solid colours include violet (*violaceo*), scarlet (*coccineo*), and crimson (*sanguineo*), which are red and purple variations; pale-red (*suaverubenti*) and flesh colour (*carnei color*), which are pastel coloured anemones; and orange (*orengiaco*).[89] Dodoens and Van de Passe II used rather different descriptive terms for the colours, but they both show that the simple words of red and purple would no longer suffice to describe the growing variation of shades and tones that could be found in the flowers.

---

84 Dodoens, *Florum*, 292.

85 Dodoens, *Stirpium historiae pemptades sex*, 430–433, including one woodcut for the *Anemone Pannonica* and one for the *Anemone trifolia*.

86 Dodoens, *Stirpium historiae pemptades sex*, 430–433.

87 Yellow anemones are rare. De Geest depicted a yellow anemone with the caption '*Anemona met een solffer cleur der Bloem*' on fol. 21 in the *Hortus amoenissimus*. However, the yellow anemone occupies the only area on the page where the paper is not severely deteriorated; the colours of the painted flowers on this page might not have been the same to early modern viewers.

88 '*In latifoliarum Anemonum floribus tanta conspicitur colorum varietas, ut verbis exprimere sit valde difficile;* [...]' in Van de Passe II, *Hortus floridus*, XV (spring garden); Savage, *Garden of Flowers*, 30. In this sentence, Van de passe II wrote the anemones as 'broad-leaved,' but the description is for the plate of three slender-leaved anemones.

89 Van de Passe II, *Hortus floridus*, XIII, XIIII, XV, XVI, XVII, XVIII (spring garden); Savage, *Garden of Flowers*, 26–36.

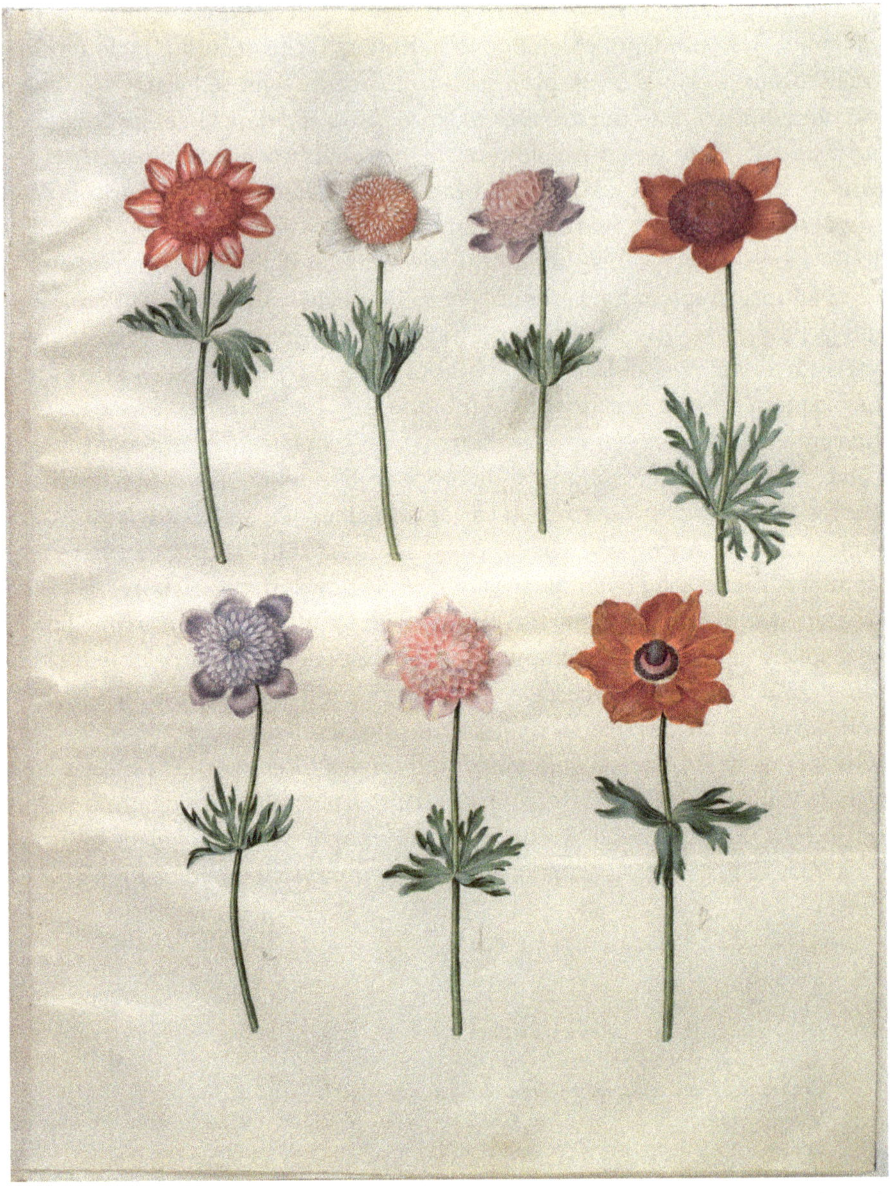

FIGURE 8 Poppy anemones from the Gottorfer Codex (1649–1659), SMK, National Gallery of Denmark, Copenhagen, KKSgb2949/35

The growing number of colours does not only refer to the varieties of one kind of flower, such as the range of red and purple anemones, but also to the number of colours that can appear on the petals of a single bloom. The eight anemones that Van de Passe II described with sentences instead of a single-colour term show a minimum of two ways that petals can have multiple colours. For the slender-leaved anemone variety 'pavo major multiplex' (*Anemone tenuifol. Pavo maior multiplex*), the compiler wrote,

> [T]he colour of which is a rather deep purple-red, with a crimson base passing into a yellowish circlet around the central portion.[90]

This gradation from one colour into another can be seen on the anemone at the bottom right corner of a parchment leaf in the Gottorfer Codex (Fig. 8), possibly depicting the same variation as the one described by Van de Passe II. Many rare and curious flowers in florilegia feature similar colour gradations, often going from one colour to white, but sometimes from one hue to a very different hue.

Yet, the most significant kind of growing colours are variegated petals, often described as *flamed, striped*, or *feathered*. The tulips that were most sought-after were flamed tulips with stripes of multiple colours on the petals (Fig. 4). Instead of being cultivated through breeding and hybridisation, the flamed tulips were a result of virus infection that degraded the petals.[91] This cause was unknown and irrelevant in the seventeenth century, as it simply generated more colours and thus rareness and curiousness in the tulips. Additionally, anemones and carnations were known to have many variegated varieties obtained through cultivation. For example, the *Hortus floridus* has two striped double broad-leaved anemones (*Anemone latifol*). One of the varieties is called 'Chalcedonica maxima many flowered' (*Chalcedon. Max. Polyanthos.*) by the compiler, writing,

> [A] broad and spreading flower, of many petals, of which the ten or twelve outer are green, remarkable for a network of scarlet veins, the

90 '[…] *cuius color Phaeniceus saturatior, sundo sanguineo in circulum flavescente circa umbilicum* […]' in Van de Passe II, *Hortus floridus*, XVI (spring garden); Savage, *Garden of Flowers*, 32.

91 De la Fuente Pedersen and Kolind Poulsen, eds., *Flowers and World Views*, 200. Anna Pavord discussed, in greater details, of the criteria that made certain colourations and/or configurations of tulips more desirable in the seventeenth century, *The Tulip*, 88–95.

> inner petals are smaller and more slender, brilliant with a somewhat pale purple and soft red colour.[92]

A hand-coloured copy of the *Hortus floridus* shows a rendition that follows this description closely (Fig. 9).[93] The original colouring has deteriorated, indicated by the grey that is often the result of tarnished white paint(s). Nevertheless, the anemone on the left clearly shows a flower with the most outer layers of petals in green, and the rest in stripes of different colours. Multiple colours within one single flower increased its rareness and curiousness, making feathered flowers something sought after for a living or pictorial curiosity collection.

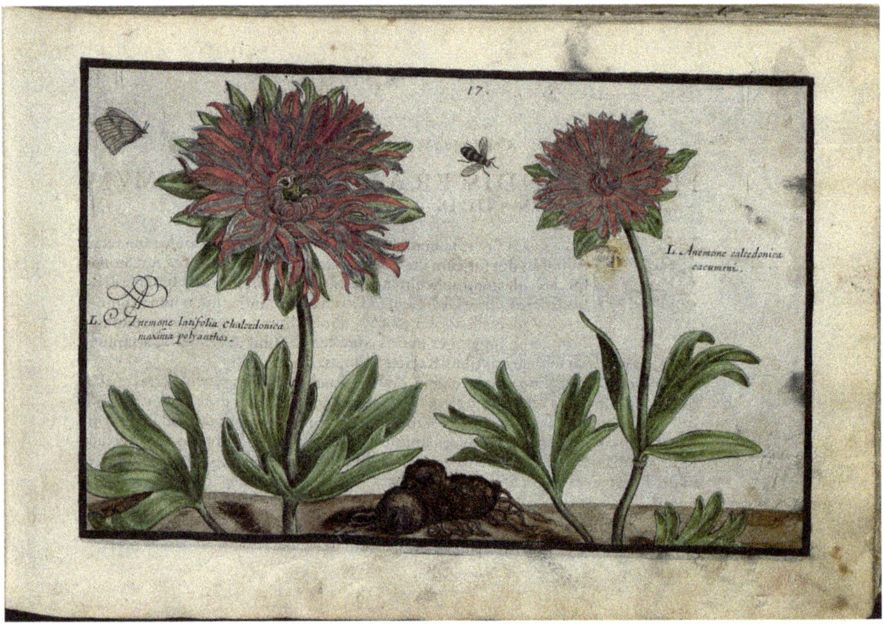


FIGURE 9 Hand-coloured double poppy anemones from a copy of the *Hortus floridus* (1614), University and State Library of Saxony-Anhalt, Halle (Saale), 143059610, https://nbn-resolving.org/urn:nbn:de:gbv:3:3-42128/fragment/page=0021

92 '[…] *florem etiam latum & extensum, multorum foliorum, quorum decem vel duodecim exteriora viridia sunt, venis coccineis intermeantibus, distincta;* […]' in Van de Passe II, *Hortus floridus,* XVII (spring garden); Savage, *Garden of Flowers*, 34.

93 Van de Passe II, *Hortus floridus*, Halle (Salle), University and State Library of Saxony-Anhalt (Universitäts- und Landesbibliothek Sachsen-Anhalt), 143059610, https://digital.bibliothek.uni-halle.de/hd/content/pageview/1664794, plate 17 (spring garden).

## 4 Conclusion

This chapter has shown that the seventeenth-century florilegium as a genre is not simply a gathering of random images of flowers. Producing a florilegium required the compiler to balance the diversity and variety of flowers and plants with decorative qualities that were rare and curious. While sometimes wild species were included, rare and curious flowers that were depicted in florilegia were usually grown and cultivated in gardens. Remote places of origin, higher petal counts, and/or unusual colours were the three major (albeit not only) characteristics of rareness and curiousness in the long seventeenth century.

It is important to reemphasise that these three major characteristics were rare and curious specifically to seventeenth-century florilegia compilers and *liefhebbers*. Other groups, such as the physician-botanists of the time, valued different kinds of rareness and curiousness.[94] Additionally, flowers with the three characteristics were rare and curious only between the second half of the sixteenth century and the end of the seventeenth century. By the eighteenth century, many flowers discussed in this chapter had become common, albeit still beloved. For example, published in 1760, *The Florist* is a collection of etchings containing sixty plates of flowers and instructions on how to colour them.[95] The small book includes essentially the same flowers as those in seventeenth-century florilegia, such as a sunflower, a double anemone, and a feathered tulip, among many more. However, the flowers were no longer rare, as described for the sunflower, 'The Petals of this noble, though common Flower, are a fine yellow, […]'[96] Other flowers, usually cultivars obtained through hybridisation, ceased to exist after a while due to not being able to reproduce.

Taking the flowers' place were different plants that became newly rare and curious to garden enthusiasts in the eighteenth century. For example, a collection of plant images now generally known as the Badminton Florilegium (1703–1705) belonged to Mary Somerset, the Duchess of Beaufort (1630–1715),

94 A prominent example is the publication that was made after the rare plants from the Hortus Medicus Amsterdam that grew many plants from Asia starting in the late seventeenth century, see Commelin, *Horti Medici Amstelaedamensis plantae rariores et exoticae*; Wijnands, *The Botany of the Commelins*.

95 Sayer, *The Florist: Containing sixty plates of the most beautiful flowers regularly disposed in their succession of blowing. …* See also Tongiorgi Tomasi, *An Oak Spring Flora*, 336–338.

96 Sayer, *The Florist*, plate 71. Even the *Neues Blumenbuch* (1680) by Maria Sibylla Merian, as mentioned in Introduction of this book already did not emphasise the rareness and curiousness of these flowers, but still beloved.

who had a renowned garden and plant collections.[97] Somerset's florilegium includes some watercolours of seventeenth-century rarities and curiosities, such as the sunflower, the crown imperial, and some crocuses and snowdrops. The focus, however, is shifted to tropical plants and succulents, many of which were among the first cultivated specimens in England.[98] In many ways, the Badminton Florilegium selected plants with similar characteristics—coming from faraway places and with unusual features—to those in seventeenth-century flower books, but valued them in very different decorative plants. Florilegia, therefore, visually documented the objects from a specific branch of plant and horticultural interest, of the cultivation of rare and curious flowers, during the long seventeenth century.

97 On the florilegium and rare plant collections of Mary Somerset, see Cottesloe and Hunt, *The Duchess of Beaufort's Flowers*; Jennifer Munroe, '"My Innocent Diversion of Gardening",' 111–123; Chapter 2 by Laird, *A Natural History of English Gardening*, 62–123.

98 Laird, *A Natural History of English Gardening*, 98–99; Rowley, 'The Duchess of Beaufort's Succulent Plants,' 1–16. See also Ruth Duthie's small book, in which she discussed how florilegium flowers continued to develop into the eighteenth and nineteenth centuries, as well as the changing perceptions towards them, in *Florists' Flowers and Societies*.

CHAPTER 2

# Growing Flowers into Rarities and Curiosities

## 1 Introduction

In 1612, the Dutch printmaker and nurseryman Emanuel Sweerts (1552–1612) compiled and published the *Florilegium Emanuelis Sweerti Septimonti Batavi …*, referred to as the *Florilegium Sweerti* in this chapter.[1] The scope of the *Florilegium Sweerti* is typical of a florilegium, which features rare and curious flowers from distant places beyond the area north of the Alps, flower varieties in different colours, and double flowers with more petals than usual.[2] However, Sweerts' flower book stands apart from most of the seventeenth-century florilegia in two major ways: the book has a comparatively heavy commercial and practical tone. First, the *Florilegium Sweerti* includes a brief advertisement at the beginning of the volume. According to the advertisement, those who were interested in purchasing the florilegium or the living flowers, decorative plants, and bulbs that appeared in the book could do so by finding Sweerts at the annual Frankfurt Fair. At other times, people could find him at the shop of the Dutch book printer Paulus Aertssen van Ravensteyn in Amsterdam.[3] This advertisement is why the *Florilegium Sweerti* is often categorised as a sales

1 Sweerts, *Florilegium Emanuelis Sweerti Septimonti Batavi.* … The reprints of the volume were renamed and now better known as the *Florilegium amplissimum et selectissimumas* …; see Chapter 4 on reprints and different editions. Bibliographic description numbered C49 in Bridson and White, *Plant, Animal & Anatomical Illustration in Art & Science*, 40. There is relatively little known about Sweerts, including what he precisely did for a living. He, however, worked for the Holy Roman Emperor, Rudolf II, to some capacity to acquire flowers and plants for the garden. Sweerts wrote in the preface of the flower books that he brought a lot of plants to the emperor, who was pleased and ordered him to make a book of these plants, see Bleiler, *Early Floral Engravings*, v–vi.

2 See Chapter 1 for discussion on the characteristics that made flowers rare and curious in the long seventeenth century.

3 Printed advertisement in Latin in the frontmatter, '*S'quis praesentis opusculi, nec non eorundem in eodem contentorum florum, plantarum & radicum copiam habere desoderat, is ea tempore nundinarum Francofurti in ipsius authoris in frontispicio libri positi officina ante curiam: extra nundinas vero Amstelodami apud Paulum Arnoldum von Ravenstein Typographum, inueniet.*'; in Dutch, '*Soo Iemant waer / die lust heeft van deze Boecken / ofte Bloemen / die daerinne begrepen staen / te coopen / one kome inder Messtyt voor het Rarthauys opten Roemer / inden Laden van desen Auteur Manuel Sweerts : Na de Messe tot Amsterdam / by Paulus Aertssen van Ravesteyn Boekdrucker / sall een jeder goet bescheyt doen.*'

catalogue of plants in the secondary literature.[4] It is important to note, however, that the book itself would have been an expensive object. It may have doubled as a sales catalogue, but not solely to be one. Sweerts himself stated in the preface that the book was made to be an image collection of rare and curious flowers.[5] Furthermore, the printmaker and nurseryman passed away shortly before the publication of the book in 1612. Thus, the printed volumes themselves might never have functioned as sales catalogues in reality.

Second, and more important to this chapter, the *Florilegium Sweerti* includes another feature that is rare to the genre of the florilegium: the essential instructions on how to plant and care for rare and curious flowers, especially the bulbous ones. The florilegium as a genre is, overall, more interested in the 'results' of gardening and horticultural activities. More precisely, flower books focus on depicting the products of the blooming flowers, rather than the process of how to grow and cultivate them. The planting instructions written by the (part-time) nurseryman of Sweerts, however, brought to the forefront the contemporary gardening and horticultural practice and the practical knowledge that were critical in supporting the production of the genre of the florilegium. Without such a practice and knowledge, there would have been no rare and curious flowers to collect, let alone depict. Without such, there would not have been the genre of the florilegium. The how-to instructions in the *Florilegium Sweerti*, as well as the other rare occasions when seventeenth-century florilegia engaged with the gardening practice through text and/or image, thus deserve some attention. Starting with Sweerts' planting and caring instructions, this chapter delves into horticultural practices in the early modern period. The chapter first discusses the gardening tasks and techniques that were described and instructed in seventeenth-century florilegia. It then turns to contemporary gardening treatises for more detailed practices to better understand *how* the seemingly simple tasks and techniques helped the flowers adapt and acclimatise to gardens in areas north of the Alps. Through the planting activities, this chapter shows how florilegium images embody the meticulous and foundational processes of growing and cultivating rare and curious flowers in the early modern period.

## 2 How to Plant According to Seventeenth-Century Florilegia

The planting instructions provided by Sweerts focus primarily on bulbs, including tubers. The instructions are concise and brief, to the point that it

4 Bleiler also questioned the use of the florilegium as a sales catalogue. The short introduction by Bleiler for the Dover Pictorial Archive Series actually has more nuanced reading of this flower book than some more authoritative pieces on the florilegium genre.

5 Sweerts, *Florilegium Sweerti*, '*Ad candidum lectorem*'; Bleiler, *Early Floral Engravings*, xi.

is easy to overlook them in the volume. The 1612 *Florilegium Sweerti* usually includes around 110 plates of rare and curious flowers. Before the first plate, there are around 40 pages of frontmatter that contain different kinds of textual information. The frontmatter is written in Latin, Dutch, German, and French, but the information each language provides is not consistent. Some pieces of information are given in all four languages, some only in Latin, while others are only in one or two languages of the four. The book opens with the frontispiece, and printed on the back is the aforementioned advertisement in all four languages. Following the advertisement is a plate with a portrait of Sweerts with an epigram in Latin; a dedication in Latin to the then-new Holy Roman Emperor Matthias, the successor to Rudolf II, who was Sweerts' most known and renowned patron; and the Latin privilege that grants the printing right of the book. The next page is the preface that explains Sweerts' motivations for putting together the florilegium, printed four times in Latin, Dutch, German, and French. A Latin poem by Damasius Blyenburgius follows, and afterwards a French congratulatory sonnet written by Augerius T.F. Clutius.[6] The planting instructions then appear in Latin and German. The last section of the frontmatter is a catalogue of all the depicted flowers and decorative plants in alphabetical order, first in Latin, then in Dutch, German, and French.

The planting instructions consist of only five short paragraphs and four steps. Each of the two languages is typeset onto one page, with the recto side of the paper sheet in Latin (Fig. 10) and the verso in German. The page states, in full,

> SOME USEFUL TEACHINGS BY THE AUTHOR CONCERNING THE PLANTING OF EXOTIC FLOWERS[7]
> Because I have been asked on many occasions in Germany, by certain eminent men and lovers of flowers and beautiful plants, to give them a good description on how one goes about lifting bulbs and resetting them, together with other matters to be of service to these men, I am pleased to give the following basic principles.[8]

6 The copy at the Oak Spring Garden Foundation has another dedication/preface in Dutch addressed to the state generals of the Dutch Republic, mostly praising God for creating flowers. This dedication/preface is not found in other copies I examined nor mentioned in bibliographical references of the title.

7 Sweerts, *Florilegium Sweerti*, '*Exoticae quaedam doctrinae authoris circa florum plantationem utiles*'. The quoted planting instructions are slightly altered translations from Bleiler's volume. Bleiler eloquently translated the instructions from the German version of the text. A few small wordings and phrasings are changed here to make the instructions read closer to the Latin version with the help of Eric Jorink. Bleiler, *Early Floral Engravings*, xii–xiii.

8 '*Quandoquidem a quibusdam viris magnatibus, qui venustissimorum & pulcerrimorum florum plantatione delectantur, saepius a me petitum, ut ijs aliquam informationem, quomodo se in*

> The first thing to be noticed is that, all flowers that have bulbs or tubers should be removed from the ground in August. However, take care that any plants that have immature seeds remain in the ground until the seed is perfectly ripe.[9]
>
> Second, bulbs or tubers which have been lifted can be kept out of the ground for three weeks or a month without harm. They can also be restored to the ground within fourteen days or three weeks. The finer the earth is prepared, the better.[10]
>
> Third, the earth can be prepared and set to order in this manner and fashion. First you must make a fairly large hole or pit; in it you must put one part of pure cow dung and one part of the best black soil and mix them together. Then you must let this stand for a good year's time, mingling them occasionally, so that weeds are destroyed and the aforementioned cow dung has become rotted. Then you will have good soil.[11]
>
> Fourth, you can also prepare good soil for yourself in this fashion. Take old tanner's bark, such as the shoemakers use, the older the better, and to it add good black earth. Put it in a pit and mix it together, according to the method described. Let it lie through the year until the bark is rotted, and then take this soil and rub it through a sieve. If you set flowers and plants in it, they will grow beautifully.[12]

The four instructions given by Sweerts could be grouped into two sets of activities. The first two concern the handling and (re)planting the bulbs and tubers and when to do so, and the last two deal with preparing the soil, or earth, so that it is of a good quality for planting.

---

*extrabendis & inferendis floribus habere debeant, intermittere non debui, quin in eorum gratiam & commodum subsequentes regulas evulgarem.'*

9 *'Primo itaque observandum, omnes flores bulbosos, siue crassas & rotundas radices habentes mense Augusto e terra esse extrahendos, cum hac tamen cautela, ut illos, quorum semina adhuc immatura, tam diu in terra relinquant donec semina perfecte maturuerint.'*

10 *'Secundo, possunt extractae radices florum, tribus septimanis, vel etiam integro mense sine corruptione extraterram conservari, & intra duas vel septimanas in terram rursus inseri, & quoterra subtilior, eo melior est.'*

11 *'Tertio, terram pro floribus inserendis aptam hoc modo praeparabis. Primo facias in terra instae magnitudinis fossam in ea ponas excrementa boum & partem optimae terrae nigrae, misceas & sinas per annum in praedicta fossa iacere, ita tamen, ut aliquoties in anno misceantur, ut superstuae & inutiles herbae expurgentur & dicta excrementa putrescant.'*

12 *'Quarto etiam hoc modo aptam terram parabis. Accipe cortices tusas de quercu quibus sutores usi sunt, quo antiquiores eo meliores, misce cum bona terra nigra & ponas simul per annum in fossa iuxta modum praescriptum, post annum per cribrum separa crassiorem terram a tenuiore, & habebis aptam terram ad inserendos flores.'*

# EXOTICÆ QVÆDAM DOCTRINÆ AVTHORIS CIRCA FLORVM PLANTATIONEM VTILES.

*QVANDOQVIDEM à quibuſdam viris magnatibus, qui venuſtiſsimorum & pulcerrimorum florum plantatione delectantur, ſæpius à me petitum, vt ijs aliquam informationem, quomodo ſe in extrahendis & inſerendis floribus habere debeant, intermittere non debui, quin in eorum gratiam & commodum ſubſequentes regulas euulgarem.*

*Primò itaque obſervandum, omnes flores bulboſos, ſiue craſſas & rotundas radices habentes menſe Auguſto è terra eſſe extrahendos, cum hac tamen cautela, vt illos, quorum ſemina adhuc immatura, tàm diu in terra relinquant donec ſemina perfectè maturuerint.*

*Secundò, poſſunt extractæ radices florum, tribus ſeptimanis, vel etiam integro menſe ſine corruptione extra terram conſeruari, & intra duas vel ſeptimanas in terram rurſus inſeri, & quo terra ſubtilior, eo melior eſt.*

*Tertiò, terram pro floribus inſerendis aptam hoc modo præparabis. Primò facias in terra iuſtæ magnitudinis foſſam in ea ponas excrementa boum & partem optimæ terræ nigræ, miſceas & ſinas per annum in prædicta foſsa iacere, ita tamen, vt aliquoties in anno miſceantur, vt ſuperfluæ & inutiles herbæ expurgentur & dicta excrementa putreſcant.*

*Quartò etiam hoc modo aptam terram parabis. Accipe cortices tuſas de quer[illegible] quibus ſutores vſi ſunt, quò antiquiores eò meliores, miſce cum bona terra nigra & ponas ſimul per annum in foſsa iuxta modum præſcriptum, poſt annum per cribrum ſepara craſsiorem terram à tenuiore, & habebis aptam terram ad inſerendos flores.*

*** Kurtze

FIGURE 10 Planting instructions in Latin in the *Florilegium Sweerti* (1612), KB, National Library of the Netherlands, The Hague, KW 371 A 11

While the practical instructions for planting are few and far between in the genre of the florilegium, another set of instructions appears in a small addendum or supplement in the *Hortus floridus* (1614) by Crispijn van de Passe II.[13] The addendum is titled, 'A short story of tulips and her practice, sowing, planting and added to it the most excellent tulips that can be found nowadays with the *liefhebbers* and that are of great value.'[14] The supplement is, at minimum, in Dutch and French. However, it is a rare addition, as the addendum is only found in a small number of volumes among the surviving copies.[15] 'A short story of tulips' includes brief instructions on planting tulip bulbs, sowing tulip seeds, and spotting diseases that can occur on tulips, a registry of tulip varieties, and an illustration of a transplanter and a description of how to use it, and so on. In particular, the instructions for the planting of tulips proceed:

> OF THE PLANTING AND PRACTICE OF TULIPS
> Of the exercise, cultivation and preparation of the soil / so it is important to note that one shall take burnt potash and tanner's bark, and sieved soil / and manure, mixed together such it will give good suitable earth. And in the autumn (or earlier) one takes all the tulips from which one does not desire the seeds; pick up and take off all cuttings that are sufficiently fed so that the tulip does not have space in the earth to grow roots, for the nature of the tulip is such that she shoots more downwards than upwards, and sometimes one sees it grows upwards with the roots, or as a faulty growth, and can be practised thus.[16]

13 Van de Passe II, *Hortus floridus*. See Chapter 1 for an introduction to this florilegium.

14 '*Een cort verhael van de tulipaenen ende haere oefeninghe, saeyinghe, plantinge ende daer by gevoecht de uutnemenste tulipaenen die nu teghenwoordich by de Liefhebbers ghevonden worden en van grooter weerden zijn.*' Translated by Henrike Scholten. The addendum is found in a hand-coloured copy of the Dutch edition of the *Hortus floridus* in Upperville, VA, Oak Spring Garden Foundation, 00004892.

15 The copy of a French edition of the *Hortus floridus* at the Oak Spring collection also has the addendum in French, 00005892. The Oak Spring copies (Dutch and French editions), however, are the only copies with this addition that I came across during the research.

16 'Van de Plantaige ende oeffeninghe der Tulipaenen. *Van oeffeninghe en bereydinge en preparatie des Eertrijck / soo is merckelijck te aenmercken dat men sal nemen gebrande Potas ende Run / ende geseefden Eertrijck / en mest sulcks onder malcanderen gedaen sal goede bequame eert geven. En in den Herfst salmen alle de Tulipanen (oft vroeger) daermen geen saet af begheert; op nemen en aflichten alle affetsels die genoech gevoet zijn overmits den Tulipa geen ruymte in de eert heeft om te schieten / dan de natuer van de Tulipa als dat sy meer naer onder als naer boven toe schieten sal / dan somtijts siet men hem wtschieten naer boven toe wortel wijs / of als een misgewas / en can op dusdanighe manier gheoeffent worden.*' Translated by Henrike Scholten, with input also from Marieke Hendriksen.

Van de Passe II's supplement differs from Sweerts instructions in details, but the tasks described in the paragraph can also be divided into the same two groups of activities. Planting bulbous flowers, according to seventeenth-century florilegia, requires handling the bulbs during a certain time of the year and preparing good soil in which the flowers can grow.

## 3 Cultivating Rare and Curious Flowers through Care and Patience

The tasks instructed by Sweerts and Van de Passe II were not necessarily complex nor difficult, and no grand gardening technology was needed in the process of cultivating rare and curious flowers. *Cultivation,* in this chapter, mostly means growing flowers and keeping them alive. While some rare and curious characteristics, such as certain colours and some double flowers, were the results of breeding new cultivars through hybridisation, this aspect and practice were rarely recorded.[17] There are occasional remarks in gardening treatises, such as on picking seeds of a doubled flower in order to ensure getting more of the same doubled flower. However, how the first flower came to have double petals is not commonly found in the text. The rest of this chapter thus focuses on the nuts and bolts of bringing flowers to bloom by helping them adapt and acclimatise to (northern) European gardens, instead of experimentations on breeding new flower varieties.

The simple tasks described in florilegia are quite contrary to the aspect of focus when discussing the cultivation of rare and curious plants, which has often been associated with grand technological developments throughout history. For example, authors, including Billie S. Britz, Mark Laird, Esther Helena Arens, and Lisa Johnson, have shown how climate-controlling structures, such as *orangeries* and hothouses, greatly assisted and enabled the growth and cultivation of plants from warmer climates in colder geographical regions.[18] In particular, Mary Somerset, Duchess of Beaufort (1630–1715), was renowned

17 Early modern naturalists also experimented with it, as discussed by Egmond, 'Experimenting with Living Nature,' 21–45. Florist societies became more documented in the late seventeenth century onward, see Duthie, 'English Florists' Societies and Feasts in the Seventeenth and First Half of the Eighteenth Centuries,' 17–35. Noël Kingsbury also briefly delves into seventeenth-century cultivars of ornamental plants in *Hybrid: The History and Science of Plant Breeding*, 329–336.

18 Britz, 'Environmental Provisions for Plants in Seventeenth-Century Northern Europe,' 133–144; Laird, 'Greenhouse Technologies and Horticulture,' 55–77; Arens, 'Flowerbeds and Hothouses,' 265–283; Johnson, 'Pieter de la Court van der Voort and Innovations in Pineapple Cultivation in Early Eighteenth-Century Gardens,' 23–41.

for her hothouse/greenhouse that cultivated many of the first tropical plants that grew in England in the second half of the seventeenth century.[19] Many of the flowers and plants in Somerset's Badminton Florilegium would not have been able to live through the cold English winter outdoors, and the climate-controlled environment in a hothouse much increased the plants' survival rate through artificial heating from a stove.[20] Such technological developments or inventions—extending to the bigger systems such as waterway designs or smaller devices including plant transportation tools—have been an important part of horticulture.[21]

However, while climate-controlling structures were developed around the same time as the thriving production period of florilegia throughout the seventeenth century, they were not a driving force behind the cultivation of rare and curious flowers. The *orangerie*, as the word suggests, was mostly constructed for citrous fruit trees and other large decorative plants and fruit trees. The hothouse with heating stoves for supplying warm air only became more easily available in the second half of the seventeenth century. This is evident from the lack of mention of such a structure in the planting instructions provided by Sweerts and Van de Passe II, both of which were published, with several reeditions printed, in the first half of the seventeenth century. This absence is especially noticeable when the addendum in the *Hortus floridus* includes a gardening tool that was novel. In 'A short story of tulips,' Van de Passe II depicted two views of a transplanter (Fig. 11). One view demonstrates how the transplanter is shut closed by a pin and another displays how the tubular device holds a tulip inside. The section of the text that follows the plate describes the basic workings of the transplanter. It also explains that the illustration was given because the transplanter could be a handy tool for gardening, but it was not yet commonly used, even though the device was already known to many people.[22] The transplanter, as a tool that was then still of certain novelty,

19 See particularly Laird, 'Greenhouse Technologies and Horticulture.'

20 See Chapter 1 for an introduction to the Badminton Florilegium.

21 See other chapters in Lee and Helphand, *Technology and the Garden*. See also 'Part IV Botanical Knowledge and Horticulture' in Fischer, Remmert, and Wolschke-Bulmahn, eds., *Gardens, Knowledge and the Sciences in the Early Modern Period.*

22 The short description for the transplanter states, '*Seer gonstige liefhebbers der edle Flora sonderlijck der seer aerdige Bloemen de Tulipaenen, mitsdien ick nu voor heb een lutsken dese exprimeren, soohebbe ick u.l. dit Instrument figuerlijck willen exprimeren, welwetende dat sulck by diverse bekent is, evenwel by de sommige nienwelingen noch niet ghebruyckelijckis, doch evenwel seer nutlijck ende prosytich ghevonden wordet.*'

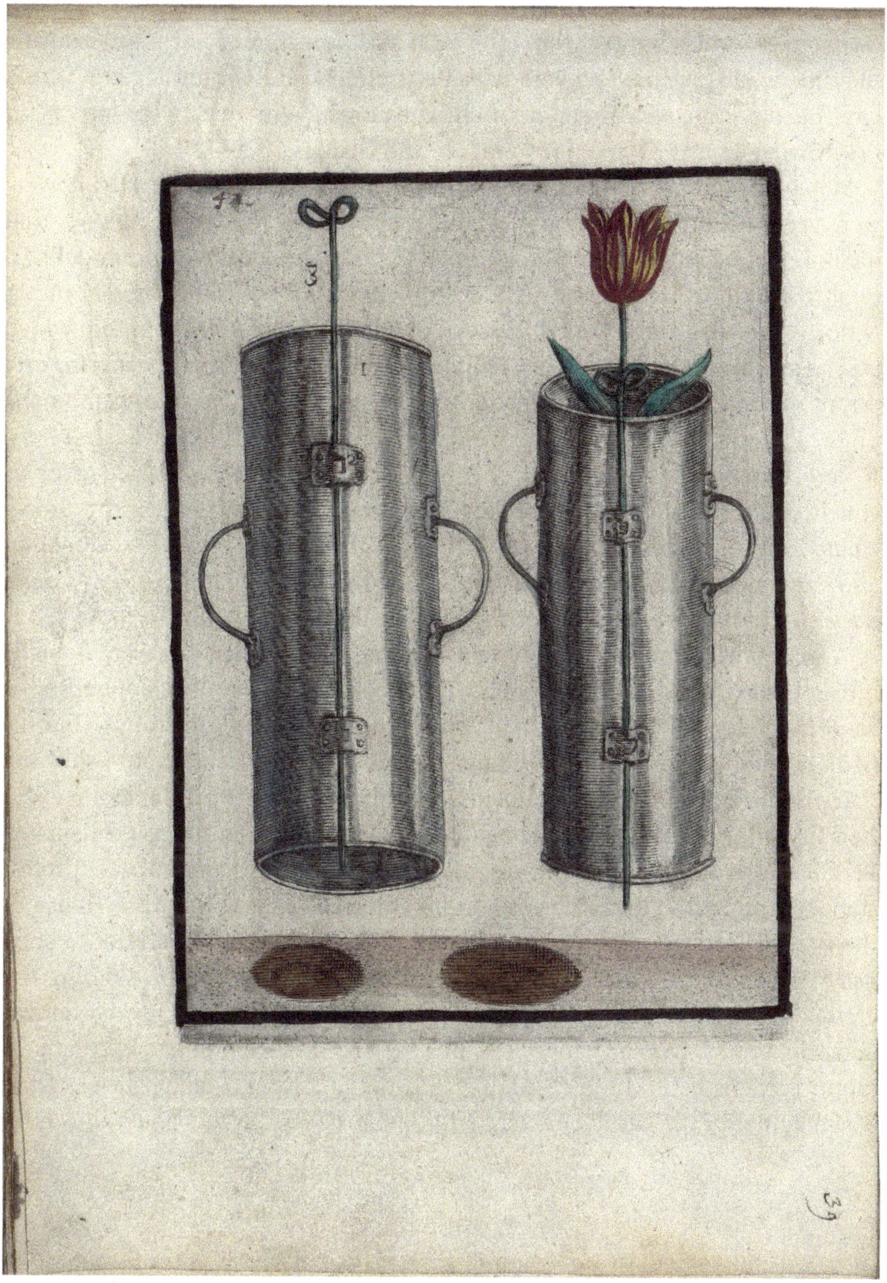

FIGURE 11 Engraving of transplanters in the addendum on tulips from a copy of the Dutch edition of the *Hortus floridus* (1614), Oak Spring Garden Foundation, Upperville, VA, 00004892

fits seamlessly into a pictorial collection of floral rarities and curiosities.[23] *Orangeries* and hothouses were rare, or at least uncommon, structures during the time when Sweerts and Van de Passe II published their florilegia. It would have been odd to leave them out if climate-control structures were important for growing living flowers that were rare and curious.

What the short instructions in florilegia emphasise is that curious flowers were grown and cultivated by performing simple but fundamental tasks that did not depend on high technology and had sustained the development of horticulture on a much longer timeline throughout history. Gardening, at times, is not glamorous work. Activities such as planting and soil prepping, which will be elaborated on in the next sections, could often be time-consuming and physically demanding. On the labour in gardens, scholars have shown how the physical works in gardening practices were performed by all levels of society. It should be noted that the type of garden discussed here is pleasure gardens, which could include flower gardens that housed the living curiosity flower collections in the seventeenth century. Such sites were a space for recreation and enjoyment, so the labour performed—no matter how intense—was not a necessity to support people's lives as farming did.

A large portion of the tasks for constructing pleasure gardens were done by hired people whose names were often not recorded, but people of higher status also engaged in the physical work of gardening. For example, historians Florike Egmond and Esther van Gelder have provided numerous accounts showing that middle-class and noble people often had demonstratively hands-on experience with gardening in the sixteenth century.[24] Garden historian Lenneke Berkhout, on the other hand, discussed the lives and works of the professional (head-)gardeners who worked for the stadtholders from the House of the Orange-Nassau in the Dutch Republic. A small section of Berkhout's book describes some techniques the gardeners used for growing and tending the greens in the stadtholders' gardens.[25] Despite writing on English gardening and only briefly, scholar Rebecca W. Bushnell further exemplified why labouring in gardens could be pleasurable to gardeners from a higher level of the social ladder in the early modern period.[26] *Gardeners*, in this book, thus refer

23 A similar transplanter also appeared in the *Florilegium renovatum et auctum* (1641) by Johann Theodor de Bry and Matthäus Merian, which will be discussed in Chapter 3.

24 See especially Chapter 1 in Egmond, *The World of Carolus Clusius*, 11–23; Chapter 4 in Van Gelder, 'Tussen hof en keizerskroon,' 213–269.

25 Chapter 6 in Berkhout, *Hoveniers van Oranje*, 222–304.

26 Chapter 3 in Bushnell, *Green Desire*, 84–107.

to anyone who performed the tasks and activities of gardening, regardless of their social status or profession.

What remains largely unexplored, however, is not *who* was engaging in the labouring in gardens, but *what* labour entailed. More precisely, what were the very tasks and activities that needed to be performed in order to grow and cultivate rare and curious flowers in early modern gardens? Because of their short and mundane nature, the instructions provided by Sweerts and Van de Passe II can be easily brushed aside. When digging deeper, however, the minute complexities of the seemingly simple tasks became more visible. The gardeners could not simply leave the bulbs and tubers to their own devices. Instead, the gardeners needed to remove the bulbs and tubers when they were ready to come out of the ground, store them away, and replant them at the right time to do so. The gardeners could not simply take any kind of soil to plant, but needed to carefully prepare soils that were suitable for the flowers they were growing. Most of the tasks could not have been done in haste, and many required the gardeners to patiently wait while managing other tasks in the garden. A gardener could only use the soil for planting after a year-long wait of the standing earth mixed with cow dung and black soil, as instructed by Sweerts, but not before.

The undisclosed labour of these instructions also underlined two critical qualities required in the gardeners in the process of growing floral rarities and curiosities: a high level of care and patience. These two words are almost omnipresent in works, both past and present, that write about gardens and gardening, as they are much integral to the processes of growing plants of all kinds. Perhaps also because of this integral nature, they are assumed and expected, sometimes acknowledged, but rarely centre-staged. To turn the spotlight onto care and patience, more significantly to the concept and practice of care, two pieces of work are useful as springboards, albeit borrowed from rather different fields of the history of medicine and the history of technology.[27] Writing about healthcare in early modern Italy, historian Sharon T. Strocchia widened the range of what constitutes medical work by looking into 'what Renaissance Italians actually did to treat and prevent illness.'[28] Through discussing many tasks, such as preparing remedies in quantity, that were not always considered

27 Alternative to care is the concept and practice of *nurse/nursing*, which is sensible when 'nursery' is also a word for growing seedlings. However, nursing is too closely associated with motherhood and the starting or young stage of one's life. The gardening instructions discuss tasks pertinent to almost the entire life cycle of a flower/plant, with a lot of steps contributing to prolonging the life of a flower, such as replanting bulbs. *Care* thus covers a longer span of a flower's/plant's life.

28 Strocchia, *Forgotten Healers Women and the Pursuit of Health in Late Renaissance Italy.*

as grand or noteworthy to be written into medical texts, Strocchia showed how these works, most often carried out by women, were instrumental in practising both domestic and institutional care for people. On a different type of care, historians Lee Vinsel and Andrew L. Russell argued for the significance of maintenance, instead of innovation, for both historical studies and future practices.[29] Through maintenance, the authors showed how the care for things is much more important in upkeeping the qualities and benefits technologies of all kinds can bring instead of their initial invention, and the dire consequences that could occur without care. While using the term *care* in very different contexts, both the works of Strocchia and Vinsel and Russell prioritised the tasks that have often escaped attention in their respective fields. Through the lens of care, the rest of this chapter aims to show the significance of basic gardening techniques in cultivating floral rarities and curiosities in the long seventeenth century.

While it is clear what the essential tasks and activities were in order to grow rare and curious flowers, what is not as apparent in the planting instructions by Sweerts and Van de Passe II is the repetitive and meticulous nature of the work. To better understand the extensive and patient care that went into growing and cultivating flowers, a gardening treatise from the end of the seventeenth century can particularly fill the gap. *De Nieuwe Nederlandse Bloem-Hof* (1700) was written by Henrick van Oosten.[30] The treatise has a heavy focus on flowers and the practical knowledge for growing them, written in great detail. There are five parts to the book. The first part is titled 'The Dutch gardener, treating of trees and shrubs.' Part two is 'The Dutch flower-garden: or, the curious florist' and is the longest part of the book. Part three is titled 'A treatise of the manuring of tulips, according to the French way.' The fourth part is 'A treatise of gillyflowers, according to the French way.' Part five is 'An account of the sowing, up-bringing, and use of orange-trees.'[31] The part on 'the curious

29 Russell and Vinsel, 'After Innovation, Turn to Maintenance,' 1–25; *The Innovation Delusion.*

30 Van Oosten, *De Nieuwe Nederlandse Bloem-Hof.* Quotes from the book for the rest of this chapter are from the English edition, *The Dutch Gardener.*

31 This is the order of the 1703 English edition. The order of the parts in the Dutch edition is different, but the orders of chapters within each part are mostly the same. The order in *De Nieuwe Nederlandse Bloem-Hof*, as printed in the preface, is as follows: *De naaukeurige Bloemist* (The curious florist), *Verhandeling der Tulpen naa de Fransche wijse* (The handling of tulips according to the French way), *Het Tractaat van de Angelieren* (An account of gillyflowers according to the French way), *De naaukeurige Hovenir* (The curious gardener), and *Het Tractaat van de Limoen en Orange-Boomen* (An account of the lemon and orange trees). Bound copies might have different orders, as each part has self-contained pagination (starting with page 1 for every part) for the Dutch edition. The pagination for the English edition is consecutive throughout the book.

florist' provides instructions for growing many of the same flowers that make frequent appearances in seventeenth-century florilegia.

Van Oosten was an experienced and reputable arborist in Leiden (*entenier tot Leiden*) and his treatise was widely disseminated throughout Europe.[32] The book was reprinted many times by several printers. The initial publication was issued in Leiden by Johannes du Vivie and Isaac Severinus, and the second printing was already in place after three years in 1703. There were at least six editions produced between 1715 and 1792, with the second, third, and fourth editions printed in Leiden and the fifth, sixth, and seventh editions published in Rotterdam by Hermanus Kentlink and Dirk Vis. The seventh edition was again republished in 1825 and 1866 in Amsterdam by Wouter Braver II.[33] *De Nieuwe Nederlandse Bloem-Hof* was translated into English in 1703 as *The Dutch Gardener: or, the compleat florist*, which was reprinted in 1711. The German translation was first published in 1706 and reprinted multiple times, and the French version was first issued in 1714. It is partly translated into Swedish as well.[34] The quoted instructions for the rest of this chapter are from the 1703 English edition, but cross-referenced with the first Dutch edition.

*The Dutch Gardener* features prescriptive instructions and practices of only one person, but its widespread dissemination indicated that the gardening tasks and techniques, as described by Van Oosten, and their variants were most likely familiar to many gardeners in the early modern period. An important note is that gardening treatises were not simply how-to instructions to get thriving plants. Writers of gardening treatises came with various motivations and goals for their text to achieve, and they do not always pertain to the actual growing of plants.[35] Yet, working as an arborist, Van Oosten was writing from the perspective of a gardener by profession, who would most likely have wanted to demonstrate his knowledge and skills in cultivating plants through writing. Taken with a grain of salt the potential exaggeration of success, his instructions can still provide a high level of information that was not too far from the practice of how flowers were grown in the long seventeenth century. Additionally, most of the described flowers were no longer that rare nor curious when Van Oosten published the treatise, even though they were still

32 Berkhout, '"De beste en kundigste enteniers van geheel Europa",' 10.

33 Mantel, '"The Leyden-gardener",' 158. Some of these editions were verified through searching in the Short-Title Catalogue Netherlands (STCN) and/or antiquarian shops listings, but not all of them were found.

34 Mantel, '"The Leyden-gardener",' 158. Mantel also mentioned a partial translation in Swedish, but did not list the book title in which the Swedish translation appeared.

35 See especially Chapter 2 in Bushnell, *Green Desire*, where the author untangled the British examples from the seventeenth century.

beloved. This decline of novelty was precisely, in part, because the horticultural practices and knowledge as documented by Van Oosten and his peers were effective.[36] The tasks and activities, executed with care and patience, multiplied many rare and curious flowers, increasing their numbers. Consequently, such multiplication reduced the flowers' novelty, despite the flowers having the same (if not more of the) characteristics that made them rare and curious in the first place.

## 4 Preparing Earth for a Good Planting Ground

Both Sweerts and Van de Passe II emphasised the importance of a nourishing environment to a flower's growth.[37] While the instructions were notably for bulbs and tubers, they are pertinent to many kinds of flowers, especially on the task of soil preparation. Sweerts dedicated two of the four paragraphs on the short page to preparing good earth so that the flowers would grow beautifully in it. Van de Passe II stated, at the beginning of the instructions, that it is important to plant the tulips in good soil, and used around one-third of the paragraph on how to achieve this proper soil. To get good earth, Sweerts advised his readers to mix 'one part of pure cow dung and one part of the best black soil' and let the mixture sit in a large hole or pit for one year, occasionally turning them to kill the weeds and rot the manure. Alternatively, black earth could be mixed with old (oak) bark for tanning leather, and, after a year of standing, the gardener could use the mixture for growing flowers after sieving. Van de Passe II did not specify the length of time needed, but also suggested using tanner's bark for the preparation of good soil. Instead of mixing with black earth, by adding tanner's bark to burnt potash and blending them together, the gardener could get good soil. Earth and soil preparation, according to the instructions, is largely equivalent to the fertilisation of soil in today's terminology. This simple task of fertilisation is essential to provide a nourishing environment for cultivating rare and curious flowers.[38]

36 More and more of these flowers started to be used for composing parterres and adding colours to plant beds, see examples including Duthie, 'The Planting Plans of Some Seventeenth-Century Flower Gardens,' 77–102; Segre, ed., *The Gardens at San Lorenzo in Piacenza*, 43–58.

37 Soil (preparation) was a topic of interest for several groups of people in the early modern period. See particularly Clericuzio, 'Plant and Soil Chemistry in Seventeenth-Century England,' 550–584; Eklund, ed., *Ground-Work*.

38 Fertilisation applies to growing all types of plants for horticulture and agriculture. See, for example, Winiwarter, 'The Art of Making the Earth Fruitful: Medieval and Early Modern Improvements of Soil Fertility,' 93–114.

There was more than one type of good earth, however, and early modern gardeners needed to consider what was the most proper and/or nourishing environment for growing each of the rarities and curiosities in the gardens. In *The Dutch Gardener*, Van Oosten regularly mentioned the quality of the planting ground. Similar to Sweerts' and Van de Passe II's instructions, a good environment was necessary for the cultivation of many flowers. The different varieties of the cornflower, for example, 'require good Ground,' otherwise they would not sprout and come out of the earth.[39] Some of the flowers were not fussy and they could adapt to different soil conditions. The autumn crocus (*Colchicum*) and the dogtooth violet (*Dens-caninus*) could both grow in any kind of ground.[40] Other flowers, on the other hand, had more specific needs than a general good ground. For example, Van Oosten's full instructions for growing the sunflower continued:

> There are some of these that must be sowed every Year, and others that keep always and encreased by dividing. They are sowed in March, in good fat Ground, which they will soon make lean; and when they are gone, you may mend the Ground doubly with Dung, to make it good again, they love a free Air and Sun.[41]

The sunflower needed a good ground that was fat. However, what was considered good earth for the sunflower may not have been the right one for other flowers. Instead of helping the cultivation of the hellebore, another flower frequently depicted in seventeenth-century florilegia, a fat ground might have hindered it, as the hellebore preferred a less heavy planting ground. The full entry for growing the hellebore is as follows,

> This Plant may be divided in the Spring, it will not be kept out of the Grounds, it loves well-dunged light Ground, and endureth all Weather.[42]

39 Van Oosten, *The Dutch Gardener*, 134; *De Nieuwe Nederlandse Bloem-Hof*, 68 (*De naaukeurige Bloemist*).

40 Van Oosten, *The Dutch Gardener*, 106; *De Nieuwe Nederlandse Bloem-Hof*, 44 (*De naaukeurige Bloemist*).

41 Van Oosten, *The Dutch Gardener*, 140; *De Nieuwe Nederlandse Bloem-Hof*, 72 (*De naaukeurige Bloemist*).

42 Van Oosten, *The Dutch Gardener*, 142; *De Nieuwe Nederlandse Bloem-Hof*, 74 (*De naaukeurige Bloemist*).

The sunflower and the hellebore needed planting grounds with rather contrasting conditions and properties. The ground condition of fat or light (and occasionally lean) is given to the majority of the instructions for the flowers discussed in Van Oosten's gardening treatise.

To change the condition of the planting grounds to provide the flowers with the most suitable environments in which to grow, gardeners could add different kinds of fertiliser to the soil. In the chapter 'The Curious Florist,' Van Oosten often did not explain how to prepare and condition the various grounds within the individual instructions for flowers.[43] Instead, some helpful passages appear in the chapter on the cultivation of trees and shrubs.[44] Van Oosten described a few ways to modify the condition of the soil by adding different kinds of excrement. Similar to Sweerts' instructions on mixing cow dung with black earth, cow manure was highly useful in fattening the soil, as it is 'the fattest and lasting long in the ground, makes it soft and light, especially if the ground is dryish,' according to *The Dutch Gardener*.[45] Sheep dung is even better for a ground that is cold because sheep manure is hotter than cow dung. Comparatively, horse manure is more suitable for flower and kitchen gardens (instead of orchards), as the dung is 'the leanest and hottest and only fit for cold and moist grounds.' Hog dung is good for almost any kind of ground, but it has to be spread thin when adding.[46] These instructions indicate that the soil in the Low Countries, where Van Oosten worked, generally had a lighter ground, which was already suitable for growing some of the rare and curious flowers. Gardeners warmed and fattened the soil through fertilisation, creating a caring environment for other flowers that needed to grow in a fatter ground, such as the sunflower that travelled across the Atlantic Ocean from a different climate zone in the warmer parts of the Americas.[47]

In addition to changing the conditions of the soil through fertilisation, gardeners sometimes needed to take more steps and seek out or prepare earth with specific properties. For example, to grow the hepatica, which is often one of the first spring flowers that appear in florilegia, the gardeners were advised to provide two kinds of grounds. Van Oosten instructed growing the seedlings

43 He usually provided preparation guidelines for flowers that needed more specific care, such as the anemone that will be discussed in the next section.

44 Part 1 in Van Oosten, *The Dutch Gardener*; Part 4 in *De Nieuwe Nederlandse Bloem-Hof*.

45 Van Oosten, *The Dutch Gardener*, 2; *De Nieuwe Nederlandse Bloem-Hof*, 2 (*De naaukeurige Hovenir*).

46 Van Oosten, *The Dutch Gardener*, 2; *De Nieuwe Nederlandse Bloem-Hof*, 2 (*De naaukeurige Hovenir*).

47 See Chapter 1 under the subsection 'Growing Distance' on the introduction of the sunflower to European gardens.

indoors (presumably in pots or small planter boxes), such as housing them in the cellar when winter hit, because they could not withstand the cold. How fat or light the soil needed to be for growing the seedlings is not clarified, but the earth must be moist until the hepatica sprouts.[48] After the seedlings reached a year old, they could be transplanted. To transplant the flower to the gardens outdoors, 'You plant them in sandy Ground mix'd with Dung (that the roots may be kept warm) that is hard and tough.'[49] This means that the gardeners needed to particularly prepare a small area in the garden with sandy earth for the hepatica, as most flowers did not thrive in such ground, according to the instructions. Furthermore, even though the ground was fertilised, the sandy texture of the earth most likely required a different method for adding nourishment for the flower to grow bigger and healthier. Instead of watering the hepatica with only water, as Van Oosten wrote, 'it is also good to water them sometimes with piss, or piss on them, yet not when they are in Flowers.'[50] With only a handful of flowers discussed in this section, there are already multiple kinds of needs of the flowers for the gardeners to meet. Providing suitable environments for the hundreds of flowers in the gardens, the gardeners' care for the living rarities and curiosities started at the ground level.

## 5 A Time to Plant and a Time to Enjoy the Blooms

The instructions by Sweerts, Van de Passe II, and Van Oosten often centre around *when* to perform *what* task in the process of cultivating rare and curious flowers. As Sweerts wrote, the gardeners should remove flowers with bulbs and tubers from the ground in August, and keep them out of the ground for around three weeks to a month before they started decaying, but could already replant the bulbs and tubers back into the ground after two to three weeks.[51] Van de Passe II did not specify the month, but similarly instructed his readers to dig the bulbs out of the ground, store them in a dry place, and replant them

48 Van Oosten, *The Dutch Gardener*, 124; *De Nieuwe Nederlandse Bloem-Hof*, 59–60 (*De naaukeurige Bloemist*).

49 Van Oosten, *The Dutch Gardener*, 125; *De Nieuwe Nederlandse Bloem-Hof*, 60 (*De naaukeurige Bloemist*).

50 Van Oosten, *The Dutch Gardener*, 125; *De Nieuwe Nederlandse Bloem-Hof*, 60 (*De naaukeurige Bloemist*).

51 See Sweerts' instructions in the previous section 'How to Plant According to Seventeenth-Century Florilegia' in this chapter.

around autumn.[52] This attention to what to do in which month or season of the year is a human effort to help the flowers acclimatise to the completely foreign environment (including climate) of European gardens. As most of the flowers were far from where they had been uprooted, the time at which to perform a task was not based on the flowers' natural growing rhythms, as they would have been in their native habitat. Rather, it was a constructed schedule for the gardeners to supply aids and care to help the flowers adapt to or thrive in different environmental conditions throughout the year.

Van Oosten's instructions for cultivating the anemone from seed to bloom offer a good example of the gardeners' extensive care around the year, sometimes multiple years, in order to cultivate living rarities and curiosities. Some of the entries in Van Oosten's treatise are short and simple, such as the aforementioned instructions for growing the sunflower and the hellebore. The sunflower needed to be sown every year, but once sown and/or planted in a suitably fertile ground, neither of the flowers needed much more care other than the presumably regular watering that may have been too obvious to be mentioned in the treatise. Growing the anemone, on the other hand, entailed many more tasks and steps, with some parts of the process painstakingly detailed. To show readers what to do at the various stages of the anemone's life, Van Oosten wrote eight sections of varying length, in the form of paragraphs with subtitles, contrasting with the sunflower and hellebore parts with one short section or paragraph for each of the flowers.[53]

Van Oosten started with collecting the seeds of the anemones with rare colours that were designedly planted for harvesting seeds, but how this first specimen came to the garden was unspecified. Once the gardeners acquired the seed, they could grow the flower from scratch. To sow the anemone, one should do as follows:

> This Seed you must dry in a Chamber; when it is well dried, lay it up in a place where it will not mould; let it lie so till March, till you sow it in the encreasing of the Moon, when you sow it, steep it in one part of Wine, and two of Water. After six Hours time, pour off the Liquor, and let it dry a little, that you may separate it, and so sow it, when all this is observed, sow it in a Box fill'd with Stones and Horse-Dung mixt with a little Earth

---

52 See Van de Passe II's instructions in the previous section 'How to Plant According to Seventeenth-Century Florilegia' in this chapter.

53 Van Oosten, *The Dutch Gardener*, 97–103 (Chap. XLIX–Chap. LVI); *De Nieuwe Nederlandse Bloem-Hof*, 37–42 (*De naaukeurige Bloemist*).

> eight or ten Days before the Full-Moon; you must always keep the Ground moist till the Seed comes up.

The properly dried seeds needed to be sowed in March. Before the gardeners could bury the seeds in the fertilised soils in a planter box, they needed to rehydrate and reactivate the seeds by soaking them in a wine and water mixture for six hours. The ground needed to be regularly watered to keep it moist for the seeds to germinate. This first stage of the anemones' life was spent indoors.[54] By simply sheltering the seedlings away from the changing weather of rain or shine, an indoor space was already sufficient to provide them with a more stable environment without needing an *orangerie* or a hothouse.

Once the seeds sprouted and grew into seedlings, they continued to spend their first year in the box and indoors, so that they could develop their 'claws,' or their tubers, away from the sun that could easily burn them and the winter frost that could easily freeze them at this early stage. After almost a year, the seedlings became more resilient and gardeners could start exposing them to the elements, as Van Oosten wrote, 'When the Spring approaches you must let them have the open Air, for a small Frost cannot hurt them.' The anemones continued to stay in the box and indoors, and after spending a little more than two years in the plant nursery,

> When they have been thus managed to the second Year, you must take them up in a dry Season, and lay them in a dry place, and when they are quite dry, keep them till September the next Year, keeping them a year out of the Ground.

According to the instructions, the seedlings had now developed their tubers and could be extracted from the earth in the planter box in the dry season, supposedly in the summer. During these years, the anemones produced leaves, but not yet flowers. It took around two years for the anemone seeds to grow into a tuber. Yet, the gardeners had to patiently wait, after extracting them from the ground, for the tubers to dry for another year.[55] After more than three years, the gardeners could finally plant the tubers outside.

---

54 Van Oosten, *The Dutch Gardener*, 98–99; *De Nieuwe Nederlandse Bloem-Hof*, 38–39 (*De naaukeurige Bloemist*). This particular section in the English translation eliminates a few passages, but the premise remains the same.

55 Van Oosten, *The Dutch Gardener*, 99; *De Nieuwe Nederlandse Bloem-Hof*, 39 (*De naaukeurige Bloemist*).

In many cases, early modern gardeners would start from this planting stage, as they would have acquired the tubers instead of the seeds for bulbous and tuberous rarities and curiosities. Planting bulbs and tubers reduced the number of years for the gardeners to wait to see their cultivations bloom, but did not decrease the level of care that was also necessary for the stages that brought them to flower. To plant the dried tubers, the gardeners needed to assess their sizes, as anemone tubers could self-divide, breaking into multiple pieces when dried. Van Oosten remarked that it was natural, but tubers of different sizes needed to be planted separately, stating,

> being small and tender, they must be put into the Ground, in the beginning of September, or sooner. You must not plant the great Claws before October, except you would have them to flower very soon.[56]

Tubers of regular or medium sizes needed to be planted in September. Van Oosten further remarked on how the local environment could be unfriendly to the anemones, and, once again, emphasised the importance of a carefully prepared ground that was warm for the flowers and could help them thrive. Most specifically,

> They must be planted eight Days before the Full Moon, but with a Southerly Wind, though it be four or five Days sooner or later. You must plant them in light Earth three Inches deep; under this good Ground you must lay a Foot deep, half perished Horse-Dung, that the Heat thereof may help the fault of our Air, they coming originally from warm Countries.[57]

About five weeks after planting, the anemones would start to show their leaves. According to Van Oosten, gardeners needed to top the soil with more horse dung when the leaves were visible to further fertilise the ground and prevent moss from overwhelming them. The gardeners should also not over-water, as that made the ground too cold for the tubers and roots at this stage of their lives. The colder months quickly followed after the tubers went into the ground, and preparing the anemones for winter was another major task to protect them from the elements and help them acclimatise and grow. Van Oosten instructed,

56 Van Oosten, *The Dutch Gardener*, 100–101; *De Nieuwe Nederlandse Bloem-Hof*, 40 (*De naaukeurige Bloemist*).

57 Van Oosten, *The Dutch Gardener*, 100–101; *De Nieuwe Nederlandse Bloem-Hof*, 39 (*De naaukeurige Bloemist*). This instruction comes before the one for planting smaller and larger sizes tubers in the chapter.

> When it begins to freeze, you must cover them against the great Frost and cutting Winds, which consume the Leaves, and make them Black, which rots when it thaws, and so the Plant perisheth. You must cover them in time with Straw or Hay, the last is better, for it covers closer, and is warmer; nor doth it easily break the Leaves. You must cover them six Inches deep, or thereabouts, and lay some Twigs at the top that it may not blow away. When it thaweth, you must give them Air, by uncovering: For if they should remain too long covered, the Leaves would grow yellow and waterish, and be killed by the lesst Cold; because in the Winter all the Strength is in the Bud; and that being gone, all is lost.[58]

Finally, the cold days passed, and the anemones bloomed. During this period, the anemones needed to stay put and not be transplanted, and the gardeners needed to pay attention to the grass that came from the covering hay and weed before the grass grew tall.[59] After several months of planting and caring, and after almost four years since sowing, the gardeners and garden visitors could finally enjoy the result of the laborious tasks and the careful labour.

Nonetheless, the work was not done. The anemones did not stay blooming, and there were more tasks to perform after they withered. As advised by Van Oosten,

> You must take up the Emonies after they have done blowing, and satisfied the Florist for his Care, let them decay till their Leaves are quite Yellow, then you must take them up before the Leaves are quite perished.[60]

Removing the tubers from the ground preserved them, so that they did not get consumed and burnt by the sun. This also gave the multiplied tubers a chance to properly self-divide, instead of being clustered and fighting with each other for the growing space underground.[61] Digging the tubers out of the earth thus prolonged the lifespan of the tubers, so that they could flower again for a few more years. The tubers needed to be washed in rainwater and ridden off the rotted roots before they could dry naturally. Once they were fully dried, they could

---

58 Van Oosten, *The Dutch Gardener*, 101; *De Nieuwe Nederlandse Bloem-Hof*, 40 (*De naaukeurige Bloemist*).

59 Van Oosten, *The Dutch Gardener*, 102; *De Nieuwe Nederlandse Bloem-Hof*, 41 (*De naaukeurige Bloemist*).

60 Van Oosten, *The Dutch Gardener*, 102; *De Nieuwe Nederlandse Bloem-Hof*, 41 (*De naaukeurige Bloemist*).

61 Van Oosten, *The Dutch Gardener*, 102; *De Nieuwe Nederlandse Bloem-Hof*, 41 (*De naaukeurige Bloemist*).

be stored and locked away until the next planting season.[62] A few months later, in September, the same planting and caring process started all over again, in order to get another opportunity to see the rare and curious anemones bloom.

With eight sections in *The Dutch Gardener*, Van Oosten demonstrated the long and laborious process of cultivating anemones with rare colours. To bring a seed to bloom, gardeners needed to sow them in spring; patiently wait for the seedlings to grow for two years before lifting, drying, and storing the tubers in summer; (trans-)plant in autumn; shield and protect the stems and leaves during the winter; and remove, wash and dry, and store the tubers in the summer, after enjoying the anemones' brief blooming season in the spring. Nonetheless, the care and patience of the gardeners could lead to blooming rarities and curiosities, but could also lead to nothing. The round-the-year plant care increased, but not guaranteed, the chance of a flower blooming. At every stage of the flowers' lives, gardeners could make mistakes with a task, and even small errors could result in damaging or killing the flowers. The gardener could have done everything right, and still had nothing to show, due to uncooperative weather conditions that prevented the flowers from blooming.

## 6 Conclusion

The meticulous care and patience of the gardeners fostered a supportive environment by preparing the earth that was most suitable for each rare and curious flower. The attentive care and patience of the gardeners tended to the flowers' various needs, according to seasons, to help them adapt and acclimatise to the (northern) European gardens. The short instructions on how to plant and care for bulbs and tubers in Emanuel Sweerts' *Florilegium Sweerti* and the advice on how to grow tulips in Crispijn van de Passe II's *Hortus floridus* constitute a snippet of this careful and patient process. Henrick van Oosten's *De Nieuwe Nederlandse Bloem-Hof*, or *The Dutch Gardener*, further textualised the repetitive tasks and the time-consuming process.

The thorough instructions on growing and tending anemones by Van Oosten, in contrast to other flowers, partially showed why Sweerts and Van de Passe II focused only on bulbs and tubers. Many rare and curious flowers in the seventeenth century were bulbous and tuberous. While roots are often eliminated, bulbs and tubers are frequently depicted alongside the blooms in florilegia. Sweerts specifically stated that the first half of his florilegium features bulbous

62 Van Oosten, *The Dutch Gardener*, 102–103; *De Nieuwe Nederlandse Bloem-Hof*, 41–42 (*De naaukeurige Bloemist*).

and tuberous flowers, and most of these flowers are depicted with their bulbs or tubers.[63] As shown in this chapter, bulbous and tuberous flowers often required extensive care and had a (much) longer nursery stage than rooted flowers. This extended growing time that went into cultivating a seed to a bulb or tuber required a high level of labour and (financial) resources, not any less than what went into tending a bulb or tuber to a bloom. Moreover, knowing that a particular flower grows out of a bulb—from their images—would be useful information to know when acquiring a specimen. It helped a gardener to spot the care the bulbous flower would need to keep it alive and bring it to bloom. Under this light, the depictions of flowers, and oftentimes their bulbs and tubers, in seventeenth-century florilegia are embodiments of the gardeners' tremendous care and patience in growing the living rarities and curiosities to bloom, instead of simply images of rare and curious flowers.

63 Sweerts, *Florilegium Sweerti*, '*Ad candidum lectorem*'; Bleiler, *Early Floral Engravings*, xii.

CHAPTER 3

# Picking Flowers for Every Season through Images and Text

## 1 Introduction

The frontispiece of the *Hortus floridus* (1614) by Crispijn van de Passe II (1594/1595–1670) (Fig. 12) depicts two allegorical figures of the sun and the moon holding up a curtain, under which is written the title of the publication in the centre of the picture plane.[1] Albeit in a more simplistic form, adorning the margins of the frontispiece are the subjects of the volume—the rare and curious flowers that were primarily grown for decoration, including species that were foreign to areas north of the Alps and garden cultivars that have double flowers and new variations of colours.[2] Presented at both ends of the primary title and occupying a prominent space close to the centre are not the subjects of the volume, but instead two portraits in medallions of the sixteenth-century physician-botanists Rembert Dodoens (1517–1585) and Carolus Clusius (1526–1609).[3] The two figures, along with their younger contemporary Matthias Lobelius (1538–1616) from the same (plant) circle in Antwerp, frequently appear on the frontispieces of intaglio-printed flower books, especially those that were first published in the first quarter of the seventeenth century. Clusius and Lobelius are depicted in full figures on pedestals, standing in front of the columns that support the architectural structure of the frontispiece of *Le jardin du roy …* (1608). This title is a florilegium compiled by the French printmaker and embroidery designer Pierre Vallet (*c.*1575–1657).[4] The head-portraits of Clusius and Dodoens similarly appear in cartouches

1 Van de Passe II, *Hortus floridus*. See Chapter 1 for an introduction to this florilegium. The two figures are also said to be Apollo and Diana in Segal and Alen, *Dutch and Flemish Flower Pieces*, vol. 2, 1012.

2 See Chapter 1 for discussion on the characteristics that made flowers rare and curious in the long seventeenth century.

3 More on the symbolic readings and visual culture of early modern frontispieces, see Tongiorgi Tomasi, 'Image, Symbol and Word on the Title Pages and Frontispieces of Scientific Books from the Sixteenth and Seventeenth Centuries,' 372–382. Bertram, Büttner, and Zittel, eds., *Gateways to the Book*.

4 Vallet, *Le jardin du roy tres chrestien Henry IV*. Bibliographic description numbered C45 in Bridson and White, *Plant, Animal & Anatomical Illustration in Art & Science*, 39–40.

at the bottom of the frontispiece of the Dutch printmaker and nurseryman Emanuel Sweerts' (1552–1612) *Florilegium Emanuelis Sweerti …* (1612).[5] These portraits visually emphasise the importance of the three physician-botanists and their herbals to the early development of the florilegium as a genre.

This chapter delves into how the first wave of the intaglio-printed flower books actively engaged with the visual and textual information from sixteenth-century herbals. The six flower books of focus are Adriaen Collaert's (*c.*1560–1618) *Florilegium* (*c.*1590), *Le jardin du roy …*, Johann Theodor de Bry's (1561–1623) *Florilegium novum* (1612), the *Florilegium Emanuelis Sweerti …*, Basilius Besler's (1561–1629) *Hortus Eystettensis* (1613), and the *Hortus floridus*. Researchers have widely recognised that the pictorial tradition of these florilegia has roots in the genre of the herbal. This close connection of flowers and plants in early modern florilegia and herbals, however, often puts flower books in a disadvantaged position. Many florilegia do not have descriptions for the depicted flowers, or the text is seemingly 'unoriginal' when there are descriptions included. Additionally, florilegia frequently arrange and order the depicted flowers according to their blooming seasons. Both the lack of (original) plant descriptions and the simple seasonal order contributed to the genre often being regarded as having little to no scientific value, especially when compared with sixteenth-century herbals.[6]

This chapter challenges the judgment based on scientific value by reconsidering the types of information and knowledge that are brought forward through the selected florilegia. It shows how several florilegium compilers emphasised the rare and curious qualities in decorative flowers and plants, as well as transferred the (visual) knowledge of rare and curious flowers from the interests of the naturalists to those of the *liefhebbers*. The first section of this chapter (re)examines the direct connections between sixteenth-century herbals and seventeenth-century florilegia through the known practice of copying in early flower books. This chapter then rethinks the role of copying through the concept and practice of excerption by focusing on *how* florilegia selected and highlighted certain visual and textual information of rare and curious flowers in sixteenth-century herbals. The last section of this chapter delves into the challenges and nuances of ordering flowers according to seasons. In so doing, this chapter argues that, while seemingly simple, seventeenth-century florilegia are an epistemic genre that explores and prioritises different kinds of (visual) plant knowledge.

5 Sweerts, *Florilegium Emanuelis Sweerti Septimonti Batavi. …* See Chapter 2 for an introduction to and more detailed discussions on this florilegium.

6 See section 'Towards a History of Knowledge' in the Introduction of this book on this issue in current florilegia studies.

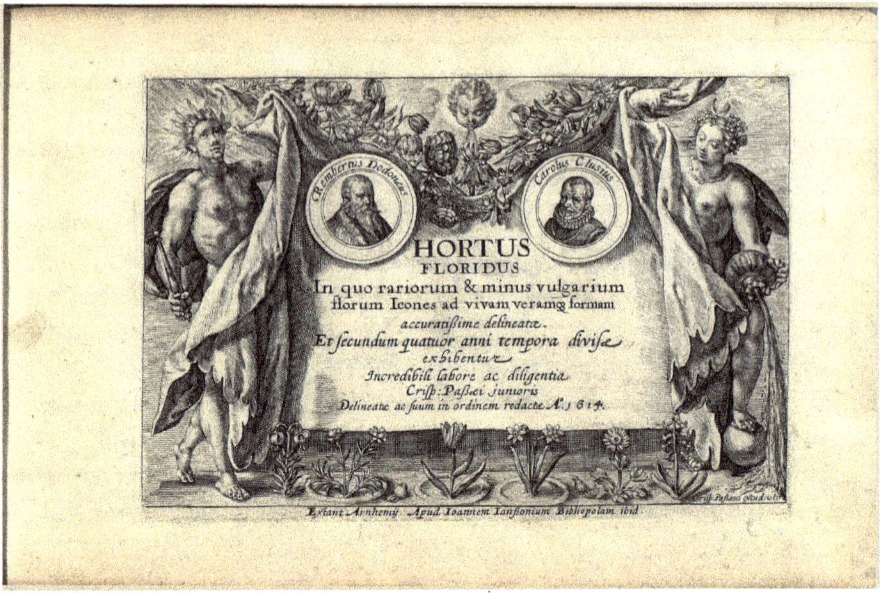


FIGURE 12 Frontispiece of the *Hortus floridus* (1614), Biblioteca Real Jardín Botánico, Madrid, Spain, A PAS, https://bibdigital.rjb.csic.es/idviewer/13646/1

## 2 Copying Flowers

While the majority of the flower depictions in intaglio-printed florilegia that were produced in the first quarter of the seventeenth century are 'new,' meaning that they first appeared in the flower books that published them, there are many depictions that are recognisable copies of the woodcuts from the herbals of Dodoens, Clusius, and Lobelius, among other sources.[7] This practice of copying herbal illustrations is particularly visible in the Flemish engraver Adriaen Collaert's *Florilegium*, the booklet that is attributed to have first used the word *florilegium* to refer to a pictorial collection of rare and curious flowers.[8] As introduced in Chapter 1, Collaert's flower book is a mixture of generic flower motifs and identifiable depictions of species, with the latter group often copies of woodcuts from the three physician-botanists' herbals. Among all, the

7 Already noted by Agnes Arber in *Herbals: Their Origin and Evolution*, 243–246. See also entries in Dackerman, ed., *Prints and the Pursuit of Knowledge in Early Modern Europe*, 196.

8 Collaert, *Florilegium*. See Introduction and Chapter 1 for a discussion on this florilegium.

most explicit examples are the two variations of the sunflowers (Fig. 2).[9] The sunflower at the top of the plate is perceptibly a partial copy of the woodcut that illustrated the sunflower description for Dodoens' 1568 *Florum* (Fig. 13).[10] Collaert's sunflower has the same composition of the head and arrangement of the leaves as the woodcut, even though it is left-facing, mirroring the direction of the woodcut, and the bottom half of the flower is eliminated. Composition, in this volume, refers to how a flower is visually composed and designed, but not how it is arranged in a picture plane.

Likewise, the bottom sunflower on Collaert's plate is modelled after the woodcut for another sunflower variety (Fig. 14) that first appeared in Lobelius' *Plantarum, seu, Stirpium historia* (1576), right next to the same sunflower woodcut that was first used for Dodoens' *Florum*.[11] Collaert's engraving is, again, a reversed image of the woodcut. Focusing only on the bloom, the engraver and florilegium compiler copied the composition of the flower head, including the star-filled pattern, and some leaves, but removed the bottom half of the stalk, as well as the roots. Additionally, the ranunculus with single flower petals, placed on the left of the top sunflower that is copied from Dodoens' herbal (Fig. 2), is copied from the woodcut for the description of the *Ranunculus Lusitanicus Clusij* in Lobelius' *Kruydtboeck* (1581); the dogtooth violet, the one (with a hanging head) on the right of the two small flowers that are sandwiched between the two sunflowers, is copied after the woodcut for the *Erythronium* in the same book.[12] Collaert only kept the blooms for his copies of the ranunculus and dogtooth violet, as he did the sunflowers. This practice of partial copying applies to most of the flowers in the *Florilegium* that have traceable models that first appeared as woodcuts for the herbals of the three physician-botanists.

Collaert's copies, despite the eliminations and mirroring effect, follow the woodcuts closely. To early modern readers and viewers, the direction of the images (left- or right-facing) did not tend to invoke different understandings of the representations.[13] Thus, what would have been most noticeable would have been *what* the engraver kept and/or eliminated from the woodcuts. In many cases, copying a flower depiction from herbal woodcut to engraving is relatively

9 The visual culture of the early modern sunflower is extensively discussed in 'Chapter 3 Representing the Sunflower in Word and Image' by Peacock, *The Look of Van Dyck*, 107–134, particularly 127–128.

10 Dodoens, *Florum*, 295.

11 Lobelius, *Plantarum, seu, Stirpium historia*, 322.

12 Lobelius, *Kruydtboeck*, 819, 232. The woodblocks that printed the two flowers are preserved at the Museum Plantin-Moretus, MPM.HB.05548, MPM.HB.05583.

13 Kusukawa, *Picturing the Book of Nature*, 64.

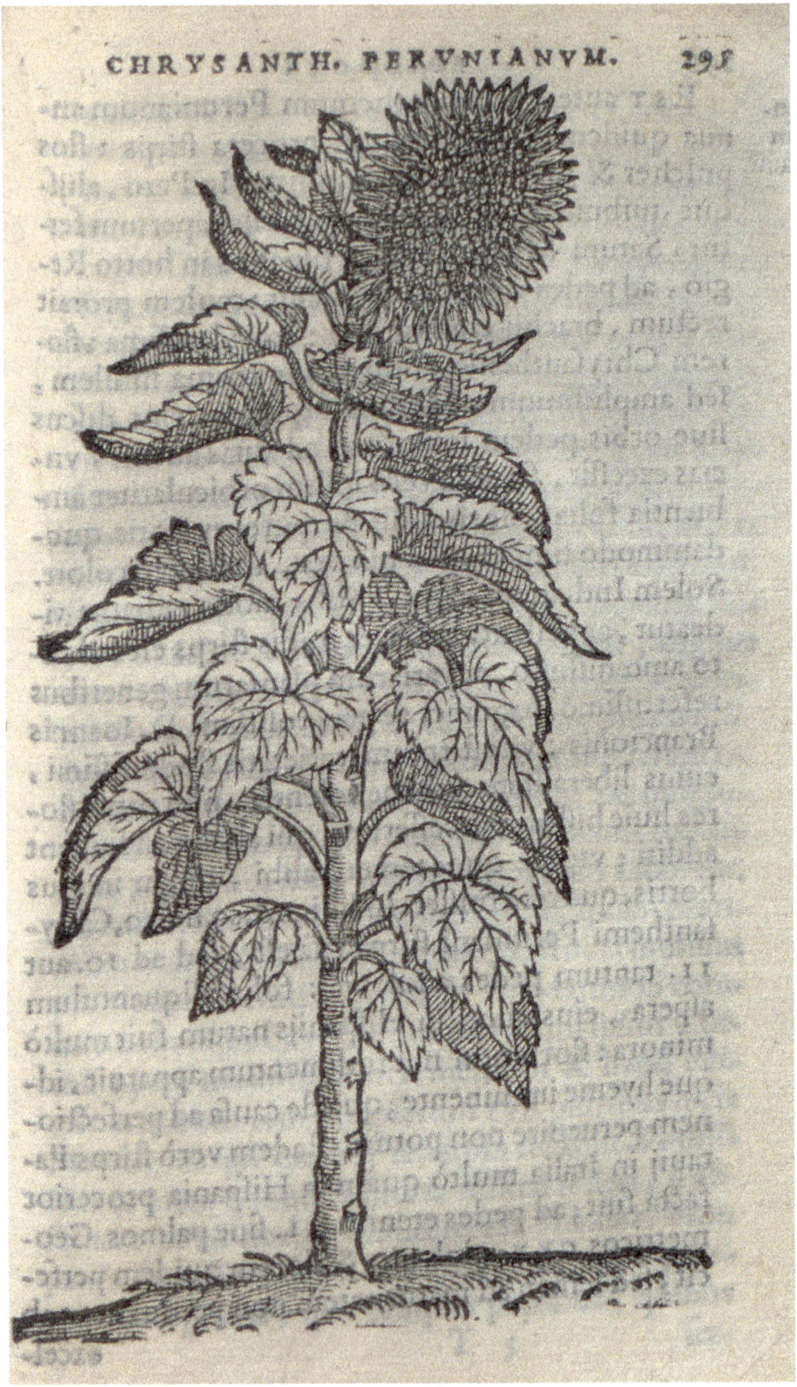


FIGURE 13 Sunflower woodcut from the *Florum* (1568), Museum Plantin-Moretus, Antwerp—UNESCO World Heritage, R 55.17

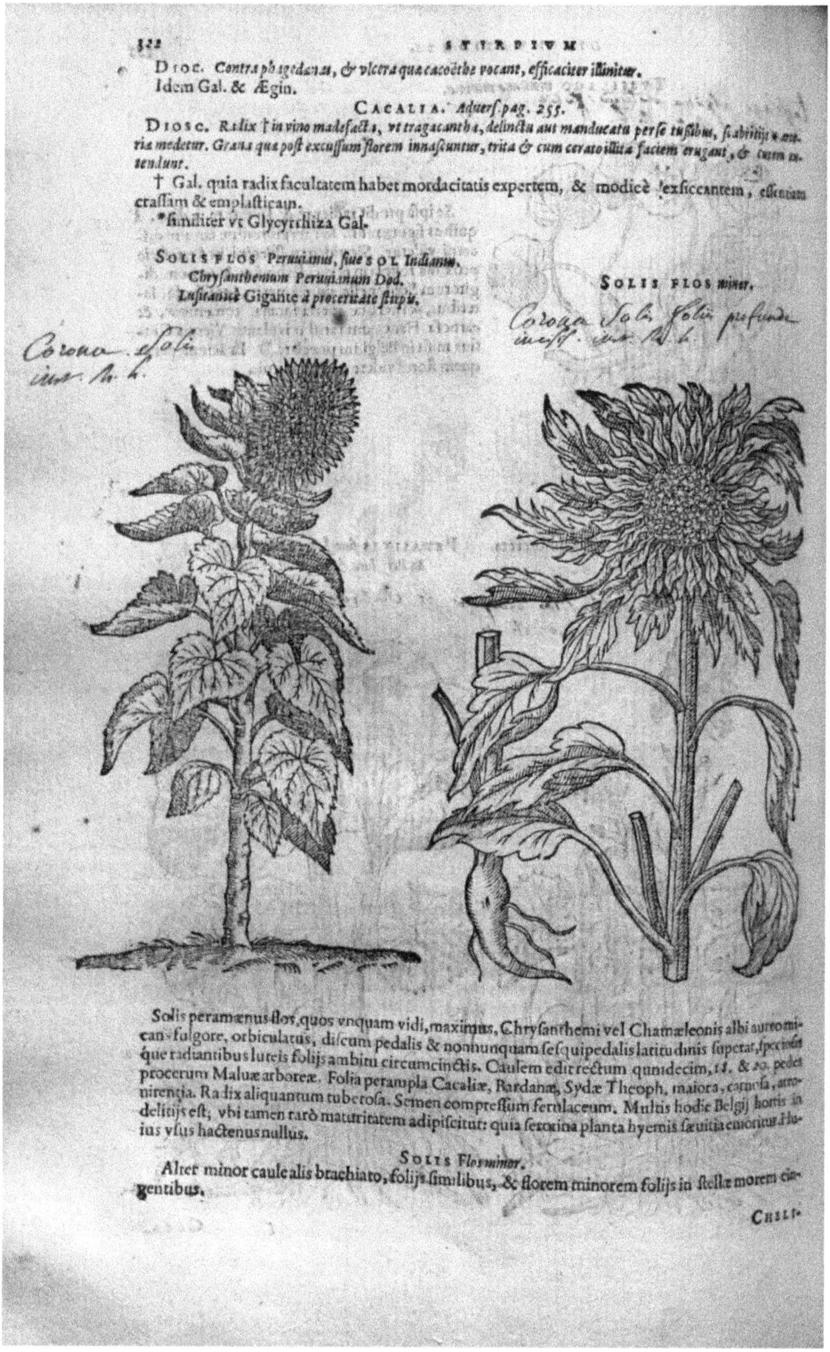

322 STIRPIVM

Dioc. *Contra phagedænas, & vlcera quæ cacoëthe vocant, efficaciter illinitur.*
Idem Gal. & Ægin.

CACALIA. *Aduers. pag. 255.*

Diosc. *Radix † in vino madefacta, vt tragacantha, delinctu aut manducata per se tussibus, scabritijs arteriæ medetur. Grana quæ post excussum florem innascuntur, trita & cum cerato illita faciem erugant, & cutem extendunt.*

† Gal. quia radix facultatem habet mordacitatis expertem, & modicè *exsiccantem, essentiam crassam & emplasticam.

*similiter vt Glycyrrhiza Gal.

SOLIS FLOS *Peruuianus, siue* SOL *Indianus.*
*Chrysanthemum Peruuianum Dod.*
*Lusitanicè* Gigante *à proceritate stirpis.*

SOLIS FLOS *minor.*

Solis peramœnus flos, quos vnquam vidi, maximus, Chrysanthemi vel Chamæleonis albi auromicans fulgore, orbiculatus, discum pedalis & nonnunquam sesquipedalis latitudinis superat, speciosisque radiantibus luteis folijs ambitu circumcinctis. Caulem edit rectum quindecim, 15. & 20. pedes procerum Maluæ arboreæ. Folia perampla Cacaliæ, Bardanæ, Sydæ Theoph. maiora, carnosa, atrouirentia. Radix aliquantum tuberosa. Semen compressum ferulaceum. Multis hodie Belgij hortis in delitijs est; vbi tamen rarò maturitatem adipiscitur: quia serotina planta hyemis sæuitia emoritur. Huius vsus hactenus nullus.

SOLIS *Flos minor.*

Alter minor caule alis brachiato, folijs similibus, & florem minorem folijs in stellæ morem cingentibus.

CHILI-

FIGURE 14 Dodoens' sunflower (left) and Lobelius' sunflower (right) from the *Plantarum, seu, Stirpium historia* (1576), Missouri Botanical Garden, Saint Louis, MO, QK41.L726 1576 [#697–698]

loosely done in florilegia. This means that there is enough resemblance for one to identify the possible woodcut as the model or image source, but the image makers took creative and artistic liberty to alter the copied woodcut in different ways. For example, the two sunflowers in Sweerts' *Florilegium Emanuelis Sweerti* ... (Fig. 15) copied the same woodcuts from the herbals of Dodoens and Lobelius. Compared with Collaert's renditions, Sweerts' sunflowers are far from identical to the woodcuts. Sweerts' sunflowers followed the tilting of the head for one depiction and a head with the star-filled pattern for the other, which are the main artistic decisions that made the two compositions distinctive from other designs of sunflowers. The placements of the leaves and the proportions of the leaves to the flower heads, however, differ somewhat drastically from the woodcuts.

The snowdrop/snowflake plate (Fig. 16) in the *Florilegium novum* presents another kind of creative license in copying flowers.[14] The flower book was compiled by the engraver and publisher Johann Theodor de Bry, who was active in Oppenheim and Frankfurt between the end of the sixteenth century and the first quarter of the seventeenth century. Similar to all the florilegia that have been introduced thus far in this book, the publication history and variations of the title are complex. Generally speaking, the first edition of De Bry's florilegium tends to include close to 80 engraved plates of rare and curious flowers. The practice of copying flowers from existing sources is highly visible in this collection, and many depictions are copies of the herbal woodcuts of the three physician-botanists. The snowdrop variety *Leucoium bulbs. pracox minus* on the left of the plate (Fig. 16) copied from the woodcut that first illustrated Dodoens' description of the flower in the *Florum*; it is referred to as the Dodoens' snowdrop from here onward.[15] The flower in the centre, captioned *Leucoum Bulbosum* (Fig. 16), is a snowflake and a copy of the woodcut for the description of the *Leucoion bulb. exaphyllon* for the same book by Dodoens.[16] Another snowdrop variety on the right of the plate, the *Leucoium bul. praecox Byzant,* is modelled after the woodcut (Fig. 17) that first appeared in the *Rariorum plantarum historia* (1601) by Clusius, called the Clusius snowdrop in this book. On the same page and next to Clusius' snowdrop is Dodoens' snowdrop (Fig. 17), printed with the same woodblock that was used for Dodoens'

14 De Bry, *Florilegium novum*. Bibliographic description numbered C48 in Bridson and White, *Plant, Animal & Anatomical Illustration in Art & Science*, 40. There is a trial/proof edition from 1611, but the publication date is generally agreed upon to be 1612 by researchers.

15 Dodoens, *Florum*, 192.

16 Dodoens, *Florum*, 194.

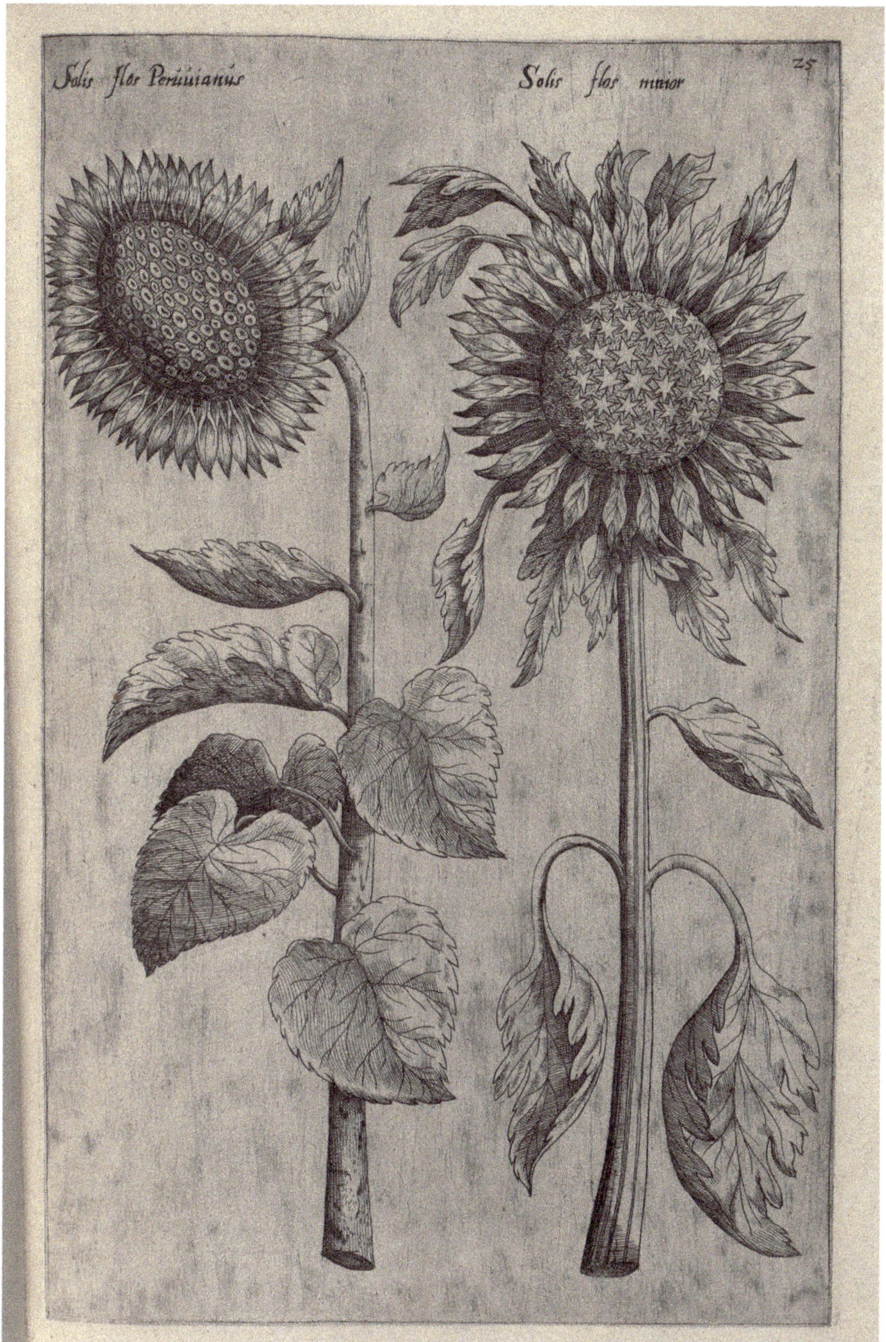


FIGURE 15 Sunflower plate from the *Florilegium Emanuelis Sweerti …* (1612), KB, National Library of the Netherlands, The Hague, KW 371 A 11

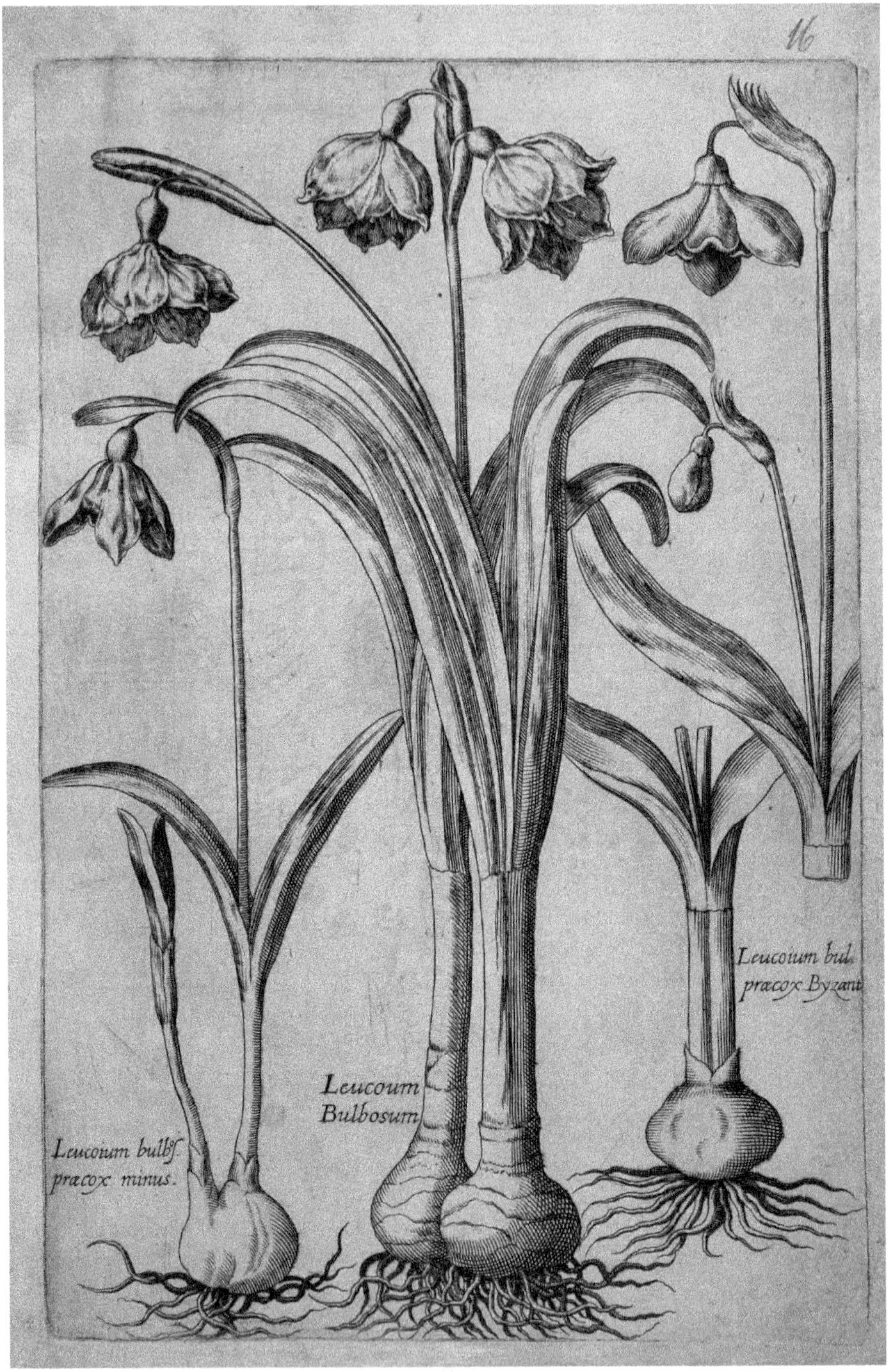


FIGURE 16 Snowdrop/snowflake plate from the *Florilegium novum* (1612), Chicago Botanic Garden, Chicago, IL, QK41.B79 1611

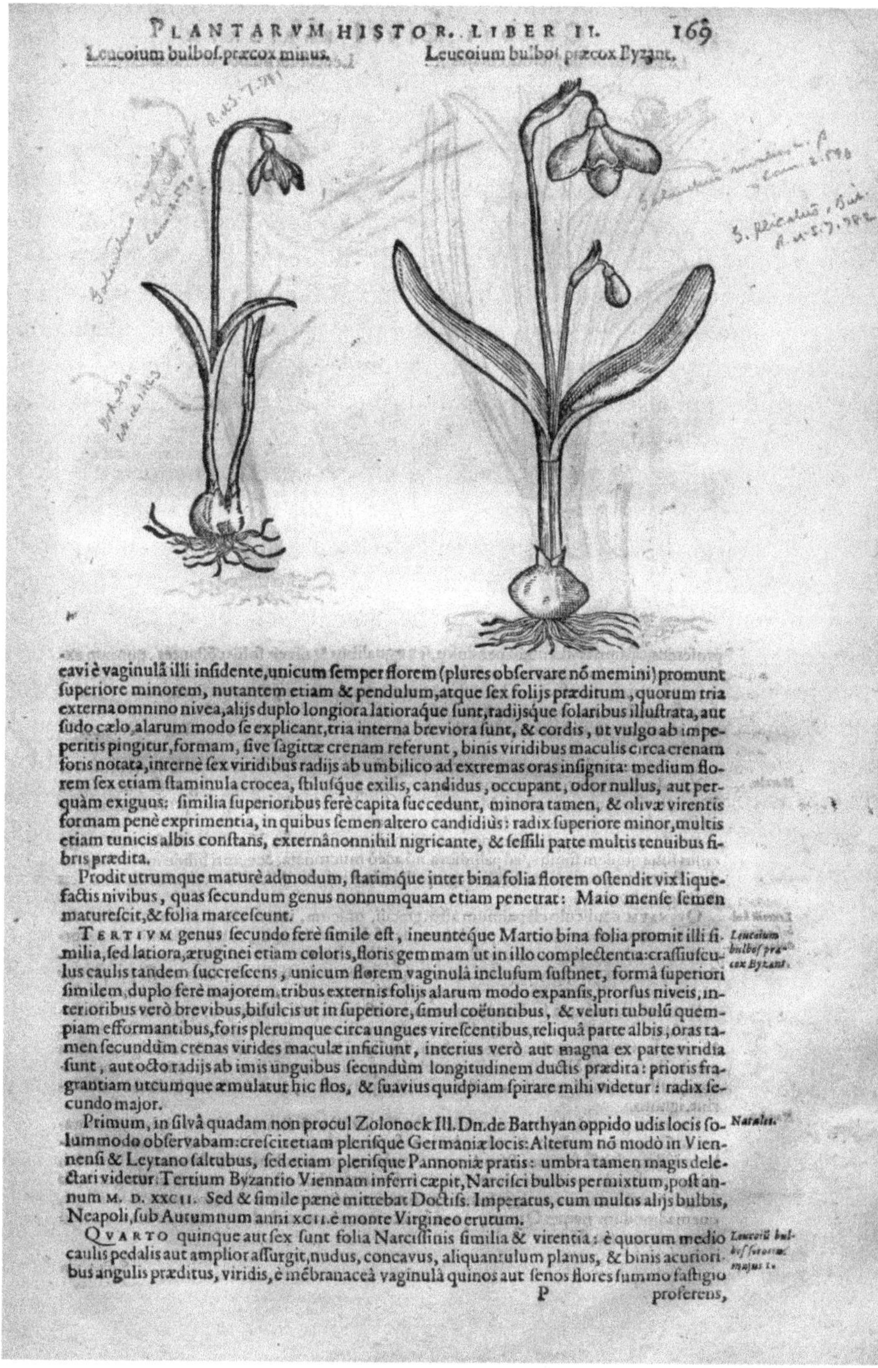

PLANTARVM HISTOR. LIBER II. 169

Leucoium bulboſ. præcox minus. Leucoium bulboſ. præcox Byzant.

cavi è vaginulã illi inſidente, unicum ſemper florem (plures obſervare nõ memini) promunt ſuperiore minorem, nutantem etiam & pendulum, atque ſex folijs præditum, quorum tria externa omnino nivea, alijs duplo longiora latioraque ſunt, radijsque ſolaribus illuſtrata, aut ſudo cælo, alarum modo ſe explicant, tria interna breviora ſunt, & cordis, ut vulgo ab imperitis pingitur, formam, ſive ſagittæ crenam referunt, binis viridibus maculis circa crenam foris notata, internè ſex viridibus radijs ab umbilico ad extremas oras inſignita: medium florem ſex etiam ſtaminula crocea, ſtilusque exilis, candidus, occupant, odor nullus, aut perquàm exiguus: ſimilia ſuperioribus ferè capita ſuccedunt, minora tamen, & olivæ virentis formam penè exprimentia, in quibus ſemen altero candidiùs: radix ſuperiore minor, multis etiam tunicis albis conſtans, externâ nonnihil nigricante, & ſeſſili parte multis tenuibus fibris prædita.

Prodit utrumque maturè admodum, ſtatimque inter bina folia florem oſtendit vix liquefactis nivibus, quas ſecundum genus nonnumquam etiam penetrat: Maio menſe ſemen matureſcit, & folia marceſcunt.

TERTIVM genus ſecundo ferè ſimile eſt, ineunteque Martio bina folia promit illi ſimilia, ſed latiora, æruginei etiam coloris, floris gemmam ut in illo complectentia: craſſiuſculus caulis tandem ſuccreſcens, unicum florem vaginulà incluſum ſuſtinet, formâ ſuperiori ſimilem, duplo ferè majorem, tribus externis folijs alarum modo expanſis, prorſus niveis, interioribus verò brevibus, biſulcis ut in ſuperiore, ſimul coëuntibus, & veluti tubulũ quempiam efformantibus, foris plerumque circa ungues vireſcentibus, reliquâ parte albis, oras tamen ſecundùm crenas virides maculæ inficiunt, interius verò aut magna ex parte viridia ſunt, aut octo radijs ab imis unguibus ſecundùm longitudinem ductis prædita: prioris fragrantiam utcumque æmulatur hic flos, & ſuavius quidpiam ſpirare mihi videtur: radix ſecundo major. *Leucoium bulboſ. præcox Byzant.*

Primum, in ſilvâ quadam non procul Zolonock Ill. Dn. de Batthyan oppido udis locis ſolummodo obſervabam: creſcit etiam pleriſque Germaniæ locis: Alterum nõ modò in Vienneſi & Leytano ſaltubus, ſed etiam pleriſque Pannoniæ pratis: umbra tamen magis delectari videtur. Tertium Byzantio Viennam inferri cæpit, Narciſci bulbis permixtum, poſt annum M. D. XXCII. Sed & ſimile pænè mittebat Doctiſs. Imperatus, cum multis alijs bulbis, Neapoli, ſub Autumnum anni XCII. è monte Virgineo erutum. *Natales.*

QVARTO quinque aut ſex ſunt folia Narciſſinis ſimilia & virentia: è quorum medio caulis pedalis aut amplior aſſurgit, nudus, concavus, aliquantulum planus, & binis acutioribus angulis præditus, viridis, è mẽbranaceâ vaginulâ quinos aut ſenos flores ſummo faſtigio *Leucoiũ bulboſ ſerotinũ majus I.*

P proferens,

FIGURE 17 Dodoens' snowdrop (left) and Clusius' snowdrop (right) from the *Rariorum plantarum historia* (1601), Missouri Botanical Garden, Saint Louis, MO, QK41.C57 1601 [#1006]

*Florum.*[17] De Bry's versions of the snowdrops are very close to the herbal illustrations in composition, with the flowers' outlines ever-so-slightly softer and less angular than the woodcuts.[18] De Bry took a major creative license, however, by cutting Clusius' snowdrop in half in order to make all the parts of the flower visible in the picture plane instead of hiding behind another flower.

Dodoens' and Clusius' snowdrops are also copied, with a greater level of artistic liberty, by Van de Passe II for the *Hortus floridus* (Fig. 5). Unlike the *Florilegium novum,* the *Hortus floridus* mostly showcases flower compositions that had not appeared before the publication of the volume. The snowdrops plate is unique in this florilegium because they are the only two depictions that are recognisable copies from existing image sources, and more specifically from the herbals of Dodoens and Clusius. The *Leucoion bulb. Triphyllon* (Fig. 5) in Van de Passe II's *Hortus floridus* was copied from Dodoens' snowdrop (Fig. 17). The variety of *Leucoion Triph. Byzant.*, on the other hand, is a copy of Clusius' snowdrop (Fig. 17). Van de Passe II did not alter the composition of Dodoens' snowdrop, keeping the artistic decision to depict the small sprout that comes out of the offshoot, which makes this illustration a distinctive composition from other representations of the same flower. Van de Passe II, however, softened his rendering of the design by giving the flower a slightly curvier stem, a rounder bulb, and a floppy leaf on one side, making Dodoens' snowdrop stiff in comparison. Clusius' snowdrop received a similar softening treatment, as Van de Passe II rounded the curvatures of the stems so that the joints where the stems turn and bend are not as sharp as in the woodcut. Moreover, the engraver cut the bulb and 'planted' Clusius' snowdrop in the ground; the bulb is, instead, placed nearby the flower. The copied sunflowers and snowdrops, along with many other copied flowers from the herbals of Dodoens, Clusius, and Lobelius, show that a portion of the florilegium visual culture came from the herbal pictorial tradition.

In addition to herbal woodcuts, seventeenth-century florilegia sometimes copy flower depictions from one another. This is most observable in De Bry's *Florilegium novum.* Aside from the herbals of Dodoens, Clusius, and Lobelius, a number of flowers in De Bry's flower book are modelled after the plates in *Le jardin du roy. …* First published in 1608, the book was produced by Vallet, based on the rare and curious flowers in the gardens of the French king Henry IV. Vallet dedicated the pictorial collection to Marie de Medici (1575–1642), who had an active embroidery circle, giving this flower book the additional functions of

---

17 Clusius, *Rariorum plantarum historia*, 169.

18 The engraving has a much bigger physical measurement than the woodcuts, so the snowdrop is also enlarged.

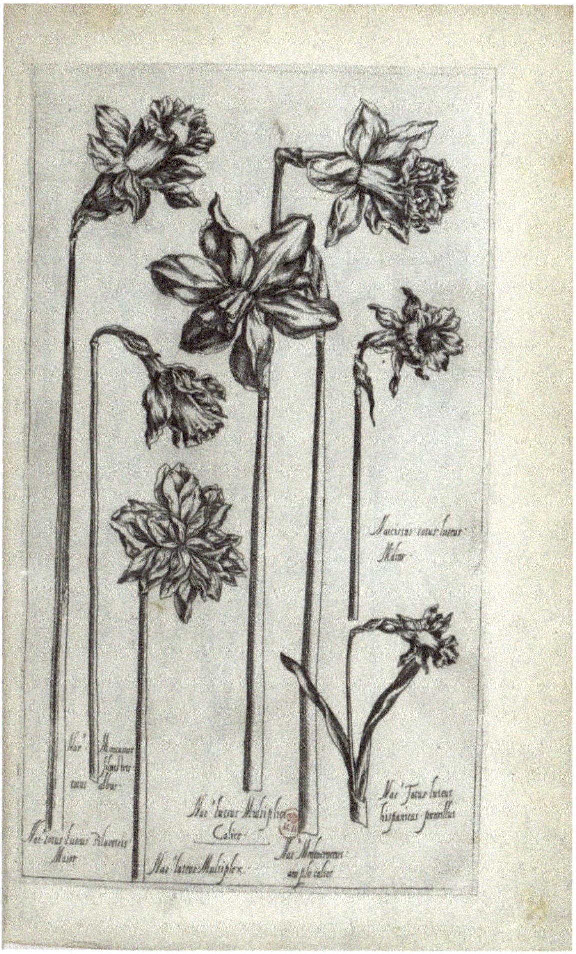

FIGURE 18
Narcissus plate from the *Le jardin du roy …* (1608), BnF, National Library of France, Paris, ark:/12148/btv1b8612033t

a pattern book and a garden catalogue.[19] One of the plates in Vallet's flower book presents seven varieties of single and double narcissi (Fig. 18). The same narcissi made an appearance four years later in the *Florilegium novum* (Fig. 19) in 1612. De Bry did not copy the plate in its entirety. Rather, he closely copied six of the blooms, excluding the *Narcissus Montanus silvestris totus albus*, and made a different visual arrangement with the flowers; he also added another variety of the *Pseudo narcissus juncifol. luteus minor*, some leaves to one of the copied narcissus, and three bulbs to his plate. The examples discussed above

19 Saunders, *Picturing Plants*, 55.

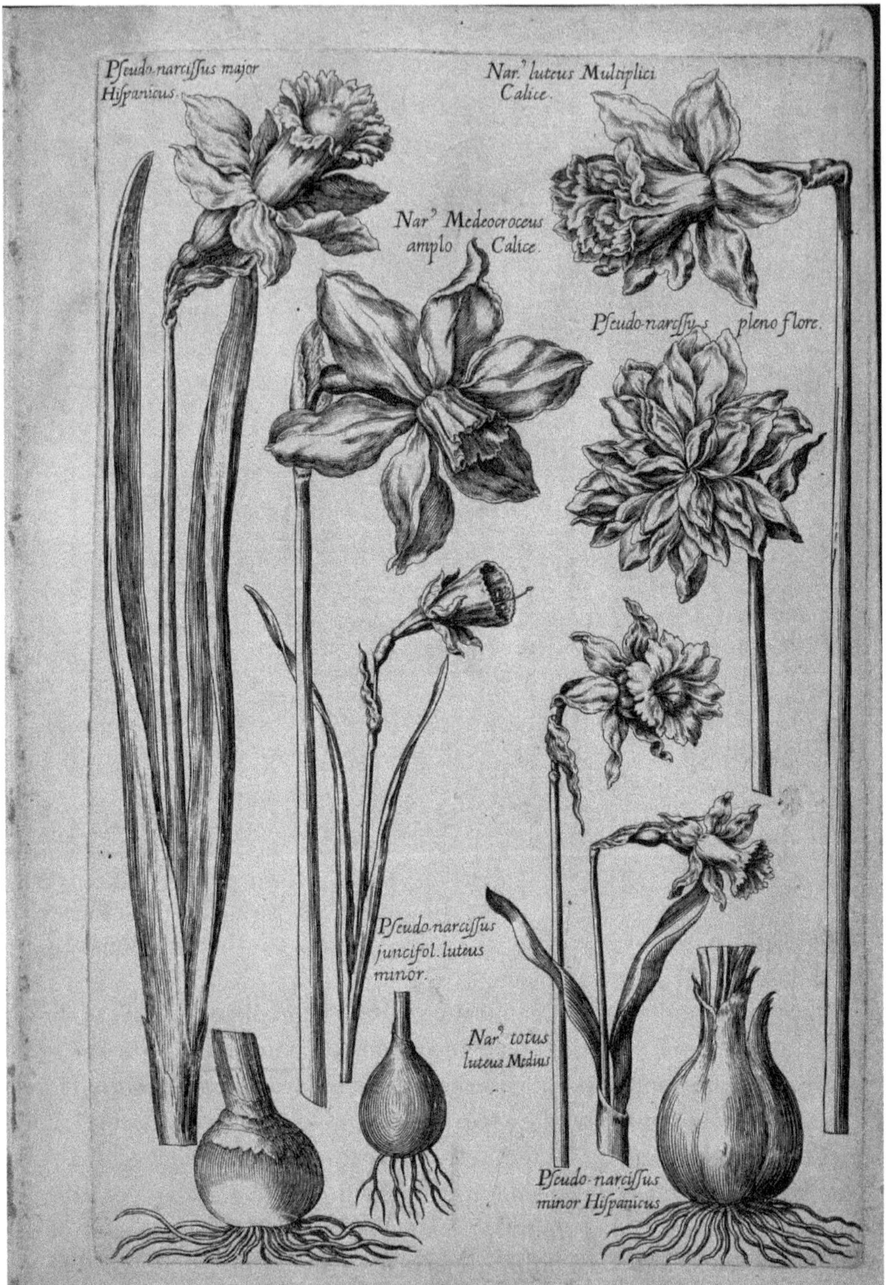


FIGURE 19 Narcissus plate from the *Florilegium novum* (1612), Chicago Botanic Garden, Chicago, IL, QK41.B79 1611

show that seventeenth-century florilegia did not hesitate to include pictures copied from published sources.

Before delving into the conceptual reasons behind this copying practice in the next section, it is important to take into account that some cases of copied flowers could simply be a result of lacking access to living specimens. As demonstrated in Chapter 1, the flowers depicted in seventeenth-century florilegia were rare during the period. While compiling a florilegium was done by people from many levels of society, owning a living collection of rare and curious flowers was only achievable for the group of *liefhebbers* with certain resources and means. Compilers of the florilegia that had clear connections to specific gardens and the rare and curious flowers growing in them, such as *Le jardin du roy …* and the *Hortus Eystettensis*, had access to live specimens.[20] Vallet could have easily drawn the flowers from life and Basilius Besler is recorded to have sent cut flowers to the contracted image makers.[21] In contrast, De Bry and Van de Passe II did not base their flower books on one specific rare flower collection. The Dutch edition of the *Hortus floridus* made explicit that Van de Passe II most likely observed and drew flowers from gardens around the Dutch Republic. Not included in other editions, the Dutch edition provides a list of *liefhebbers* and the provinces in which they were located. This edition also includes a poem that praises the pleasant flowers in their gardens, which were now made accessible to more people through Van de Passe II's efforts.[22] As the compiler wrote that snowdrops were only grown in curiosity gardens in the Low Countries, it is possible that the gardens Van de Passe II visited did not have the snowdrops in their living flower collections, or at least not profuse enough for him to draw from life.[23] Such reasoning can easily apply to De Bry's copying of flowers as well. The rare flower collections in *liefhebbers'* gardens around Oppenheim where De Bry worked were most likely (much) smaller than that of the French king.[24] Copying flowers, on a practical level, could be

20 Besler, *Hortus Eystettensis*. See Chapter 1 for an introduction to this florilegium.

21 It is not certain how the image makers for these two projects worked to produce the depictions, but broadly agreed that the images were drawn from live specimen, see Saunders, *Picturing Plants*, 55; Segal and Alen, *Dutch and Flemish Flower Pieces*, vol. 2, 1015. See Chapter 5 for more discussions on (not) depicting flowers from life.

22 Printed on the verso side of the titlepage in Van de Passe II, *Den Blom-hof*.

23 The description for the plate in the Latin edition states that the flower '[…] abounds in Italy; but is not to be found here except in the gardens of the curious.' (*In Italia frequens est, hic nisi in hortis curiosorum minime invenitur.*) in Van de Passe II, *Hortus floridus*, 4 (winter garden). English translation from Savage, *Garden of Flowers*, 212.

24 De Bry had a little garden, but it is unlikely that he would have had a large number of rare flowers and plants growing in it, see Van Groesen, 'The De Bry Collection of Voyages,' 110.

a means to pictorially acquire a rare specimen for an image collection of rare and curious flowers.

## 3 Excerpting Flowers, Extracting Knowledge

Portraying the physician-botanists on the frontispieces, however, goes beyond the fact that the illustrations in their works occasionally served as models for florilegia to copy rare and curious flowers that were hard to access in real life. Van de Passe II's positioning of Dodoens' and Clusius' portraits could be seen as an act of paying homage to the two physician-botanists, as researchers Sam Segal and Klara Alen wrote.[25] Art historian Jaya Remond similarly argued that the portraits are 'discursive and validating tools' to link the images in the *Hortus floridus* to the plant knowledge of the time.[26] The interest in the rare and curious flowers that commonly appear in seventeenth-century florilegia was already on the rise in the sixteenth century. Dodoens, Clusius, and Lobelius were among the first European authors to publish descriptions and depictions of these flowers, along with other rare and common medicinal and utilitarian plants. Dodoens' herbal is one of the most translated books in the early modern period, and Clusius was a major contributor who described many plants that were rare and exotic to Europe. While Lobelius is less studied nowadays, he was highly active in publishing herbals in Antwerp and London and was often cited by his contemporaries. The trio's works were translated, usually from Latin, into several languages, including Dutch, English, and French, and were widely read in the Low Countries and abroad.[27] The three physician-botanists were thus sensible authoritative figures to pay homage to and to present as validating figures. What remains underexplored, however, is not *why* these intaglio-printed florilegia included herbal materials, but *how* the compilers engaged with sixteenth-century herbals, aside from copying certain woodcuts. A more nuanced understanding of early modern copying practice, as well as the concept and practice of excerption and extraction, is useful for unpacking the *how* and reconsidering the epistemic choices made by the compilers.

25 Segal and Alen, *Dutch and Flemish Flower Pieces*, vol. 2, 1012.

26 Remond, 'The Pictorial Idioms of Nature,' 273–296; especially 289.

27 On the publications of the three physician botanists' herbals and botanical treatises, see especially De Nave and Dirk Imhof, eds., *Botany in the Low Countries*. On Dodoens, see Gilias, Van Tilburg, and Van Roy, eds., *Rembert Dodoens: Een zestiende-eeuwse kruiden-wetenschapper*. On Clusius' dealings with rare and curious flowers and plants, see notably Chapters IV, V, and VI in Egmond, *The World of Carolus Clusius*.

Scholars from both fields of history of art and science, broadly speaking, have shown that copying as a concept and practice is integral to many artistic and scientific developments in the early modern period.[28] In particular, contributors to the book *Making Copies in European Art 1400–1600*, edited by Maddalena Bellavitis, researched the various meanings of making copies of artworks, from copying as a way to imitate and emulate artistic techniques and styles to copies as witnesses to lost or destroyed works. Regardless of the reasons for making copies, 'copies almost always exhibit deliberate variations, additions, subtractions, and reimaginings,' as Peter M. Lukehart wrote for the volume's introduction.[29] Such alterations require making decisions on what to retain, eliminate, add, and change, no matter how small the decisions are. While focusing on different kinds of images (and texts), the contributions in the recent special issue of 'The practice of copying in making knowledge in Early Modern Europe' in the journal *Word & Image*, edited by Sietske Fransen and Katherine M. Reinhart, further reinforce how copying is a process of deliberate decision making.[30] Copying text and images was not a mindless process. Instead, 'copying was also employed as a way to prioritise, confirm, and validate knowledge—a practice that was especially prevalent within collaborative scientific environments.'[31] Copying, in this regard, is close to the practice of *excerption* in acquiring information and knowledge.

The genre of the florilegium, in both the literal and figurative meaning of 'gathering flowers,' pictorially assembled a collection of flowers and excerpted the flowers with the most sought-after qualities from authoritative sources of plant studies of the time. As shown in the previous section, even though the sunflowers and snowdrops have identifiable features showing them to be partial copies of herbal woodcuts, the copies are modified and reiterated. The elimination of leaves and roots for Collaert's sunflowers, in particular, invokes a more *literary* interpretation of the term *florilegium* as a device for collecting

28 A few pieces that are not directly relevant to the images and objects that are discussed in this chapter, but shaped the thinking of this chapter with interesting questions on copying are Preciado, ed., *Retaining the Original*; Lisa Pon, *Raphael, Dürer, and Marcantonio Raimondi: Copying and the Italian Renaissance Print*. Closer to natural historical images, see Egmond, 'Curious Fish: Connections Between Some Sixteenth-Century Watercolours and Prints,' 245–272; Egmond and Kusukawa, 'Circulation of Images and Graphic Practices in Renaissance Natural History,' 29–72.

29 Bellavitis, ed., *Making Copies in European Art 1400–1600*, 7.

30 Fransen and Reinhart, 'The Practice of Copying in Making Knowledge in Early Modern Europe,' 211–222. See also articles from the same issue.

31 Fransen and Reinhart, 'The Practice of Copying in Making Knowledge in Early Modern Europe,' 217.

excerpted sentences. As historian Ann Blair has dissected, the term *florilegium* in its literary use stands for a genre that selects 'the best passages, or 'flowers' from authoritative sources,' instead of summarising a piece of written work. It is often used for medieval collections of sentences, but the term was likely first used by the Italian humanist Aldus Manutius (1449–1515) before the first quarter of the sixteenth century for a Latin translation of a collection of Greek epigrams.[32] The text that was excerpted or extracted from a body of text and compiled into the new collection changed and shifted according to the interests of focus and the time period, as Blair shows throughout her book, but this practice of selecting and collecting memorable text already existed in antiquity and continued through the Middle Ages and into the early modern period. A major difference between medieval excerpt collections and early modern ones, generally speaking, is that the former ones tend to highlight the 'flowers,' picking passages that were deemed most worthy for the person collecting the text. Early modern note-taking collections, on the other hand, tend to be expansive, acquiring as many pieces of textual information as possible.[33] This line of difference is artificial, but it is useful here as a way to distinguish what people focused on for their text collections.

The pictorial genre of the florilegium in the seventeenth century is similar to the medieval mindset of bringing 'flowers' forward in this regard. This is not to claim that Collaert had a deliberate intent when using the word *florilegium* for his booklet. Rather, it is to emphasise that the pictorial genre of the florilegium, at least among the printed ones, was practised similarly to the literary genre in its excerption of existing works.[34] Returning to Collaert's sunflower plate, the compiler and image-maker excerpted the blooms of the sunflowers, the ranunculus, and the dogtooth violet—keeping only the 'flower' of the flowers—instead of copying the whole composition of the woodcuts for the herbals of Dodoens and Lobelius. The blooms were the main part of the flowering plant that possessed the rare and curious features of unusual petal counts and/or colours. Excerpting the blooms and visually eliminating the roots and some leaves resulted in undivided attention paid to the flowers.

32 Blair, *Too Much to Know*, 34–35.

33 Dana Jalobeanu more specifically discussed how seventeenth-century naturalists and natural philosophers used some of these note-taking devices investigated by Blair in 'The Toolbox of the Early Modern Natural Historian,' 107–123.

34 Another related word for botanical publications, *flora*, which refers to books or catalogues that document and/or describe plants of a particular geographical regions in the wild, similarly had different literary uses in the early modern period, see Berrens, 'The Meaning of Flora,' 237–249. Writing botanical texts was also anthological to a high extent, see Knight, *Of Books and Botany in Early Modern England*, 46–49.

Moreover, Collaert's *Florilegium* excerpted the rare and curious flowers from the whole corpus of botanical woodcuts in the works of Dodoens, Clusius, and Lobelius. The herbals of the three physician-botanists contain hundreds of, sometimes more than one thousand, woodcuts of plants, among which include images of grains, medicinal plants, mushrooms, and trees. Flowers, especially rare and curious ones, constitute only a small portion of them. Dodoens, Clusius, and Lobelius shared many of the woodcuts in their works, largely due to them having the same printer-publisher, Christophe Plantin (1520–1589) and his successors of the Plantin Press (*Officina Plantiniana*).[35] Plantin commissioned a team of image makers to produce a large stock of botanical woodblocks, with Pieter van der Borcht (1535/40–1608) attributed to be the draftsman and Arnold Nicolai (active 1555–*c*.1568), Gerard Janssen van Kampen (active 1564–1583), Cornelis Muller (active 1564–1572), and Antoon van Leest (active 1566–1584) as the recorded woodcutters for the blocks.[36] A large number of these botanical blocks were reused and recycled to print the illustrations for different titles and editions of the three physician-botanists' works.[37] Without counting the repeated appearances of many of the illustrations, the entire stock still amounts to more than 3,000 images of plants. Out of this vast number of plant images, Collaert picked those of the rare and curious flowers and excerpted the depictions into his small florilegium.

In addition to the excerptions of the blooms from selected woodcuts and the rare and curious flowers from the whole stock of plant images, the first wave of the printed florilegia extracted the botanical knowledge of the selected flowers from the herbals of Dodoens, Clusius, and Lobelius. The Latin edition of the *Hortus floridus* is a prominent work for demonstrating this practice of extracting botanical knowledge in a flower book. This version of the flower book is perhaps unique within the genre in the way that it provides a high level of textual information, considerably more than other editions of the title and other titles in the genre. Among all the entries, Van de Passe II's snowdrops remain the most straightforward example of such practice in text, especially within the context of the copied flowers. The Latin description for the *Leucoion Triph. Byzant.* in the *Hortus floridus* states,

35 Voet, 'Christopher Plantin as a Promoter of the Science of Botany,' 39–45.

36 See 'Chapter 8 Illustration' in Voet, *The Golden Compasses*, vol. 2, 194–243.

37 On the use and reuse of the Plantin botanical stock, see Chen, 'A Woodblock's Career,' 20–63.

> [...] *tribus externis folijs alarum modo expansis plane niveis, interioribus vero brevioribus simul coninuctis,* [...][38]

The original sentence by Clusius is written as,

> [...] *duplo fere majorem tribus externis folijs alarum modo expansis, prorsus niveis, interioribus vero brevibus,* [...][39]

The passages inform, in essence, that the three snow-white petals in the outer ring are larger than the three inner ones. While not identical, the choice and arrangement of words for describing the excerpted picture of Clusius' snowdrop shows that Van de Passe II closely consulted the herbal for selecting visual and textual information to be included in his flower book. Even though Dodoens' and Clusius' names are not always cited explicitly in every entry, most of the descriptions in the *Hortus floridus* are easily recognisable as summaries or reiterations of the plant descriptions by the two physician-botanists.[40]

For his florilegium, Van de Passe II extracted particularly the information on the flowers' appearances, including their colours, and excluded their virtues and/or utilitarian functions found in the herbals. For example, even though Dodoens' 1568 *Florum* features flowers and decorative plants that have relatively few medicinal and utilitarian functions, these properties are still regularly documented. For the entry of the sunflower, Dodoens mostly described the large appearance of the flower. He also briefly touched on the culinary use of cooking the sunflower head like an artichoke and its possible aphrodisiac effect after consuming the plant, which was noted by his contemporaries in Italy.[41] Van de Passe II extracted much of Dodoens' information on the appearance of the sunflower, addressing particularly the massive size of the plant;

38 '[...] (with) three outer petals/leaves extended in the way of wings, clearly as white as snow, but shorter inner (ones), conjoined together [...]' in Van de Passe II, *Hortus floridus*, 4 (winter garden). English translation by Ada Migliazza.

39 '[...] (with) tree external leaves/petals two times bigger, expanded in the way of wings, of the precise colour of snow, while the inner ones are shorter [...]' in Clusius, *Rariorum plantarum historia*, 169. English translation by Ada Migliazza.

40 Van de Passe II may have read other authors of plants as well, indicated by his mentions of figures such as the ancient authors of Pliny the Elder and Dioscorides and the Italian naturalists Pietro Andrea Mattioli. However, these could simply be extracted from Dodoens' works since the physician-botanist often discussed in his herbals what these authors wrote.

41 Peacock, *The Look of Van Dyck*, 111.

he forwent completely its practical use, albeit limited to begin with during this period.

The florilegium compiler, however, did not simply extract information of what was already known. He sometimes confirmed, commented on, and/or disagreed with the cited authors as well. A major example regards the categorisation of an 'anemone.' Van de Passe II placed the two flowers on plate 18 as two varieties of the slender-leaved anemone (*Anemone tenuifol*). For the 'Flower of Adonis, Eranthemum of Dodoens' (*Flos Adonis Eranthemum Dod.*), the text states,

> The herb, which, Dodonaeus calls *Eranthemum*, and Mathiolus wrongly the flower of Adonis, since that flower very well known from the tales of the poets, seems to be a species of Anemone.[42]

Van de Passe II is most likely the outlier for categorising the plant as an anemone variety, but this commentary shows that the compiler read and digested the sources in the process of extracting the plant information.

The *Hortus Eystettensis* presents yet another way of extracting information by compiling and citing sources for each and every entry. The *florilegium*, in its literary use, sometimes also works as a reference genre, and the *Hortus Eystettensis* has a comparable collection of references in its printed text.[43] For example, Besler included two kinds of sunflowers, one with a single giant head, the *Flos Solis maior*, and another with multiple heads of smaller flowers, the *Flos Solis prolifer*.[44] Similar to Van de Passe II's entries, Besler's description of the giant sunflower focuses primarily on the appearance of the plant. At the very beginning of the entry, Besler pointed out that the sunflower had multiple names, including the *Flos Solis major*, *Helianthemum*, *Chrysanthemum Peruvianum*, and the *Helenium Indicum*. These were not names in the vernacular, as Besler also provided the flower's German name of *Groß Indianisch Sonnenblumen*, but the Latin names used by different botanists or naturalists of the time who had previously described the flower. The *Chrysanthemum*

42 '*Herba, quam Eranthemum nuncupat Dodonaeus; & Mathiolus non recte Adonidis florem, cum iste flos poetarum fabulis notissimus, Anemones species videatur* [...]' in Van de Passe II, *Hortus floridus*, XVIII (spring garden); Savage, *Garden of Flowers*, 36.

43 Blair, *Too Much to Know*, 124–126.

44 Besler, *Hortus Eystettensis*, quintus ordo. fol. 1, 2 (summer garden); Littger and Dressendörfer, *Florilegium: The Book of Plants*, plates 204, 205. This Taschen reproduction is the most accessible and widely circulated secondary source, with clearly marked plate numbers and index of plants; this book refers to the plate numbers as given in this reproduction.

*Peruvianum,* which term is also adopted by Van de Passe II, was named by Dodoens and indicated the physician-botanist's thought that the flower was a kind of chrysanthemum from the Americas. The Italian naturalist Pietro Andrea Mattioli (1501–1577), according to Besler, called the flower *Helenium Indicum* instead.[45]

Early modern botanists and naturalists often named the plants according to how they saw fit, as this was a moment in time before the Linnaean classification system developed and became widely applied for standardised naming.[46] A way to achieve some level of concordance was not to standardise plant names for each plant. Instead, botanists and naturalists often compiled the names in several vernacular languages and Latin variations to accompany their description of a plant.[47] One significant feature of the Lobelius' *Kruydtboeck*, for example, is the extensive index of plant names in Dutch, Latin, German, French, Portuguese, and English, compiled by the physician-botanist from different sources.[48] This provided readers of herbals and botanical treatises with translations of plant names as reference points to identify and organise plants when reading works from different authors and European regions. The many names of the sunflower listed by Besler are not far from the efforts contemporary botanists and naturalists made to identify plants in text. The difference is that Besler did not add new Latin or vernacular names to the mix.

The names Besler and other florilegia compilers used for referring to the flowers in the captions or descriptions thus work as strong indicators of the authors and works they were reading, even if no explicit names of the botanists are given. At the bottom of the giant sunflower description, Besler provided a list of contemporary authors and their works, as well as the page numbers on which the sunflower entries can be found. For example, readers can find the description of the sunflower in 'Dodon. Fol. 264.,' which refers to page 264 in Dodoens' folio-size herbal, the *Stirpium historiae pemptades sex*. The

45 For a more streamlined account on who called the sunflower what in the sixteenth and seventeenth centuries, see Peacock, *The Look of Van Dyck*, 110–114.

46 Anna Pavord delved more into the different systems on a longer historical timeline in *The Naming of Names*. Many of these naming practices are currently under scrutiny for their connection to the history of racism and othering, see Mabry et al., 'Building an Inclusive Botany', 544–557.

47 See especially discussions by Florike Egmond and Philippe Selosse on the issues of communicating different naming systems in the early modern period. Egmond, 'Names of Naturalia in the Early Modern Period,' 131–162; Selosse, 'The Role of Carolus Clusius (1526–1609) as Translator in the Emergence of a Taxonomic Terminology in Botany,' 349–368.

48 Wille, 'Botanical Collectors and Collections in the Low Countries,' 80.

works referenced for Besler's *Flos Solis maior* include the herbals of the familiar figures of Dodoens and Lobelius, as well as the books of the Swiss botanist Caspar Bauhin (1560–1624), French botanist Paul Reneaulme (1560–1624), and the Spanish botanist Nicolas Monardes (1493/1508–1588), among others. The *Hortus Eystettensis* directs its readers to further readings on the rare and curious flowers from the recent scholarship of the period, functioning similarly to a reference book, albeit a highly extravagant one. These aforementioned examples show the several ways in which the printed florilegia extracted the visual and textual knowledge from sixteenth-century herbals through copying, excerption, and referencing, which reinforced the reasons for portraying Dodoens, Clusius, and/or Lobelius on the frontispieces.

A reprinted edition of De Bry's *Florilegium novum*, however, presents a clear change of direction in the book genre from which to excerpt, by copying plates from a horticultural treatise instead of a herbal. All of the printed florilegia discussed so far were reprinted multiple times, which will be further discussed in Chapter 4. De Bry's *Florilegium novum*, in particular, grew larger in size (plate counts) with every new version. Some of the versions may have been issued more similarly to supplements, including new plates but not those that initially made for the *Florilegium novum*.[49] The most significant reedition, however, is the *Florilegium renovatum et auctum*, which was published in 1641, almost twenty years after De Bry's passing, by his son-in-law and engraver Matthäus Merian (1593–1650).[50] For disambiguation, this chapter refers to the 1612 publication as De Bry's florilegium and the 1641 reedition as Merian's florilegium. The 1641 florilegium of Merian is significant in two ways. First, Merian did not simply add more depictions of rare and curious flowers, as De Bry's previous reprints and/or supplements had. Instead, Merian connected the flowers and plants in the book to the garden of the Frankfurt Burgomaster Johannes Schwindt (1580–1648), as stated,

> We have renewed and enlarged this florilegium, but also, for a better adornment of this work, we have treated first that famous Garden of Schwindt, celebrated by many, which can be visited here.[51]

49 See Chapter 4 for more information and discussion on reprints and reeditions of the intaglio-printed florilegia.

50 De Bry and Merian, *Florilegium renovatum et auctum*. Bibliographic description numbered C85 in Bridson and White, *Plant, Animal & Anatomical Illustration in Art & Science*, 44–45.

51 '*Florilegium hoc renouauimus & auximus, sed etiam ad maiorem operis ornatum, Hortum illum Schuuindianum a multis celebratum, qui hic uisitur, praemisimus.*' in De Bry and Merian, *Florilegium renovatum et auctum*, 13. English translation by Ada Migliazza.

This makes Merian's version of the flower book more similar to the *Hortus Eystettensis*, serving partially as a catalogue of Schwindt's rare and curious flower and plant collections.

Second, Merian excerpted the plates from the book *De florum cultura*, first published in 1633 by Italian Jesuit Giovanni Baptista Ferrari (1584–1655).[52] Merian explicitly announced, in the preface, that more than 30 plates included in the *Florilegium renovatum et auctum* were diligently reproduced from Ferrari's 1638 reprint of *De florum cultura*.[53] Similar to all the examples discussed in this chapter, Merian's copies altered and modified their visual sources. The plates in *De florum cultura* often include a ribbon that inscribes the name of the depicted flower, serving as an ornate caption that becomes part of the composition of the images. Merian's plates removed the ribbons, keeping only the flowers for his copies. In a few cases, Merian combined images that were initially printed on multiple plates in Ferrari's books into one picture plane, compiling the visual sources into one plate.

These excerpted plates show that the florilegium progressively aligned its (visual) interests in contemporary horticultural development instead of the herbal tradition. Even though the early modern categorisations of book genres were not strict, *De florum cultura* is clearly not a herbal. The book, as its title suggests, focuses on the (horti-) culture of flowers. Its several chapters cover some general how-to instructions on gardening and the practical growing of flowers, as well as the ways of identifying, arranging, preserving, and even creating rare and curious flowers that were being practised throughout Europe during this period. While it may not strictly be a horticultural treatise in definition, *De florum cultura* presents the cultivation and use of decorative flowers, especially the rare and curious ones, where their medicinal or utilitarian qualities take a backseat, if at all present. By extracting the plates from Ferarri's books, Merian's version of the florilegium further aligned the pictorial genre with the interest and knowledge of horticulture in the seventeenth century. All in all, the excerption, extraction, and referencing of herbals and horticultural treatises in florilegia does not make the genre a diluted version of the works that have been considered more scholarly. Rather, they are deliberate choices that emphasise and prioritise different kinds of (visual) knowledge of rare and curious flowers.

---

52 Ferrari, *De florum cultura libri IV*.

53 De Bry and Merian, *Florilegium renovatum et auctum*, 13.

## 4 Putting Flowers in Seasonal Order

The biggest departure from the herbal tradition that showcases how the florilegium prioritises different kinds of knowledge through the excerpted information on rare and curious flowers is the seemingly simple ordering of flowers according to their blooming seasons.[54] Before diving into the seasonal order for organising plates and leaves in seventeenth-century florilegia, it is important to emphasise that this is not a standardised nor universal system across the genre. Sweerts, for example, divided his florilegium into two parts, with the first featuring bulbous and tuberous flowers, and the second rooted ones.[55] Additionally, early modern books always have a layer of customisation, and inserts that are not part of the printed corpus of a title are often found.[56] However, added materials do not tend to alter the sequence of the reading experience. Compared with other book genres, surviving copies—in this section, it means the physical book item from the same title—of a printed florilegium title can vary drastically from one to another. Some copies include so many additions, such as plates from other titles, that the intended order is largely altered, if there was an intended order with which to begin.[57] Manuscript florilegia, on the other hand, are usually bound long after the completion of individual sheets. While the final order(s) of leaves in the bound volumes can often be seasonal, this system might not have been the intended one during the initial owners' lifetimes.[58] The content and order of each and every florilegium,

54 Almost every piece of secondary literature on the genre writes about the same observation, that florilegia arrange their plates in seasonal order, but usually undermines this system by stating that it is not a 'botanical' or 'scientific' approach for organising plants.

55 Sweerts, *Florilegium Emanuelis Sweerti …*, '*Ad candidum lectorem*'; Bleiler, *Early Floral Engravings*, xii.

56 There are some debates of whether or not early modern books were sold bound or unbound. Most likely both options would have been available. See especially Reynolds, 'Worthy to Be Reserved,' 350.

57 The surviving copy of Merian's florilegium at The Newberry Library in Chicago is a prominent example. The order is almost the reversal of most copies of the same title, and plates from other books (to be identified) extracted and added to this copy. I was only able to see selected parts of this copy, based on the photos and videos shared with me by Suzanne Karr Schmidt through email correspondences, but they already show a clear impression of the copy. De Bry and Merian, *Florilegium renovatum et auctum*, Chicago, The Newberry Library, 318931, https://i-share-nby.primo.exlibrisgroup.com/permalink/01CARLI_NBY/i5mcb2/alma99427513880 5867.

58 The Gottorfer Codex by the German painter Hans Simon Holtzbecker (see Chapter 1 for an introduction of this florilegium), for example, has a few sets of numbers on the sheets, indicating that the collection experienced different ordering systems, see De la Fuente Pedersen and Kolind Poulsen, eds., *Flowers and World Views*, 114–118.

even when the plates are from the same printed title, can thus fluctuate and vary.

Despite the inconsistencies within the genre, ordering flowers according to their blooming seasons remains the most common approach in organising the rare and curious flowers in seventeenth-century florilegia, and the *Hortus Eystettensis* and the *Hortus floridus* cemented this seasonal order. Both flower books were issued as a collection, releasing the plates at once in 1613 and 1614 for their first editions. Their later editions have changes mainly in the text and practically no growth in plate numbers.[59] Van de Passe II's plates and the accompanying text are numbered; Besler's plates are numberless, but the plant descriptions are given numbers. Both collections have indexes that list the names of the depicted flowers and the number of the folio or page for locating the depictions within the volumes. Although the order may be disrupted by additions or rearrangements in a handful of surviving copies, such disruption is comparably lower than the other titles of the printed florilegia, most likely due to the clear numbering and indexing in these two flower books. Going through the *Hortus Eystettensis* and the *Hortus floridus*, therefore, means going through a pre-determined sequence of the pictures provided by the compilers. Additionally, both florilegia include sub-titlepages that clearly mark the seasons of spring, summer, autumn, and winter. Each sub-titlepage is followed by the flowers, including decorative plants, that bloom during that season, functioning similarly to a book chapter. These two collections unquestionably applied a seasonal order for organising the depicted flowers that would become the common system for the genre of the florilegium.

The following object descriptions and plate counts for both florilegia are specifically based on the copies—the 1613 *Hortus Eystettensis* and the 1614 *Hortus floridus* in Latin—that are now preserved at the Royal Botanic Garden in Madrid (Biblioteca del Real Jardín Botanico).[60] An ornate engraving marks the beginning of every season in the *Hortus Eystettensis*. The engraved plate is used four times, giving all the sub-titlepages the same image, but changing the printed text so that they reflect the season in focus by printing the text and

59 See Chapter 4 for more information and discussion on reprints and reeditions of the intaglio-printed florilegia.

60 Besler, *Hortus Eystettensis*, Madrid, Royal Botanic Garden (Biblioteca del Real Jardín Botanico), Af BES, https://bibdigital.rjb.csic.es/idurl/1/10908; Van de Passe II, *Hortus floridus*, Madrid, Royal Botanic Garden (Biblioteca del Real Jardín Botanico), A PAS, https://bibdigital.rjb.csic.es/idurl/1/13646. The sub-titlepage for the autumn garden in the *Hortus floridus* is printed as 1616. It is highly possible that the year 1616 is a typo, as it has shown up on every copy I have consulted thus far, when all the other details point to them being the 1614 version.

image separately.[61] A cartouche contains the announcing text of the season, which states 'The Spring Order of Plants of the Eichstätt Garden' (*Plantarum Horti Eystaettensis Classis Verna*) for the chapter on the spring flowers. The cartouche is framed by the allegorical figures of flora and panacea (personification of medicine), each sits on top of a column; a selection of rare plants and animals (from the Americas) and some shells fill the space around the columns; an alembic and other distilling instruments occupy the bottom of the picture plane.

In the chapter on the spring garden, Besler included 134 plates, each plate (in the entire volume) depicts between one to six images of different species and/or variations of rare and curious flowers. The flowers that bloomed in this chapter include crocuses, anemones, ranunculi, hyacinths, narcissi, snake-head fritillaries, tulips, crown imperials, lilies, roses, peonies, irises, and many more. The summer garden is the most substantial chapter of the volume, containing 184 plates of flower depictions. Exhibiting their blooming flowers in this chapter are the carnations or gillyflowers, tagetes, bindweeds and morning glories (both considered variations of the *Convolvulus* in the early modern period), poppies, orchids, rose champions, hollyhocks, hibiscuses, pot marigolds or calendulas, sunflowers, campanulas, irises, martagon lilies, delphiniums, columbines, nigellas, stocks (*Matthiola incana* L.), and foxgloves, to name a few. The chapter on the autumn garden drastically reduces the number of plates to 42. Many of the flowers in this chapter are species from the Americas that were grown for decorative purpose during this time due to their rarity. One can find variations of cacti, artichoke buds (the edible portion, before they bloom), aloes, autumn crocuses (*Colchicum*), and cyclamens, as well as the tobacco flower, cannas, chilli peppers, a tomato plant, among others. The winter garden, expectedly, is the most barren chapter, consisting of only 7 plates. While small in number, the plates depict several variations of hellebores, snowdrops, pasqueflowers (*Pulsatilla* L.), and crocuses.

In contrast, the *Hortus floridus* has plain and simple sub-titlepages for each of the seasons. The spring garden begins with the index of flowers, following the general preface of the book. Likewise, the rest of the seasons are marked with simple sub-titlepages, with typeset text and decorative *fleurons*, but no images.[62] The indexes of flowers follow, and an engraving of an enclosed garden

---

61 For an introduction on the separate processes of printing text and images, see Gaskell, *A New Introduction to Bibliography*, 118–124, 154–159.

62 A *fleuron* is a typographic element of a small decorative design. It is casted in the same manner as a lead type for printing text. *Fleurons* can be set in the same way as types and printed together with text. This is a much simpler process than printing engravings or

with a geometric parterre appears before the flower depictions for every season. There are two versions of the garden plates. One shows a woman reaching for a tulip while a man looks into the garden; the flowers that are blooming in the plant beds are from the spring and summer chapters. The other garden plate shows a man watering a flower, and two others walking in the garden; the same plant beds contain flowers from the autumn and winter chapters instead. This careful differentiation of seasons in the two garden plates, however, does not correspond to the chapter divisions in most copies, as the Madrid copy has the autumn/winter plate for the spring chapter.[63] The spring garden is the most luxuriant in the *Hortus floridus*, containing 41 plates of flowers, with each plate in the book depicting one to four variations of a flower. Blooming in spring are the hepaticas, crocuses, narcissi, primroses, auriculas, irises, hyacinths, a crown imperial, anemones, tulips, snake-head fritillaries, columbines, ranunculi, orchids (*Satyrium* L.), and so on. The chapter on the summer garden is smaller, consisting of 19 plates with blooming peonies, irises, martagon lilies, lilies, gladioli, roses, campanulas, poppies, rose champions, carnations or gillyflowers, and nigellas, among others. The autumn garden grows larger to 27 plates. The flowers in this chapter include a clematis, two sunflowers, a canna, a few tagetes, two hollyhocks, a tobacco flower, cyclamens, and several autumn crocuses. The winter garden has 12 plates, with blooming hellebores, snowdrops, a winter aconite, hepaticas, crocuses, dogtooth violets, and a few more. Neither the *Hortus Eystettensis* nor the *Hortus floridus* have balanced distributions of flowers throughout the books, with some chapters much thicker than others. This reflects the natural ebb and flow of seasonality and creates an experience of visiting a garden throughout the year when going through the pages.[64]

Ordering flowers according to their blooming seasons may seem to be a simple and intuitive system of presentation, but this practice was not-at-all the default for publications on plants. Information related to seasons is not

etchings that requires a different press. Besler's sub-titlepages were printed with two different presses, one for the plate and one for the text. More on *fleurons*, see Vervliet, *Vine Leaf Ornaments in Renaissance Typography*.

63 A copy of the Dutch edition at Utrecht University has the right correspondence of garden plates and seasons, but the winter chapter shows the garden plate after the flowers instead of before. Van de Passe II, *Den Blom-hof*, Utrecht, Utrecht University Library Special Collections (Bijzondere Collecties Universiteitsbibliotheek Utrecht), R qu 851 obl (Rariora), https://utrechtuniversity.on.worldcat.org/oclc/66465813.

64 Many of the secondary literature that observed the seasonal order in florilegia have statements similar to this view of a 'virtual' garden. It is one layer of significance for this order, but not the only layer, as will be further discussed in the following paragraphs.

difficult to find when reading through the plant descriptions in herbals. For example, in the *Stirpium historiae pemptades sex* (1583), Dodoens wrote that the snowdrop, the variation of Dodoens' snowdrop (Fig. 17), emerges early in February, preceding almost all the other flowers.[65] Another variation, however, does not flower until the end of April, together with the narcissus.[66] This attention to flowering months or season-related information is not essential, nor consistent. Dodoens' entry for the sunflower has no mention of its flowering month(s).[67] Similarly, Lobelius did not mention at which time of the year the sunflower blossomed, even though he pointed out the effect of the local seasonal climate on the plant by stating that the sunflower has difficulty reaching maturity because it cannot survive the harsh winter in the Low Countries.[68]

Moreover, none of the herbals excerpted or referenced by florilegia compilers arranged their plants according to growing or flowering seasons. Early modern herbals or botanical treatises do not have a universal method of ordering plants. Instead, early botanists organised and arranged the plants they described using different systems according to what they saw as most suitable, if not placing plants in alphabetical order. For example, one book in the bibliography is the *Hortus medicus et philosophicus* (1588) by German physician-botanist Joachim Camerarius II (1534–1598), who might have been involved in the curation of the rare plant collection in the garden of Eichstätt and even the production of the *Hortus Eystettensis*.[69] Camerarius II ordered the plants alphabetically, and two variations of the sunflower, the *Flos solis* and the *Flos solis Matth.* can be found on pages 60 and 61, under the section for plants with names starting with the letter F, and in between descriptions of the *Flos Indicus* and *Foeniculum maius siue dulce Italicum*.[70]

65 Dodoens, *Stirpium historiae pemptades sex*, 229.

66 Dodoens, *Stirpium historiae pemptades sex*, 231.

67 Dodoens, *Stirpium historiae pemptades sex*, 263–264.

68 Lobelius, *Plantarum, seu, Stirpium historia*, 322.

69 Camerarius, *Hortus medicus et philosophicus*. Camerarius may have been the person who started dividing flowers according to their blooming seasons for the *Hortus Eystettensis*, see Barker, *Hortus Eystettensis: The Bishop's Garden and Besler's Magnificent Book*, 28. An important image collection that is often mentioned in conjunction with the *Hortus Eystettensis* but somewhat difficult to access is the Camerarius Florilegium, recorded to be at Erlangen University Library (Universitätsbibliothek Erlangen-Nürnberg), but does not appear in the library catalogue. A digital facsimile can be viewed through a CD-ROM by the Harald Fischer Verlag, *Camerarius-Florilegium* (Erlangen: Harald Fischer Verlag, 2004). Additionally, some useful descriptions can be found in Wickert, *Das Camerarius-Florilegium*.

70 Besler's citation only lists page 61.

The ordering of plants shares both similarities and differences among the three most referenced figures for the genre of florilegium. The publications of the works of Dodoens, Clusius, and Lobelius tend to follow a similar trajectory, compiling several of their titles into a big herbal close to the end of their careers; this chapter thus only examines their masterworks for this reason.[71] Each of Dodoens' six parts in the *Stirpium* focuses on a group of plants, and each part is further divided into more books (*liber*) and chapters (*caput*). The first part concerns plants that have long been used for medicine, most of which descended from antiquity and were already written about by figures such as the Greek physician Galen.[72] The second part is an updated version of the *Florum*, describing flowers for garlands and wreaths and fragrant and umbelliferous herbs.[73] A majority of rare and curious flowers and plants referenced in florilegia are categorised under this part, but many come from other parts of the herbal as well. Part three groups plants with different uses, including rooted plants, purgative herbs, trailing plants, harmful and poisonous plants, and also ferns, mosses, and fungi.[74] Part four focuses on grains, beans, marsh and aquatic plants, and other pertinent herbs.[75] The fifth part describes vegetables and cardoons.[76] The sixth and last part deals with fruits and trees.[77] Within these larger divisions of parts, Dodoens ordered the plants loosely according to the perceptions of the plants' natural affinities.[78]

Clusius's *Rariorum plantarum historia* categorised some plants he described in similar groups to Dodoens and others in very different ones, and he presented them in a different order. The main portion of the volume is his work on the history of plants, which is divided into six books. The first book starts with trees, bushes, and shrubs.[79] Book two deals with plants with bulbous or tuberous roots, a large part of which are elegant and eye-pleasing flowers that are suitable for garlands and wreaths.[80] Book three continues with rooted plants

71 Similar orders have already existed Dodoens, *Cruijdeboeck* (1554). Dodoens' later works, including the *Florum*, are updates of this 1554 Dutch herbal and compiled again as into the *Stirpium historiae pemptades sex* (1583), before being translated back to Dutch as the *Cruydt-boeck* (1608).

72 Dodoens, *Stirpium historiae pemptades sex*, 1.

73 Dodoens, *Stirpium historiae pemptades sex*, 155.

74 Dodoens, *Stirpium historiae pemptades sex*, 321.

75 Dodoens, *Stirpium historiae pemptades sex*, 479.

76 Dodoens, *Stirpium historiae pemptades sex*, 603.

77 Dodoens, *Stirpium historiae pemptades sex*, 729.

78 Dodoens, *Stirpium historiae pemptades sex*, 4; Arber, *Herbals: Their Origin and Evolution*, 174.

79 Clusius, *Rariorum plantarum historia*, 1.

80 Clusius, *Rariorum plantarum historia*, 127.

that are also beautiful and often fragrant.[81] Book four focuses on flowers that are elegant and pleasant, such as species of *Scabiosa* and *Aster*, many of which also have strong scents.[82] The fifth book describes plants that are narcotic or poisonous, and other plants that have leaves similar to these toxic plants.[83] The sixth book discusses miscellaneous plants that were difficult to be placed into a convenient order for Clusius, including umbelliferous plants, grasses, reeds, marsh and aquatic plants, and so on.[84] There are also several smaller books as addendums that focus solely on mushrooms and local-specific plants of a region following his overall history of plants. Within the books on flowers, both Dodoens and Clusius grouped the flowers more-or-less based on their morphological information, usually indicated by the names of the genus, and moved from one to the next. Clusius, for example, started his book two with descriptions of several kinds of *Lilium*, before moving to the many *Tulipa* he studied.

In contrast, the *Plantarum, seu, Stirpium historia* by Lobelius does not have dividers such as parts of books. The big herbal is usually a combination of two books. The first title, the 1576 *Plantarum*, was published and printed by Plantin and is of focus here. The second part, the *Nova stirpium adversaria* (1571), has a completely different origin, but was almost always sold together with the 1576 *Plantarum*.[85] The physician-botanist started with the group of *Graminis* (grass), followed by several kinds of *Tritici* (wheat), and continued to introduce one kind of plant after another without breaks until the end of the title. Lobelius noted the similarities among the leaves, roots, flowers, fruit, and other parts of the plants he studied. The grouping and ordering of plants are heavily based on morphology in his herbal.[86] Dodoens, Clusius, and Lobelius had varying degrees of interest in grouping and categorising plants according to their uses and morphological information, and their categories and the order of presenting these groups differ from one another. These examples show that grouping flowers according to their blooming seasons and placing them in a seasonal order throughout the year was not a consideration nor practice in presenting plant information in early modern herbals and botanical treatises.

81 Clusius, *Rariorum plantarum historia*, 282.
82 Clusius, *Rariorum plantarum historia*, j.
83 Clusius, *Rariorum plantarum historia*, lxxxiij.
84 Clusius, *Rariorum plantarum historia*, clxxx.
85 The book was an earlier work by Lobelius and the French physician-botanist Pierre Pena in London under the title *Stirpium adversaria nova* (1571). The compilation of the two titles resulted in some complications in bibliographical descriptions, see Voet, *The Plantin Press, 1555–1589*, vol. 3, 1406–1411.
86 Lobelius, *Plantarum, seu, Stirpium historia*, 5; Arber, *Herbals: Their Origin and Evolution*, 176–177.

This is not to imply, however, that flower books present their flowers in a month-by-month manner, showing a sequence that fully reflects what was blooming in a garden throughout the year. For example, Van de Passe II included several varieties of narcissus, most of which appear in the spring garden.[87] The plate containing the two variations of the large narcissus (*Narcissus omnium maximus* and *Narcissus maximus duplicat follis*) is the fourth entry of the spring garden; the two rush-leaved narcissi (*Narcissus iuncifol*) and their accompanying text are numbered as entry 41 in the spring garden. The rest of the narcissi in the spring garden occupy entries 19 to 24. This ordering is not because the rushed-leaved narcissi bloom later in the spring, as the description states that they flower in April, or even March if the winter is mild.[88] Appearing on plate 21 are two varieties of the slender-leaved narcissus (*Narsissus tenuifol*). The narcissus with a white somewhat double flower (*flore albo pleniore*) 'is found nowhere except in gardens, and usually appears later, namely in the month of May.'[89] While the two plates of narcissi are both placed under the order of the spring garden, they are not in sequence of their recorded flowering months. What this grouping demonstrates is that the compiler was attentive to the general time of the year when one could encounter the depicted flower in a garden.

Furthermore, albeit subtle, the seasonal orders of flowers were not identical among florilegia, suggesting that the compilers arranged the orders based on their own observations or studies of what was blooming when in the gardens. As previously shown, Besler and Van de Passe II gave different weights for the four seasons. While both feature the least number of plates, unsurprisingly, in the winter section, summer is the most blooming season in the *Hortus Eystettensis*, whereas spring is in the *Hortus floridus*. The two compilers did not group their flowers identically, either, despite both placing them in a seasonal order. Besler and Van de Passe II both included a variation of the double columbine with starlike petals, called *Aquilegia stellta flore violaceo* in Besler's florilegium and *Aquileia stellato purpureo flore* in Van de Passe II's flower book.[90]

87 'The great autumn narcissus' (*narciscus autum: maior luter flore*) and 'The unknown narcissus of Clusius' (*narciscus incognitus Clusius*) are numbered as plate 15 and plate 25 in the autumn garden in Van de Passe II, *Hortus floridus.*

88 '*In Belgio Aprili mense florem promit; Et Martio si Hyems clemens fuerit.*' in Van de Passe II, *Hortus floridus,* XLI (spring garden); Savage, *Garden of Flowers*, 82.

89 '*Narcissus quisequitur nusquam nisi in hortis reperitur, seriusque Maio nempe mense, prodire solet* […]' in Van de Passe II, *Hortus floridus,* XXI (spring garden); Savage, *Garden of Flowers*, 42.

90 Besler, *Hortus Eystettensis*, secudus ordo. fol. 8 (summer garden); Littger and Dressendörfer, *Florilegium: The Book of Plants*, plate 171. Van de Passe II, *Hortus floridus*, plate 37.

Besler placed all the varieties of the depicted and described columbines in the order of the summer flowers, whereas Van de Passe II's columbines appear in the spring garden. This could simply be because columbines flowered in the months that one might have considered to be early summer and another late spring. This could also have been because of the geographical difference, with Besler in Eichstätt in the south of Germany and Van de Passe II working in the Dutch Republic, which slightly affected the blooming times of the columbines. It nevertheless shows that the florilegium compilers had similarities and differences in their ordering of flowers, comparable to the physician-botanists, but the florilegium compilers differed their orders in the details of the organisation instead of the major groupings. Placing flowers in seasonal order in florilegia is, under this light, a system that deliberately brought forward a different kind of plant knowledge through attentive observations of rare and curious flowers and decorative plants.

## 5 Conclusion

This chapter has shown how seventeenth-century florilegia extracted and brought into focus the rare and curious flowers from the vast amount of plant knowledge that was available at the time. The intaglio-printed flower books that were first published in the first quarter of the seventeenth century especially exhibit the tight connection to sixteenth-century herbals through copying, excerpting, and referencing the visual and textual information mostly from the works of Rembert Dodoens, Carolus Clusius, and Matthias Lobelius. The flower books highlight, notably, the visual and textual morphological details of the blooms, which are the main plant part with feature(s) that made a flower rare and curious in the long seventeenth century. The organising system of grouping and placing flowers in seasonal order further shows the compilers' attentive observations of the flowers in their local garden settings, as the collections do not have identical groupings of flowers from different seasons. Similar to the *florilegium* in the medieval literary practice of 'picking flowers' out of a text, the compilers of these printed flower books emphasised and prioritised the visual and textual information of flowers in a published form. By reorganising and adding to the extracted plant knowledge from herbals, compilers of these printed florilegia produced epistemic objects for *liefhebbers* that centred the period's understanding and knowledge of rare and curious flowers.

CHAPTER 4

# From Describing Flower Colours to the Colouring of Flowers

## 1 Introduction

The *Hortus Eystettensis,* a big florilegium compiled by the Nuremberg apothecary and botanist Basilius Besler (1561–1629), was first published in 1613 and reprinted several times.[1] Being the biggest printed collection in the seventeenth century on rare and curious flowers, the florilegium included a large number of flowers and decorative plants that had travelled to areas north of the Alps, usually from the Mediterranean, Asia Minor, and the Americas, as well as cultivars with new and/or unique colours or with (many) more petals than their native form.[2] Shortly after the 1613 first printing with flower descriptions in Latin, which is the version discussed in this book thus far, Besler issued another set of copies with only the plates and an updated preface in 1613.[3] The plates of this textless version are printed on higher-quality paper, with the edition intended to be a deluxe version and more valuable as a book object than the version with Latin text.[4] A third version has been identified to have been issued in 1627, but only two copies survived and both of them have significantly fewer plates. One surviving copy has 96 plates and the other 23, and both are without text. The fourth edition was issued in 1640, a decade after Besler's death in 1629, with the same content and plate number as the 1613 editions, but without the flower descriptions and with a different preface.[5] A fifth and final edition was printed 100 years later in 1713, and this time the Latin text was fully restored.[6]

1 Besler, *Hortus Eystettensis*. See Chapter 1 for an introduction to this florilegium.

2 See Chapter 1 for discussion on the characteristics that made flowers rare and curious in the long seventeenth century.

3 Hofmann and Zöhl, *Hortus Eystettensis: Studien zur Entstehung des Kupferstichwerks und zum Exemplar des Andrea Vendramin*, 26–30.

4 Barker, *Hortus Eystettensis: The Bishop's Garden and Besler's Magnificent Book*, 35.

5 Besler, *Hortus Eystettensis* (1640). Bibliographic description numbered C80 in Bridson and White, *Plant, Animal & Anatomical Illustration in Art & Science*, 44.

6 A comparison of a small number of plates from the 1613 and 1713 editions verified that the types, as expected, are set differently, but the contents of the text were not altered. The original copperplates were thought to have been melted for the minting bureau after five editions, until most of them resurfaced in the Austrian National Library (Österreichische

© JESSIE WEI-HSUAN CHEN, 2025 | DOI:10.1163/9789004735149_006

Besler himself never reused the text after the initial publication, but continued to reprint the plates. In the posthumous 1640 edition, the preface stated that the flower descriptions were removed because they were not well-received and did not reflect the standard of serious plant studies of the time.[7] It should be noted that the production of the first Latin edition was highly complicated, so there might have been practical reasons for not including the flower descriptions and referenced works. This edition was produced after the destruction of the garden due to the Thirty Years' War. There might not have been enough resources to reproduce the full volume with both images and text. The text was brought back in the eighteenth century, when it would have been much more outdated. The return of the text may not have been produced for the purpose of studying, either. Rather, it may have been an act of yearning for an older time when the garden of Eichstätt was in its most glorious state, as the florilegium and its text received criticism in the eighteenth century because it did not adhere to the epistemic values of the time.[8] Regardless of how the text was received, what is more important to consider here is that this removal and re-adding of text changed the information available to readers when going through different editions of the *Hortus Eystettensis*.

Similar to the *Hortus Eystettensis*, the other four intaglio-printed florilegia that were first published between 1608 and 1614 saw changes in images and text in their reprints and different editions, which will be further discussed in this chapter. Generally speaking, the reprints and later editions have two major trajectories, sometimes including both. First, many titles saw a growth in the number of plates and flower depictions in the reprints and later editions. Second, there is often a decrease in morphological descriptions and/or an increase of textual details for the colour information of the depicted flowers in the reprints and later editions. The growing number of plates was a result of the active cultivation and horticultural practices, in addition to the continuous introduction of (decorative) plants from distant places.[9] Not only were there ever more new species of rare flowers and plants, but also (colour) varieties that are now considered insufficient to be listed as their own species. The adding of new depictions was not accompanied by new morphological descriptions, except for the colours of the flowers that continued to be described.

Nationalbibliothek) in Vienna in 1998, see Hofmann and Zöhl, *Hortus Eystettensis: Studien zur Entstehung des Kupferstichwerks und zum Exemplar des Andrea Vendramin*, 39.

7 Besler, *Hortus Eystettensis* (1640), section '*Spectatori.*'

8 Hofmann and Zöhl, *Hortus Eystettensis: Studien zur Entstehung des Kupferstichwerks und Zum Exemplar des Andrea Vendramin*, 35–36; Daston, 'Epistemic Images,' 24–27.

9 See Chapter 2 for a discussion on growing and cultivating rare and curious flowers in the seventeenth century.

The provided text on colours also shifted from the format of flower descriptions that aligns with the practice of early modern physician-botanists in Latin to the vernacular languages (most often in Dutch, French, and English) in a writing format that is closer to how-to instructions for artistic practices.

This chapter examines such changing textual information and how it worked with the same florilegium images throughout the several reprints and editions in the seventeenth century. Within this volume, a *reprint* means a second, third, or more printing of a title that has little to no changes in content from the initial publication. An *edition*, on the other hand, refers to a later print run that includes apparent differences in images and/or from the first publication of a title. This chapter focuses on the diachronic developments, instead of the chronological changes, of the genre on a micro level. While noticeable, the shifting image-text relations and image-text proportions in these printed florilegia do not alter the primary function of a flower book as a picture book. As mentioned in the general introduction, seventeenth-century florilegia usually have little text, and it is most often in the form of a preface and captions. As captions are usually engraved or etched onto the plates, they are part of the images and will be regarded as images in this chapter. The points of analysis here are the textual information addressing the plates, but outside the boundary of the picture planes. The small amount of text in florilegia means that the flower descriptions and colouring instructions that made it to the pages and correspond to the images bear even more weight in their working with the images to deliver different facets of information on the rare and curious flowers.

The heuristics of delivering textual information related to plates shifted from and/or between the approaches that were common in early modern herbals and how-to treatises. The flower descriptions and colouring instructions hence need to be examined and reconsidered to better understand how florilegium compilers used text and for what purpose. In her article 'Image, Text and *Observatio*: The *Codex Kentmanus*,' historian of science Sachiko Kusukawa examined 'the inter-relationship between image, text and object in the *Codex Kentmanus*,' a collection of plant depictions and corresponding descriptions of the flora in the botanical garden at Padua and studied by the German physician-botanist Johannes Kentmann (1518–1574).[10] The reasons for producing and for the use of the codex are much different from those for the genre of the florilegium, with the codex being Kentmann's personal experience and records of what he learned from his contemporaries regarding the

10 Kusukawa, 'Image, Text and *Observatio*,' 445–475. More on the codex, see Alex, ed., *Die schönsten Pflanzen-Aquarelle der Renaissance.*

studied plants. However, what Kusukawa underlined is how image and text worked closely together in the study of natural objects in the early modern period, as she concluded:

> [I]mages were used as record-keeping, enabled people to study objects over time and space, and to accumulate and incorporate information from elsewhere. No doubt, others found other uses for images too. The fact that Kentmann named the textual part *observationes* also underscores the fundamental importance of the verbal text as part of the study of nature in this period.[11]

Similarly, in his recent book *Aesthetic Science*, Alexander Wragge-Morley discussed how the early scientists of the Royal Society of London utilised different kinds of rhetoric to 'picture' natural things. For example, Wragge-Morley showed how the English naturalist John Ray (1627–1705) largely used verbal descriptions to picture the internal features that are not visually depicted in his plate of cetaceans. To Ray, the external features on the plate alone was not sufficient, as many internal features of the animals were what distinguished the species.[12] The text in seventeenth-century florilegia displays a similar practice of 'verbal picturing,' of textually describing the morphological and colour information that could not be depicted, without an aim of competing with the works of the naturalists of the time.

Colour terms are especially important for verbally picturing florilegium images in prints to present colourful images of flowers through the black-and-white medium of intaglio printing. Despite its immediate association with the hand-colouring practice, most copies with hand-coloured plates are not the editions with colouring instructions; hand-coloured copies are most often the Latin or textless editions without any how-to text for colouring practice. As Doris Oltrogge demonstrated in her book chapter, 'Writing on Pigments in Natural History and Art Technology in Sixteenth-Century Germany and Switzerland,' the use of colour terms was inter-connected between naturalists and art practitioners.[13] The colour *cerulean*, for example, is sometimes used to describe flower colours in florilegia. Cerulean is related to the ancient term *caeruleum*, which covers two blue pigments, one is processed from the mineral

11 Kusukawa, 'Image, Text and *Observatio*,' 475.

12 Wragge-Morley, *Aesthetic Science*, 117–119.

13 Oltrogge, 'Writing on Pigments in Natural History and Art Technology in Sixteenth-Century Germany and Switzerland,' 47–69.

of azurite, and one is the artificial Egyptian blue.[14] The use of the word *cerulea* for describing flower colours, however, does not mean to colour the flower with the two specific pigments (although one could), but to invoke an association with the colour blue upon reading. It should be stressed that colour terms in florilegia do not aim to communicate precise shades of hues, but to direct readers to a (wide) range of colours within a hue. Through a combination of verbal picturing and shared colour terms, this chapter emphasises *how* florilegium images and their changing texts worked together to deliver information on the rare and curious features of decorative flowers and plants.

## 2 Modifying and Reprinting Seventeenth-Century Florilegia

The five major intaglio-printed florilegia that were first published between 1608 and 1614 include *Le jardin du roy …* (1608) by Pierre Vallet (*c.*1575–1657), the *Florilegium novum* (1612) by Johann Theodor de Bry (1561–1623), the *Florilegium Emanuelis Sweerti …* (1612) by Emanuel Sweerts (1552–1612), the *Hortus Eystettensis* by Besler, and the *Hortus floridus* (1614) by Crispijn van de Passe II (1594/1595–1670). All five titles saw several editions and reprints, but each of them had a different trajectory. Additionally, every title has some reprints and (re)editions that are better known and with a higher survival rate, while some versions fell into obscurity. This section does not attempt to provide a full bibliography of each title. Rather, it gives an overview of how the genre developed through the most relevant (re)editions where significant changes in the included textual information occurred.

The *Florilegium novum* by De Bry grew bigger in the size of the collection by publishing supplements and addendums throughout the years.[15] There are at least three collections of added plates recorded, including the *Amplificatio sive dilatatio Florilegii nuper coepti auctique* (1613), the *Augmentatio uberior Florilegij* (1614), and the *Anthologia magna* (1626), with the last one produced by De Bry's son-in-law Matthäus Merian (1647–1717).[16] The full collection of

14 The colour *cerulean* is the pigment example provided by Oltrogge, and I applied it here to the florilegium captions, see Oltrogge, 'Writing on Pigments in Natural History and Art Technology in Sixteenth-Century Germany and Switzerland,' 53–54.

15 De Bry, *Florilegium novum*. See Chapter 3 for an introduction to this florilegium.

16 De Bry, *Amplificatio sive dilatatio Florilegii nuper coepti auctique*; there are very few copies of *Amplificatio* documented, and WorldCat records only two copies in the world, both located in Germany; bibliographer Claus Nissen mentioned this addition in the essay, but did not describe it in *Die botanische Buchillustration*, 75. Warner, 'The Augmentatio, 1614, of De Bry's *Florilegium Novum*,' 29–32; there is not yet entry on the *Augmentatio* on

these plates that De Bry accumulated was republished as the *Florilegium renovatum et auctum* in 1641 by Merian.[17] As discussed in the previous chapter, this edition shifted the function of the flower book more towards the spectrum of serving as a garden catalogue by connecting the plates and their depicted flowers to the specific garden of the Frankfurt Burgomaster Johannes Schwindt (1580–1648). Additionally, Merian's version turned the flower book more towards the visual culture of horticulture by excerpting the plates from the horticultural treatise *De florum cultura* (1633, 1638) by the Italian Jesuit Giovanni Baptista Ferrari (1584–1655).[18] Neither De Bry's *Florilegium novum* nor Merian's *Florilegium renovatum et auctum* include flower descriptions or colouring instructions. The prefaces, in German or Latin, and the captions of flower names in Latin compose the majority of the textual element.

Sweerts' *Florilegium Emanuelis Sweerti …* developed similarly to Besler's *Hortus Eystettensis*.[19] The number of plates remained mostly the same throughout the several reprints in 1620, 1631, 1641, 1647, and 1655, but the text in the frontmatter was somewhat streamlined and reduced.[20] While the first edition, as discussed in Chapter 2, included several dedications and praises of or by different people, the later editions removed most of this information. The preface, the planting instructions, and the registration of flowers remained. The three parts, most notably in the 1647 version, are printed consistently in the four languages of Latin, Dutch, German, and French, as opposed to the 1612 first edition with some parts only in Latin or a couple of the languages, such as the planting instructions that were initially only in Latin and German.[21] The *Hortus Eystettensis* by Besler, as discussed at the beginning of the chapter, removed the flower descriptions in the several reprints for the rest of the seventeenth century.

---

WorldCat. De Bry and Merian, *Anthologia magna*; bibliographic description numbered C66 in Bridson and White, *Plant, Animal & Anatomical Illustration in Art & Science*, 43.

17 De Bry and Merian, *Florilegium renovatum et auctum*. See Chapter 3 for an introduction to and discussion on this florilegium.

18 Ferrari, *De florum cultura*.

19 Sweerts, *Florilegium Emanuelis Sweerti Septimonti Batavi*. See Chapter 2 for an introduction to and more detailed discussions on this florilegium.

20 The most accessible edition is the 1647 reprint, *Florilegium amplissimum et selectissimum*.

21 See Chapter 2 for description of the frontmatter in the 1612 edition. The streamlining took place gradually throughout several editions. I had access to the 1612, 1620, and 1647 editions. The frontmatter of the 1620 edition already eliminates the praises, but only includes planting instructions in Latin and German. Bibliographic descriptions numbered C58, C71, C87, C90, C94 in Bridson and White, *Plant, Animal & Anatomical Illustration in Art & Science*, 42, 43, 45.

The reprint of Vallet's *Le jardin du roy* … turned in a different direction, bringing the artistic practice of hand-colouring to the foreground.[22] The 1608 version has a small section of flower descriptions that is written in Latin by Jean Robin (1550–1629), the director of the Royal Gardens.[23] According to Robin, he did not describe every flower in the florilegium, but only the rare(st) ones that were exotic to France and had not been described by other physician-botanists in herbals, with a total of 12 entries, a small portion in a collection of 73 plates and more than 200 depictions.[24] Vallet's flower book was reprinted several times in 1623, 1638, 1662, and 1665[?].[25] Already in 1623, the Latin flower descriptions by Robin were replaced with another type of description in the frontmatter.[26] Written in French, a new section is titled '*Aux amateurs des plantes, et de l'enlumineure.*' (To lovers of plants, and of illumination.) The short paragraph provides essential instructions on how to prepare the papers before colouring the plates, as well as how to make watercolour paints by mixing gum arabic with various kinds of pigments.[27] Instead of featuring only around a dozen flowers, as with Robin, the 266 entries covered all the flowers that were depicted in the volume, albeit some being extremely short, informing readers what colours the flowers were and/or which colours to apply.

The several editions of the *Hortus floridus* by Van de Passe II are more translations than reprints, as they were issued within a couple years from one another.[28] The publication history for this florilegium is highly complex, more so than for the other titles in an already complicated genre, because the surviving copies often included two parts that were printed in different years.[29] Broadly speaking, the first was the 1614 Latin edition that has been frequently discussed

---

22 Vallet, *Le jardin du roy tres chrestien Henry IV*. See Chapter 3 for an introduction to this florilegium.

23 Saunders, *Picturing Plants*, 55; Tongiorgi Tomasi, *An Oak Spring Flora*, 40.

24 Total number of plates is around 76 in most copies, including a portrait of Vallet.

25 Out of all the reprints, I could only access a copy of the 1623 edition through a digitised copy from the Biodiversity Heritage Library, Washington D.C., Smithsonian Libraries and Archives, QK41.V18 1623, https://doi.org/10.5962/bhl.title.62899. Bibliographic descriptions numbered C61, C78, C104, C109 in Bridson and White, *Plant, Animal & Anatomical Illustration in Art & Science*, 42, 44, 46.

26 The title was also updated to *Loys XIII* instead of *Henry IV*. Vallet, *Le jardin du roy tres chrestien, Loys XIII*.

27 Vallet, *Le jardin du roy tres chrestien, Loys XIII*, '*Aux amateurs des plantes, et de l'enlumineure.*'

28 Van de Passe II, *Hortus floridus*. See Chapter 1 for an introduction to this florilegium.

29 Provided that the years are accurate to reality. The sub-titlepage for the autumn garden in the *Hortus floridus* is printed as 1616. It is highly possible that the year 1616 is a typo, as it has shown up on every copy I have consulted thus far, when all the other details point to them being the 1614 version. More on the complex publication history of this title, see

in this book thus far. In the same year, the Dutch edition emerged. A significant addition to the Dutch version of *Den Blom-hof*, on top of the changing textual information, is that Van de Passe II added insects and little creatures (such as snails and a mouse) to the plates, making it the only printed florilegium that altered the state of the plates in a significant manner.[30] There are a handful of new plates of tulip and carnation varieties that were not in the Latin edition as well. A French edition, *Jardin de fleurs*, appeared also in 1614, and the English edition, *A Garden of Flowers*, was printed in 1615.[31] A new Latin edition with the plates with insects and little creatures was issued in 1617, merely three years after the title's initial publication.

In general, there are two sets of information: flower descriptions that are exclusively in Latin and colouring instructions, which are combined with simple morphological descriptions for some of the flowers, in Dutch, French, and English. Some rare copies of the Dutch and French editions have the short addendum on tulips and their cultivation that is discussed in Chapter 2. While there are anomalies, surviving copies in the respective editions have the following characteristics. The 1614 Latin edition has flower descriptions but no colouring instructions. The 1614 Dutch edition has colouring instructions in Dutch. The French edition has the colouring instructions translated into French, but also has the same Latin flower descriptions for each of the plates. The English edition has the translated colouring instructions in English and no flower descriptions in Latin.[32] Except the Latin edition, all the versions of the *Hortus floridus* provide textual information for colouring flowers.

The several directions in which seventeenth-century florilegia developed in their reprints and later editions boil down to three kinds of working dynamics. First, images and text do not correspond to one another, with only captions that are part of the plates offering essential textual information. Second, images correspond to flower descriptions with morphological and other kinds of information (such as places of origin), often on a one-to-one basis. Third, images correspond to colouring instructions on a one-to-one basis. By working

Savage, 'The Hortus Floridus of Crispijn Vande Pas the Younger,' 181–206; Segal and Alen, *Dutch and Flemish Flower Pieces*, vol. 2, 1012–1015.

30 Van de Passe II, *Den Blom-hof*. When changes are made to a printing plate, such as when a printmaker added things to a plate, the impressions become a new state. The prints from the original plate are the first state; if revision is done and more prints are pulled from this revised plate, they are impressions from the second state.

31 Van De Passe II, *Jardin de fleurs*; *A Garden of Flowers*.

32 The English edition is harder to find. So far, I have only come across this version at the Upperville, VA, Oak Spring Garden Foundation, 00005583.

with different types of text, the same plate may deliver different kinds of messages and/or present different pictures of the depicted flowers.

## 3 Leafing through a Flower Book

To further complicate the matter, there are multiple ways in which the images and text are ordered and arranged in a flower book when there is a one-to-one correspondence. This affects how the visual and verbal information work together to present a picture of a rare and curious flower. The orders of the pages can differ from what is discussed below, as they heavily depend on how a book owner had their copy bound. The rest of this section describes the common order through copies that are bound in the supposedly intended manner by the florilegium compiler, as indicated by the page and plate numbers.

As mentioned, seventeenth-century florilegia usually have a bulk of frontmatter before the appearance of the first plate. For example, the 1613 first edition of the *Hortus Eustettensis* starts with the frontmatter: a preface introducing some background information of the collection to the readers; a portrait of Besler and a poem praising him, written by a certain Georg. Remus; the sub-frontispiece that marks the beginning of the spring garden, as discussed in Chapter 3.[33] The spring garden section starts with the many pages of the index of spring flowers, first listed in Latin and then in German. A plain sub-title page in text, printing '*Primus ordo collectarum plantarum vernalium*' (The first order of the collected spring plants) reminds readers of the content of the following plates.[34] When readers turned the page to the first plate of the volume, they would see a spread with the plate of flower depictions on the left side and the text of lower descriptions on the right side (Fig. 20). The text is printed on the verso side of the next plate, which means that all the visual and textual information of the depicted and described flowers is contained within the boundary of the large opening spread. The order of sub-frontispiece, index, sub-titlepage, and plates of the flowers from the season repeats for the summer, autumn, and winter garden for the rest of the book. The deluxe version, on the other hand, does not have text printed on the back side of the plates. Instead of seeing the descriptions on one side of the spread, when opening

33 See section 'Putting Flowers in Seasonal Order' in Chapter 3 on the order of the book and the sub-frontispieces that divide the seasons.

34 This type of sub-titlepage with different numbers (eg. *secondus*, *tertius* ...) appears periodically throughout the season. Every season has multiple of such sub-titlepages, except the winter garden.

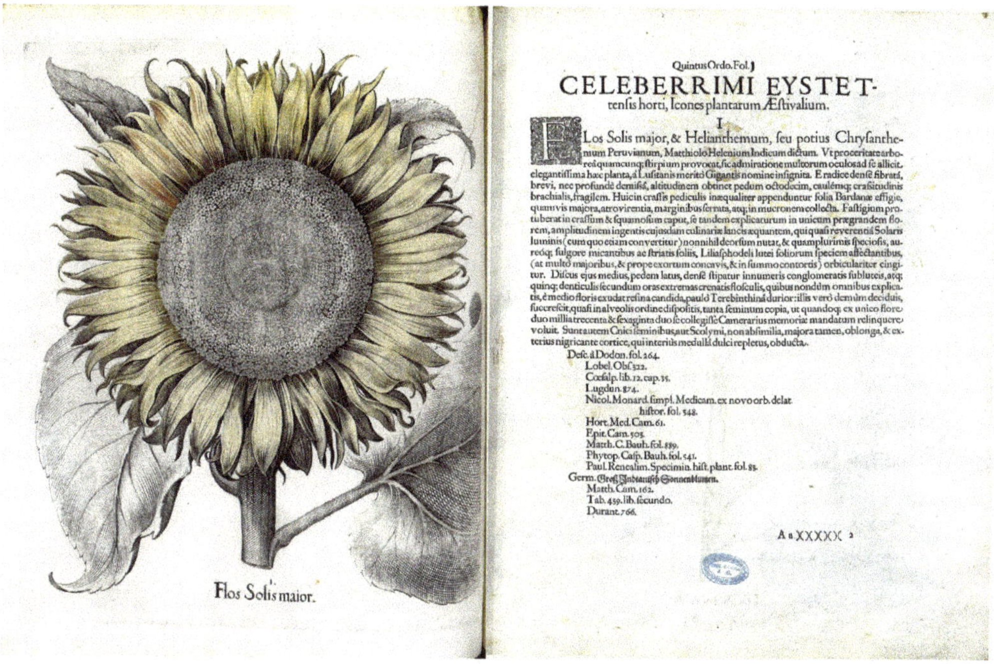


FIGURE 20 Sunflower spread from the *Hortus Eystettensis* (1613), Biblioteca Real Jardín Botánico, Madrid, Spain, Af BES, https://bibdigital.rjb.csic.es/idviewer/10913/138; https://bibdigital.rjb.csic.es/idviewer/10913/139

a bound book and after the frontmatter, the reader would see a blank page within a spread.[35]

The *Hortus floridus* has a similar image-text arrangement as the *Hortus Eystettensis* for the flower descriptions on the spread. The frontmatter starts with the frontispiece, followed by a preface by Van de Passe II, and a page of two short poems by Arnoldus Buchelius (1565–1641) and Jacobus Valentinus vander Voort. An index of flowers in the spring garden appears next before the sub-frontispiece of an enclosed garden.[36] The plates of the flowers in the spring garden follow. For the summer, autumn, and winter gardens, there are usually

35 I have not been able to access a copy of the deluxe version. The consulted copies during research are all the edition with text. A digitised copy appears to be the deluxe version due to its lack of text. This copy is bound in the way that viewers see a blank page on the left and an image on the right of a spread, as far as the photos can show, Erlangen, University Library of Erlangen-Nürnberg (Universitätsbibliothek Erlangen-Nürnberg), H62/CIM.M 12, https://gateway-bayern.de/BV009173203.

36 The poems and the index are printed on the same sheet of paper, and they often switch order depending on how a copy is bound.

a sub-titlepage, an index of the flowers of the season, and a sub-frontispiece, but the order of the three components is not consistent, before the images of the flowers in that season appear. Unlike the *Hortus Eystettensis*, the flower descriptions are printed on the verso side of the previous plate. As a result, when opening a copy of the Latin edition, readers see the text on the left side of the spread and the plate of the same flowers on the right (Fig. 21). The Dutch and English editions, in contrast, give all the textual information, including the colouring instructions, at the beginning of the flower book, together with the frontmatter.[37] Each entry of the colouring instruction has a header with numbers and flower names that correspond to the pertinent plates. There is no text printed on the back of the plates, so readers would only see the images when going through the book (Fig. 22).[38] As the French edition has both flower descriptions in Latin and colouring instructions, readers also see text on the left when leafing through the volume.[39]

Separating and juxtaposing plates and descriptions/instructions are the two dynamics for images and text to work together in these flower books. Flower descriptions and colouring instructions are both supplements to the flower depictions. Intaglio-plates and text were printed separately in the early modern period. To print a sheet of paper with text on one side and images on another, as the pages in the Latin versions of Besler's and Van de Passe II's florilegia, the piece of paper would need to undergo two separate printing processes.[40] The plate would be printed using a rolling press and the type would be printed using a handpress, requiring more cost and labour especially for producing flower books that juxtapose images and text.[41] For the most part,

37 Every copy I have been able to consult during research has a different compilation of frontmatter. Some copies have the Latin preface, others have missing pages. The colouring instructions are, however, consistently included in all the copies I have studied.

38 With the exception of the spread that contains the plate with two hepatica varieties (plate 1, spring garden), as viewers tend to see the last entries of the colouring instructions on the left side of the spread.

39 Occasionally, copies are bound in the way that the spreads alternate, with one spread showing two plates facing each other, and the next spread showing two sets of flower descriptions, such as the copy in Utrecht, Utrecht University Library Special Collections (Bijzondere Collecties Universiteitsbibliotheek Utrecht), Rariora oct 109 Obl, https://utrechtuniversity.on.worldcat.org/oclc/66465814.

40 For an introduction on the separate processes of printing text and images, see Gaskell, *A New Introduction to Bibliography*, 118–124, 154–159. More specifically on composing text, see Febvre and Martin, *The Coming of the Book*, 61–68.

41 Secondary literature on book history usually discusses the additional costs that was required for adding illustrations to text in early modern book printing. While adding text to plates most likely would not have been as expensive as the other way around, this

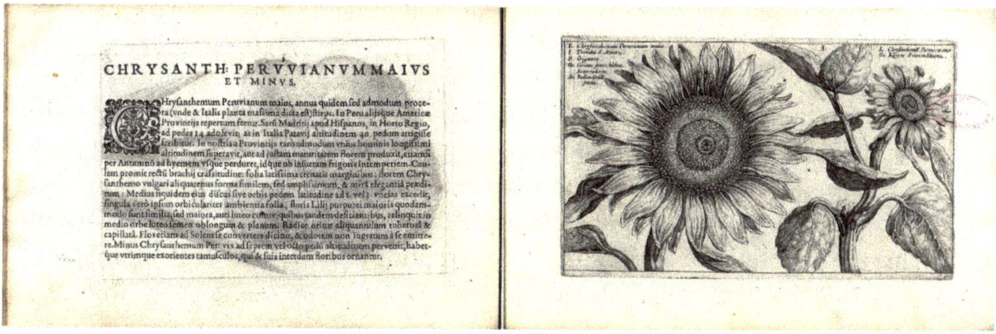


FIGURE 21 Sunflower spread from the Latin edition of the *Hortus floridus* (1614), Biblioteca Real Jardín Botánico, Madrid, Spain, A PAS, https://bibdigital.rjb.csic.es/idviewer/13646/146; https://bibdigital.rjb.csic.es/idviewer/13646/147

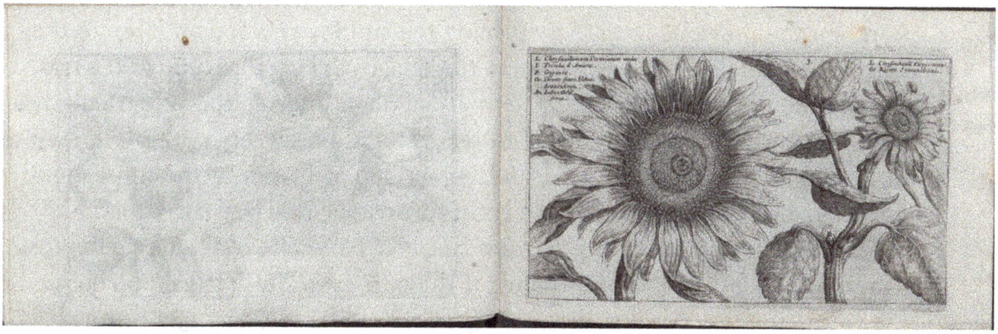

FIGURE 22 Sunflower spread from the Dutch edition of the *Hortus floridus* (1614), Utrecht University Library Special Collections, Utrecht, R qu 851 obl (Rariora)

images and text are printed separately in seventeenth-century florilegia. With the genre as a whole, it is more common for a reader to see a blank page on one and the image on the other side of a spread. Vallet's *Le jardin du roy …*, De Bry's *Florilegium novum*, and Sweert's *Florilegium Emanuelis Sweerti …*, as well as their reprints and different editions have mostly the same arrangement of images and text. They have the frontmatter with prefaces and additional materials, such as the planting or colouring instructions, at the beginning of the books, before the plates make an appearance. These titles and variations of florilegia do not have textual information that is next to the images within a spread. Textless florilegia are further explored in Chapter 5, and the remainder

combination pf printing images and text still required more effort than printing only the plates.

of this chapter will zoom in on the verbal information in the flower books of Besler, Van de Passe II, and Vallet.

It became noticeable why Besler and Van de Passe II put in additional effort to print the text on the verso side of the plates when looking into the sources from which the *Hortus Eystettensis* and *Hortus floridus* excerpted some of their visual and/or textual information. Both methods of arranging descriptions and depictions in close proximity of one another and having all the images following the text were commonly used for presenting visual and textual information in sixteenth-century herbals and botanical treatises. For example, the woodcut of the sunflower in the Flemish physician-botanist Rembert Dodoens' small herbal, the 1568 *Florum*, is printed on page 295 (Fig. 13).[42] The description starts on the back of the woodcut on pages 296 and ends on page 298. Similarly, the sunflower woodcut appears on page 264 in Dodoens' *Stirpium historiae pemptades sex* (1583) and sits next to the text block of the flower's description, which starts on the previous page.[43] By placing the sunflower woodcut close to the description, readers can easily turn to the image when going through the text. This method of delivering textual and visual plant information is particularly prevalent in herbals that provide an image for almost every plant described in the book.[44] The descriptions and depictions of plants in the most cited European herbals of the sixteenth century are usually laid out in this manner.[45]

On the other hand, in the *Hortus medicus et philosophicus* (1588), a botanical treatise cited by Besler in many of the entries in the *Hortus Eystettensis*, the German physician-botanist Joachim Camerarius II (1534–1598) provided 56 depictions for only a selection of plants among those he listed and/or described. The woodcuts are placed towards the end of the book after the main body of descriptions. On the top of each woodcut, the plant name and a page number are provided to assist the readers in locating the entry within the main text. For example, an image of the *Nasturtium Indicum*, a flower that occasionally appears in florilegia, is one of the 56 provided woodcuts.[46] Readers can find the description of the flower on page 105 in the first book of the herbal on medicinal plants. It takes the readers more effort to consult the woodcuts, as

42 Dodoens, *Florum*, 295.

43 Dodoens, *Stirpium historiae pemptades sex*, 264.

44 See particularly Kusukawa, *Picturing the Book of Nature*, 113–114.

45 Including the big herbals of the German physician-botanists Hieronymus Bock, Otto Brunfels, Leonhart Fuch, Netherlandish physician-botanists Rembert Dodoens, Carolus Clusius, and Matthias Lobelius, Italian physician-botanist Pietro Andrea Mattioli, and the English physician-botanist John Gerard.

46 Woodcut numbered XXXI towards the end of the book in Camerarius, *Hortus medicus et philosophicus*.

one may need to flip back and forth through the book if wanting to compare the text and image(s) of a plant. However, this method keeps the reading flow smooth when moving through the entries of plants, which are in alphabetical order in Camerarius II's treatise. For intaglio printing, this method especially reduced the production cost and labour, but still allowed the inclusion of textual information as supplements. It is thus sensible that Robin's flower descriptions are not typeset next to their respective plates, but are printed as part of the frontmatter. As Besler's and Van de Passe II's florilegia are the two collections with the heaviest references to sixteenth-century herbals, having the one-to-one description and depiction side-by-side within a full spread aligns with how the textual and visual information was presented in their source materials of herbals.

## 4 Describing Flowers and Their Colours

Even though the juxtaposition of the Latin descriptions and depictions of flowers delivers the visual and textual information in a manner that is similar to sixteenth-century herbals, there is a clear change of image-text proportion that flipped the dynamic in seventeenth-century florilegia. Instead of images being created to support and enrich the text, the role is reversed with the text generated to support and enrich the images. Generally speaking, herbal illustrations tend to visualise, clarify, add to, and/or complement the plant descriptions in the sixteenth century, most often through depicting morphological details in a lifelike manner for herbals and botanical treatises.[47] For example, the sunflower description and woodcut in Dodoens' 1568 *Florum* demonstrate this dynamic of an image supporting the text. In addition to stating that the flower is unusually tall, Dodoens wrote,

> It has a straight stalk as thick as an arm. The leaves are very broad; the flower is similar in shape to a 'Chrysanthemum' but is much larger. The diameter of its disc or circumference is afoot across and it weighs two to three ounces. The petals surrounding this disc are somewhat similar

47 For an overview of the visual strategies of herbal woodcuts, see Landau and Parshall, *The Renaissance Print*, 259–245; Swan, 'The Uses of Realism in Early Modern Illustrated Botany,' 239–249.

> to those of the great purple lily, but are larger and of a golden yellow colour.[48]

The accompanying woodcut (Fig. 13) visualises the details of the description closely. The picture itself is small, measuring approximately 13.1cm tall. This miniaturised sunflower, however, provides visual information that closely adheres to the description, as the woodcut presents a tall flower with a thick stem, large leaves, and a flowerhead that has a big disc. The woodcut does not replace the description, but instead provides the readers with a more concrete representation of the flower's appearance within one picture.

The flower descriptions in florilegia, on the other hand, tend to supplement information that is not easily presented in visual form. The *Hortus Eystettensis* showcased the same type of the large sunflower as Dodoens', named *Flos solis maior*, on the plate (Fig. 20), whereas the *Hortus floridus* depicts the greater sunflower (*Chrysanthemum Peruvianum maius*) together with the lesser variety (*Chrysanthemum Peruvianum* minus) on one plate (Fig. 21). The depictions of the greater sunflower have rather different stylisations and details, but they both present the flower with a frontal view of the head. Both images show the head as the most prominent feature of the flower and the focal point of the images, with the one for Besler's florilegium measuring around 35.3cm and the one by Van de Passe II measuring around 13cm in diameter, a few leaves that are mostly cut-off and hence more suggestive than informative, and a little bit of a thick stem.

The texts by Besler and Van de Passe II differ in writing style and details, but offer similar basic information. In addition to morphological information, Besler's entry includes the various names in different languages according to how the sunflower was known in several regions in Western Europe, as well as the Latin names given to the flower by different sixteenth-century physician-botanists or naturalists who wrote about the sunflower.[49] Furthermore, Besler briefly remarked on how Camerarius II once collected 2,362 seeds from a sunflower head, before listing several herbals and botanical

---

48 '[…] *caulem promit rectum, brachij crassitudine: folia latissima: florem Chrysanthemo aliquatenus forma similem, sed amplissimum. Medius siquidem eius discus siue orbis pedem latitudine ad duas aut tres uncias excessit, singular vero ipsum orbiculariter ambientia folia, floris Lilij purpurei maioris quodammodo similia, sed maiora, auri luteo colore.* […]' in Dodoens, *Florum*, 296; translation from De Nave and Imhof, eds., *Botany in the Low Countries*, 144 (catalogue entry 102).

49 Besler, *Hortus Eystettensis* (1613), quintus ordo. fol. 1 (summer garden); or see section 'Excerpting Flowers, Extracting Knowledge' in Chapter 3 on the names listed in the *Hortus Eystettensis*.

treatises that discussed the sunflower for reference.[50] The majority of the description is on the large sunflower's form and exterior structure, stating,

> From a root that is densely fibrous, short, and not reaching deep, [the plant] develops a height of eighteen feet and a fragile stalk of the thickness of an arm. *From the thick pedicels hang some leaves that have the shape of the burdock, extremely big, dark-coloured, serrated at the edges, and gathered into a narrow point. A thick and scaly head bulges out of the apex, which eventually will grow out into one enormous flower, whose remarkable width is comparable to that of a kitchen plate,* and which bends considerably forwards as if in reverence towards the light of the Sun (around which it also revolves), and *which is surrounded in the manner of a circle with very many splendid leaves, twinkling of a golden gleam, and marked with lines, that look like the leaves of the yellow lilies, (but much bigger and nearly hollow at the base and twisted at the top). Its middle disc, [which is] one foot wide, is densely crowded with very many yellow conglomerates, and with little flowers notched by five small teeth alongside the external margins,* and when those [flowers] are not yet all bloomed, from the centre of the flower exudes a white resin, a bit harder than the Turpentine: but when at last those [are] falling down, under [them] grows a great multitude of seeds, as if in small cavities set in order.[51]

Van de Passe II's entry, on the other hand, is largely excerpted and extracted from Dodoens' entry. The florilegium compiler reiterated the place of origin of the flower, the gardens throughout Europe in which a large specimen had been

---

50 Besler, *Hortus Eystettensis* (1613), quintus ordo. fol. 1 (summer garden).

51 '[...] *E radice dense sibrata, brevi, nec prosunde demissa, altitudinem obtinet pedum octodecim, caulemq; crassitudinis brachialis, fragilem. Huicin crassis pediculis inaequaliter appenduntur folia Bardanae effigie, quam vis majora, atrovirentia, marginibus serrata, atq; in mucronem collecta.* Fastigium protuberatin crassum & squamosum caput, se tandem explicaturum in unicum praegrandem florem, amplitudinem ingentis cujusdam culinariae lancis aequamtem, *qui quasi recerentia Solaris luminis (cum quo etiam convertitur) nonnihil deorsum nutat, & quamplurimis speciosis, aureoq;* fulgore micantibus ac striatis foliis, Liliasphodeli lutei foliorum speciem affectantibus, (at multo majoribus, & prope exortum concavis, & in summocontortis) orbiculariter cingitur. Discus ejus medius, pedem latus, dense stipatur innumeris conglomeratis subluteis, atq; quinq; denticulis secundum oras extremas crenatis flosculis, *quibus nodum omnibus explicatis, e medio floris exudat resina candida, paulo Terebinthina durior: illis vero demum deciduis, succrescit, quasi in alveolis ordine dispositis, tanta feminum copia* [...]' in Besler, *Hortus Eystettensis* (1613), quintus ordo. fol. 1 (summer garden). This quote was translated in collaboration with Ada Migliazza.

grown, and remarks on the specimens' heights, among other small snippets of information. The main portion that describes the morphological details proceeds,

> It puts forth a straight stem, of an arm's width; very broad leaves notched at their edges; a flower in some degree similar in shape to that of the corn marigold, but very much larger and endowed with a wonderful gorgeousness. *Although its middle disc or orb exceeds a foot in width by two or three inches, surrounding it are separate petals, in a measure similar to those of the flower of the greater purple lily, but larger, of a golden yellow in colour.* At length losing these petals, the oblong, flat seed is left behind in the middle yellow orb. From the fibrous, lumpy roots vert little is produced.[52]

The depictions of the greater sunflower in Besler's and Van de Passe II's flower books visualise descriptive details that more-or-less correspond to the italicised sentences in the entries.

In both of the depictions and descriptions by Besler and Van de Passe II, what the images do not or cannot visually convey is supplemented by the accompanying verbal descriptions, with the visual qualities and appearances of the petals becoming much more emphasised. The engraved flower heads, while not in the exact sizes as described in the text, both give a much closer idea of the largeness of the flower in real life and are much more detailed than that of the woodcut in Dodoens' herbals. By zooming in on the blooms, the images emphasise the flower's large head, which was commonly remarked on as a curious feature of the flower, in addition to its foreign origin and golden colour.[53] However, having the flower head occupying the majority of a picture plane means that there is no space, in both depictions, to visually present other parts of the flower. Instead, such morphological information of the non-flowering elements, including the shapes of the leaves, the form of the roots, and the colour of the flower, are described in more detail through text.

52 '[...] *Caulem promit rectu brachij crassitudine: folia latissima crenatis marginibus: florem Chrysanthemo vulgari aliquatenus forma similem, sed amplissimum, & mira elegantia praeditum:* Medius siquidem eius discus sive orbis pedem latitudine ad 2. vel 3. uncias excedit, singula vero ipsum orbiculariter ambientia folia, floris Lilij purpurei maioris quodammodo sunt similia, sed maiora, auri luteo colore; *quibus tandem desicientibus, relinquit in medio orbe luteo semen oblongum & planum. Radice oritur aliquantulum tuberosa & capillata.* [...]' in Van de Passe II, *Hortus floridus*, 2 (autumn garden); Savage, *Garden of Flowers*, 152.

53 See Chapter 6 for a more extensive discussion on the visual techniques and materials that affect the flower sizes, in terms of their physical measurements, in florilegia.

The additional information on the place of origin, gardens that had the specimens, and the seed-counting of Camerarius II, further enriched some interesting facts and/or anecdotes about the flowers beyond the picture planes. However, the non-visual senses, including the smells and tastes of the flowers and leaves that were regularly discussed in herbals, were often left out in florilegium text. The examples above show how florilegium compilers used verbal descriptions to supplement visual information (or other types of information) for the depictions, with image and text working together to present a more comprehensive picture of the flowers beyond the picture planes.

The flower descriptions in the first 1608 publication of Vallet's *Le jardin du roy*... also supplement visual information for the plates, but selectively. As mentioned, there are only 12 descriptions of the flowers that were considered rarest and most curious by Robin, provided as part of the frontmatter at the beginning of the volume. The flower names are at the top of every entry, but there are no corresponding plate numbers for quickly locating the described flowers in the volume. For instance, the short entry for the *Martagum Pannonicum* corresponds to the plate that depicts a flower with the caption *Martagum Pannonicum flore suave rubente* (Fig. 23) at the bottom right corner. Similar to most plant entries in sixteenth-century herbals, Robin wrote about several aspects of the flower, including its colour:

> This very elegant kind of mountain lily does not differ much from the *Hemerocallis* yellow flower, but [it differs] only in regard to the kind of flowers that are of a reddish colour, this plant [was] imported from Pannonia together with many other foreign [ones].[54]

The *Martagum Pannonicum* is a kind of martagon lily (*Lilium martagon* L.). The term 'mountain lily,' which invokes a much different group of flowers today, seems to have often been applied to what would now be categorised as martagon lily in the early modern period, as shown in Robin's description and the following examples in the next section. This short description indicated that Robin expected his readers to be familiar, to an extent, with the depicted flowers in the volume. The morphological information is mostly presented as a comparison to another flower that he did not describe, that the martagon

54 '*Non multum ab hemerocallide flavo flore differt, perelegans hoc lilij montani genus, sed sola florum varietate qui spadicei sunt coloris, ex Pannonia delata haec planta cum quamplurumis alijs peregrinis.*' in Vallet, *Le jardin du roy tres chrestien Henry IV*, '*Joannes Robinus botamicus regius et iatrici horti celeberrimae scholae parisiensis curator.*' Quote translated by Ada Migliazza.

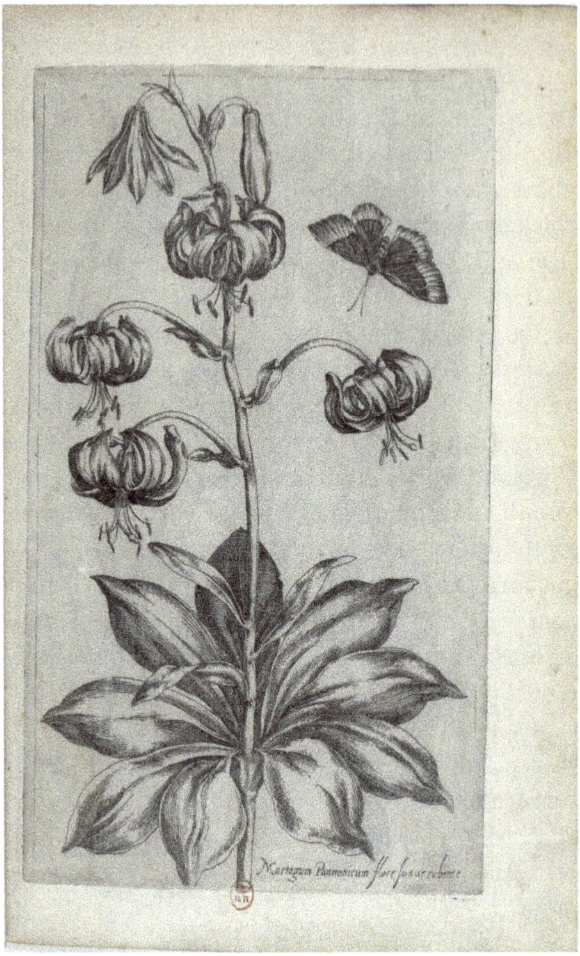

FIGURE 23
*Martagon Pannonicum* from the *Le jardin du roy* … (1608), BnF, National Library of France, Paris, ark:/12148/btv1b8612033t

lily is similar to the yellow *Hemerocallis*, which may have been a kind of yellow pompone lily (*Lilium pomponium* L.) that also has curled petals.[55] Instead of yellow, the colour of the *Martagum Pannonicum* petals is in the range of red, also as indicated in the caption of *flore suave rubente* (sweet red flowers)

55 Folio 63 in the *Hortus amoenissimus* by Franciscus de Geest has an orange lily that has a similar form as the Vallet's *Martagum Pannonicum*. The caption writes '*Hemerocallus Tursicus Turxse crol-lelie*' and the flower is identified by Sam Segal as the *Lilium pomponium* L. in Menghini, ed., *Hortus Amoenissimus Omnigenis Floribus, Plantis, Stirpibus*, 76.

on the plate.[56] This means that readers would have needed some idea about what a *Hemerocallis* was, perhaps by encountering them in the Royal Garden.[57] Another piece of important information is that the flower came from Pannonia in eastern Europe, making it exotic to the Royal Gardens in France. The separation of image and text required more work from the readers to connect the description to the plate. Although to a lesser degree, Robin's description likewise supplies verbal information for presenting a multi-faceted picture of the *Martagum Pannonicum*.

## 5 Describing How to Colour Flowers

In a similar manner to Robin's flower descriptions, the colouring instructions in the later editions of Vallet's and Van de Passe II's flower books supplemented information despite the separation of images and text, zooming in ever closer on the colour information. The term *colouring instruction* and the hand-colouring practice warrant clarification here. The term *instruction* and its iterations in different European languages was not explicitly used in *Le jardin du roy* … nor in the *Hortus floridus*. Van de Passe II used the 'description of the Colours', '*Beschrijvinghe van de couleuren*', and '*Description des couleurs*' for the prefaces of the English, Dutch, and French editions to inform the readers what the entries are. In most cases, the instructions that are discussed below may not seem to be remarkably pedagogical on the surface. Vallet offered only a short paragraph on what the materials for watercolours are, and Van de Passe II excluded such information altogether. Teaching readers the mechanics of how to colour, however, is not the intention of these colouring instructions in florilegia. More precisely, readers did not consult a flower book to learn, from scratch, how to make watercolours and paint the flowers. Instead, readers were expected to have some level of practical knowledge and/or experience on painting. Hand-colouring was a common pastime among learned people.[58]

56 There is some uncertainty with what word to use for the translation of the Latin word *spadicei*. The possibilities, however, all point to the spectrum of red and its tints. The colour of the flower may have been similar to what will be considered pink or magenta today, as a digitised plate from a hand-coloured copy has the colouration of magenta. Vallet, *Le jardin du roy tres chrestien Henry IV*, Paris, National Library of France (Bibliothèque nationale de France), btv1b21000678, https://gallica.bnf.fr/ark:/12148/btv1b21000678.

57 Robin listed a yellow *Hemerocallis flore luteo* in Robin, *Catalogus stirpium tam indigenarum quam exoticarum quae Lutetiae coluntur*, 30.

58 Goedings, '*Afsetters en meester-afsetters*,' 23–25.

Many *liefhebbers* would have learned the basic handling and working with early modern watercolours, the most common technique for hand-colouring (further discussed in Chapter 6), by taking lessons from a painter and/or through art treatises that were written more specifically as how-to texts.[59]

For example, a well-known French treatise for hand-colouring that is slightly later than Vallet's flower book is *Les lecons royales* … (1686) by the French miniature painter Catherine Perrot (1620?–1685). The book is a manual showing readers how to hand-colour a series of prints on flowers and birds by the French printmaker and miniature painter Nicolas Robert (1614–1685), from whom Perrot learned how to paint.[60] Perrot wrote about what pigments to get in order to assemble a palette that was suitable for colouring the prints; how to prepare the pigments to turn them into watercolour paints; how to preserve the colours so that they can last for multiple sessions of painting; and other practical matters. To instruct the readers on colouring the mountain lily (which may be a martagon lily) on plate 16 in Robert's series, Perrot wrote,

> The mountain lily is purple and calls for an initial application of a very light crimson lake solution mixed with a little bit of umber. To darken it, use the same three colours mixed together. Paint the dots on the top with the same mixture made as dark as possible. The lily's bud must be painted in the same manner as the flower.[61]

The entry briefly describes what colour the flower petals are and the paints to use for mixing the several shades of red-purple. The descriptions of colours in the later editions of Vallet's and translations of Van de Passe II's florilegia are

---

59 Art treatises, however, are not purely practical, see Remond, 'Artful Instruction,' 101–134. A study that shows a clear example of the pros and cons of using text to record and disseminate practical knowledge is Hagendijk, 'Learning a Craft from Books,' 198–235.

60 More on Perrot, see Lavezzi, 'Catherine Perrot, peintre savant en miniature, Les Leçons Royales de 1686 et de 1693,' 229–245; McKeown, 'Self-Portrait of an Artist,' 38–46.

61 *'Le Lys de Montagne couleur de Pourpre, s'ebauche d'une eau de Laque de Carmin & un peu de terre d'Ombre melez ensemble fort claire, & se rembrunit des trois mesmes couleurs melangez, les points de dessus se font du plus brun de ce melange, le bouton du Lys se fait comme la Fleur.'* in Perrot, *Les lecons royales*, 49. Translations from Tongiorgi Tomasi, *The Royal Lessons by Catherine Perrot*, 70. The described plate, albeit from a later edition and in reverse orientation of the original, can be seen in Nicolas Robert, *Variae ac multiformes florum species expressae ad viuum et aeneis tabulis* (1665), Providence, John Hay library, Book 255281, image 25, https://library.brown.edu/projects/rome/books/255281/; Bulckens, 'A Flower Book: Some Remarks on Nicolas Robert's "Variae Ac Multiformes Florum Species".'

comparable with how Perrot instructed on the colouring of the mountain lily, but without the preliminary preparation and handling of the painting materials. Despite not being written specifically to instruct on the hand-colouring practice, *colouring instruction* is a useful term here for distinguishing the text from the aforementioned flower descriptions that align more with the language and style of sixteenth-century herbals.

As mentioned, the 1623 and later reprints and editions of Vallet's *Le jardin du roy* ... replaced Robin's Latin descriptions with colouring instructions in French in the frontmatter. There are no plate numbers to link the instructions to the plates, but the names of the flowers are given on top of every entry in the order of the plates, which have numbers etched on the top right corner. For example, the instructions start with the *Corona Imperialis maior*, followed by the *Tulippa bombisima, Tulippa Persica, Corona Imperiallis, Tulipa varie, Tulipa varie, Tulipa varie*, and so on.[62] The seven flowers correspond to those that are depicted on the plates numbered 1 through 3. Plate 1 shows a crown imperial and two tulips, with the captions showing *Corona Imperialis Maior, Tulipa bombicina flore Rubro*, and *Tulipa persica*; plate 2 displays another variety of the crown imperial with the caption *Corona Imperialis*; plate 3 depicts three varieties of tulip, captioned *Tulipa vario flore*. While not straightforward, through the flower names that more-or-less correspond to one another, readers can find the colouring instructions for each of the depicted flowers in the collection.

Vallet's colouring instructions offer more detailed colour information for different parts of the flowers—including blooms, leaves, and stems—but with few (if any) other morphological details or other types of information. The colouring instruction for the same image of the martagon lily described by Robin is listed as the *Martagon panonicum* in the frontmatter of the 1623 edition. The *Martagon panonicum* has a short entry, with the full text stating 'Everything is the same as above, except it's a little more winey.' (*Tout de mesme que dessus, excepte qu'il est un peu plus vineux.*).[63] The most important information is the winey colour of the flower, which pushes it towards the spectrum of red, as described by Robin. The entry Vallet was referring to is the instruction for colouring another kind of martagon lily (or pompone lily), the *Martagon montanum longiore spica*. The entry is written as follows:

62 Vallet, *Le jardin du roy tres chrestien, Loys* XIII, 1, *'Aux amateurs des plantes, et de l'enlumineure.'*

63 Vallet, *Le jardin du roy tres chrestien, Loys* XIII, 2, *'Aux amateurs des plantes, et de l'enlumineure.'* This quote is translated with the help of Jaya Remond and V.E. Mandrij.

> Lay down a thick coat of white laque, and end it with a thick laque, the dots are the same laque with a hint of black, the stamens are orange, the heart green, the stem is purple and green.[64]

Unlike Robin's description, the characteristic of travelling from afar is removed, and the rare and curious quality of the flower's colour(s) becomes the most important aspect of the martagon lily in the instruction. At the beginning of the section on illumination, Vallet wrote in a short paragraph on colouring materials a kind of *laque* that is extremely red is the carmine, which usually refers to lake pigments made of the carmine insect.[65] The florilegium compiler described the red petals by referring to the word *laque*. The different plant parts of a flower are given in greater details, such as including the colours of the stem and the dots and speckles on the petals, and the petal colours of each of the flower's varieties are further indicated.

While not consistent, the colouring instructions in Van de Passe II's *Hortus floridus* similarly became more detailed with the colours on different parts of a flower, with the text working more and more within the boundary of what is already depicted in the images. For example, Van de Passe II did not include the *Martagon Pannonicum* in the *Hortus floridus*, but he described and depicted *Lilium montanum* (Fig. 24), which may be another kind of martagon lily or pompone lily, that is close in appearance. The description is much longer and more extensive than Robin's entry for the *Martagon Pannonicum*, but it has the same structure, with information on morphology and place of origin. Living as a wildflower in German woods and mountains, the *Lilium montanum* is instead a garden flower in the Dutch Republic. Among the descriptions of the morphology of the flower, including the form of the stems, leaves, flowers, and bulbs and roots, the sentences on the flower's colours state,

> The flowers hang downwards from their own stems, with the cleft of a lily, but smaller than in others, faded purple in colour and brilliant with purple speckles, the stamens in the centre being of the same colour. The

64 '*C'est d'une grosse laque blanchette qu'il faut coucher, & l'acheuer de grosse laque, les poincts sont de la mesme laque auec un peu de noir, les estamines orangees, & le Coeur*[?] *vert, la tige est de pourpre & de verd.*' in Vallet, *Le jardin du Roy tres chrestien, Loys XIII*, 2, '*Aux amateurs des plantes, et de l'enlumineure.*' This quote is translated with the help of Jaya Remond and V.E. Mandrij.

65 Carmine most likely refers to cochineal in the seventeenth century, which replaced carmine as the colourant for insect-based red lake, see Eastaugh et al., *Pigment Compendium*, 91; Phipps, 'Cochineal Red,' 4–48.

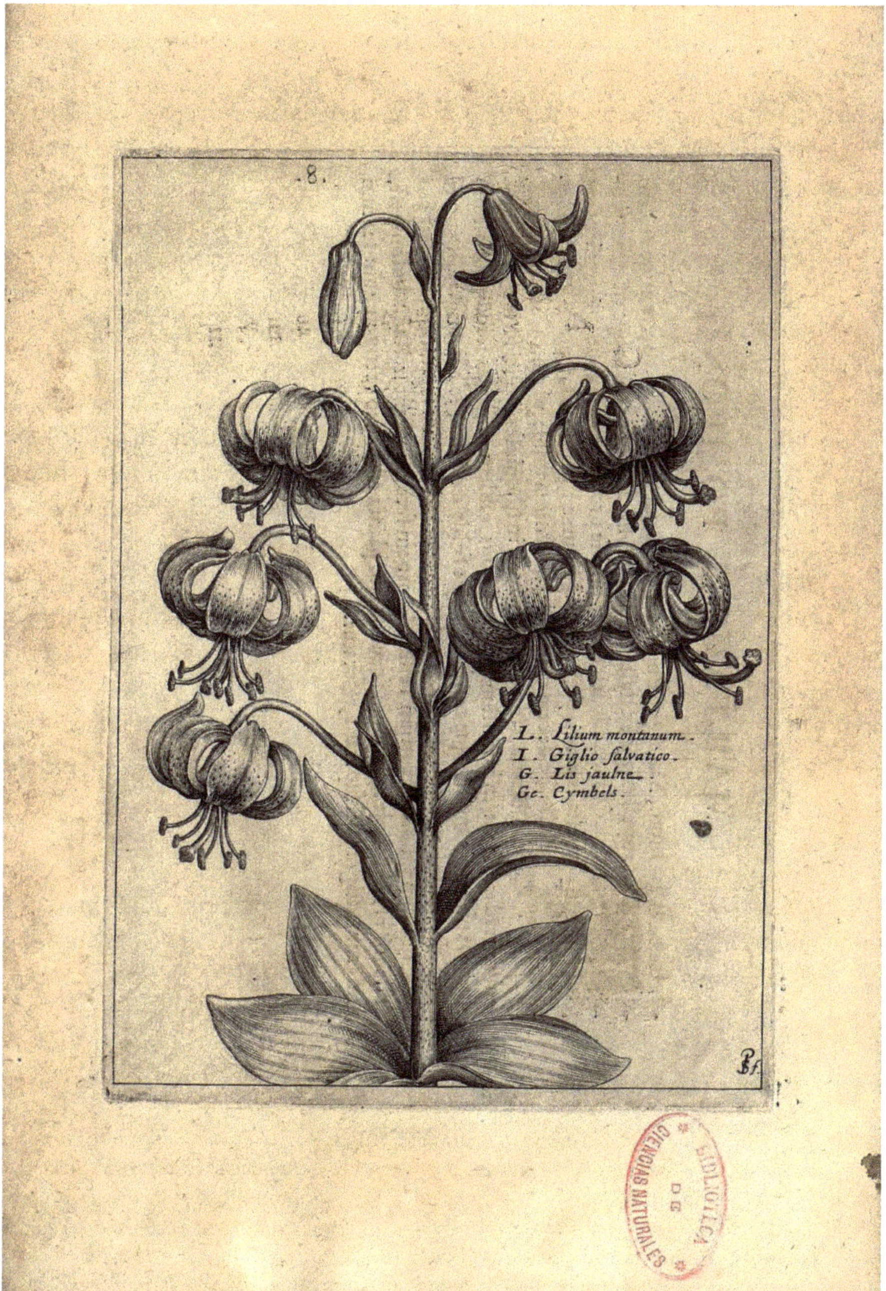


FIGURE 24 *Lilium montanum* from the *Hortus floridus* (1614), Biblioteca Real Jardín Botánico, Madrid, Spain, A PAS, https://bibdigital.rjb.csic.es/idviewer/13646/111

> petals are reflexed and turned backward almost in a circle. The bulbous roots, like many kernels joined together, are of a golden or yellow colour.[66]

The colouring instruction for the same *Lilium montanum* (Fig. 24) in the English, Dutch, and French editions of the *Hortus floridus*, in contrast, is much shorter. The full entry states,

> The mountaine Lilley hath his flower of a withered purple colour, & shadowed accordingly, and garnished with certaine sad purple spots, the steale hath certaine spots, but more darklt appearinge.[67]

Similar to the entries in Vallet's 1623 edition of *Le jardin du roy* …, Van de Passe II removed the flower's place of origin and most of the morphological information. The instruction shows the readers where on the flower the colours should be applied—the tinted purple and the darker purple speckles on the petals, and some spots in even darker purple on the stems—functioning solely within the boundary of the depiction.

In some cases, Van de Passe II provided both the colours and the paints to use, in a closer fashion to how Perrot instructed the readers. For example, the sunflower plate (Fig. 21) in the *Hortus floridus* includes two varieties. The text in the frontmatter is divided into two components: a simple flower description and a comparatively extensive colouring instruction that applies to both the greater and lesser varieties of the sunflower. The full entry is as follows:

> THE MARY GOLDE OF THE SONNE.
> This great and faire sonne marigolde cometh late, not before September, and althoughe this be a very great flower, yet it groweth yearely up, of a smale seede, and must be new sewen every yeare.

66 '[…] *flores e suis pediculis deorsum dependent, Lilij diuisura, sed minores, colore purpureo obsoleto, purpureusque aliquot maculis splendentes, & staminibus eiusdem coloris in medio; quorum foliola retrorsum reflexa in circulum feru recuruantur: Radices bulbosae veluti ex multis nucleis conjunctae, aurei vel flavi coloris* […]' in Van de Passe II, *Hortus floridus*, 8 (summer garden); Savage, *Garden of Flowers*, 118.

67 Van de Passe II, *A Garden of Flowers*, number 8 (summer garden); '*Dit berch Leliken heeft zijne bloemen van coleur / als verstorven purpur / nae zijnen eysch gediept / met eenige schoone donckere purpure plerckens verciert / den steel heeft gelijcke plerckens doch wat flauwer voortcomende.*' in *Den Blom-hof*, number 8 (summer garden).

> THE LESSER MARIGOLDE OF PERU.
> This smaler is in all things like unto the former, but is of a shorter growth, and there ariseth of these three or foure of them upon one steale and of one seede.
>
> The leaves that stand rounde about are of a faire masticott coloure, and if the masticot be not or a high colour, it must be tempered with a little lack, made shyninge, and shadowed with sad yellow, the innermost must be of a berry yellow there must be regard had in toppinge the stars with light masticott, the laggs must also be topt with the former coloure, the cowne within is the saddest of all, the leaves that come after the yellow leaves, must very evidently appeare, because they are greene, these leaves and the steale must be shadowed with sad yellow and ashcoloure, and topt with white and masticott.[68]

The first component offers some information on why the sunflower is placed in the autumn garden, with Van de Passe II writing that the flowers bloomed in September or later in the year, and that the flower was fast growing and needed to be sown every year. The larger text block elucidates the major visual aspect that is missing within the picture plane: the colour of the flowers, and to a lesser extent the colour of the leaves. Van de Passe II describes several kinds of yellow through the materials. Massicot (*masticott*) is a tinted yellow pigment made from roasting lead. The berry yellow most likely refers to a golden wash made from buckthorn berries that was common to a colourist's palette of the

68 Van de Passe II, *A Garden of Flowers,* number 3 (autumn garden); '*FLOS SOLUS OFT CHRYSANTHEMUM. Sonne blom, dese seer schoone ende groote blom / ende is een spade blom comt in September voort / ende hoe wel dese blom soo groot wort / soo wast sy van een cleynachtich sadeken en moet oock alle jare op een nieuw ghesadet worden / etc. CHRYSANTHEMUM PERUI: MINUS. De cleyne is in alles ghelyck de groote van ghedaente / van sy is corter van ghewas ende syrupt met dry oft vier cleyne bloemkens / uit / ende wassen uit eenen sade. Die bladeren die rontom staen zyn schoon Masticot coleur / oft so den Masticot niet hooch van verbeen is / moet met wat lack getempert worden / blinckende / ende wat schytgeel ghediept zyn / het binnenste met besien aengeleght / dan men moet wel acht nemen datmen de steerkens met lichte masticot verhoocht. De horenkens oft den tweeden rant / met het voorghenoemde coleur verhoocht moet zyn / de middel-cruyn is het bruynste van allen. De bladeren die achter de geele bladeren comen die moeten wel af steerken want het groene bladers zyn / den steel ende de bladeren moeten met asch / en schytgeel ghediept zyn / ende met wit ende Masticot ghehoocht.*' in *Den Blom-hof,* number 3 (autumn garden).

time.[69] The sad yellow stands for a yellow deeper in colour. While not explicitly instructed, a sad yellow can be achieved by adding a red lake to the listed yellows, in the same way, one can bring a massicot that is too light to a deeper colour.

The terms for the materials in these colouring instructions have a double function for describing the colours of the flowers and instructing on the pigments to use for colouring the flowers. Functioning solely within the boundary of the prints, the colouring instructions no longer work with the images to present multifaceted pictures of the rare and curious flowers. Rather, they work together to present verbal pictures of colourful flowers. Vallet and Van de Passe II used the language and writing style of artistic practice to provide another means of supplementing the crucial information on colours that made many flower varieties rare and curious. By textually colouring the images, the colouring instructions enriched the black-and-white images with vivid colours in the reprints and later editions of *Le jardin du roy …* and *Hortus floridus*.

## 6 Conclusion

Flower descriptions and colouring instructions were useful ways for florilegium compilers to supplement information to what could not have been depicted within the picture planes of a black-and-white medium. The shifting image-text relations and proportions throughout the printing and reprinting of seventeenth-century florilegia show a general trend of the genre to place increasing emphasis on colours. While the morphological information became less prominent and important in the later editions of *Le jardin du roy …* and the *Hortus floridus*, colour information remained present. Catering to the widely read contemporary *liefhebbers*, florilegium compilers utilised writing styles and terms that were similar to early modern herbals and/or hand-colouring treatises for verbal picturing. By juxtaposing plates with flower descriptions, presenting information similar to sixteenth-century herbals, the images and text presented multifaceted pictures of rare and curious flowers beyond the picture planes. The colouring instructions gave vividly colourful pictures of the depicted flowers, despite requiring more work from the readers to connect the images and text.

As discussed in the previous chapter, florilegia, by and large, did not claim to be authoritative studies of flowers and decorative plants. Even the 12 entries by

69 The *schytgeel* in the Dutch edition is also known as the *shitgeel*, see Eastaugh et al., *Pigment Compendium*, 328.

Jean Robin of the flowers that were previously undescribed were intended to simply give readers a better idea of what the flowers were. Returning to the different editions of the *Hortus Eystettensis* discussed at the very beginning of this chapter, the re-adding of flower descriptions in the 1713 version can be understood through this lens. The inclusion of the then-outdated text, 100 years after its initial publication, was not to offer authoritative information for the studies of the depicted flowers, but to present more comprehensive pictures of the depictions. Plates in intaglio-printed florilegia can function on their own, but flower descriptions and colouring instructions enabled the depictions to offer more beyond the inked lines.

CHAPTER 5

# Keeping Flowers between the Pages

## 1 Introduction

The Rijksmuseum in Amsterdam houses an album with 194 watercolours of flowers and plants on paper. The frontispiece of the album contains the phrase '*Plusieurs especes de fleurs dessinees d'apres le naturel*' (Several kinds of flowers drawn after nature), but the album is better known as the *Bloemenboek*.[1] The majority of the 194 watercolours depict rare and curious flowers of the seventeenth century, including flowers and decorative plants that travelled from the Mediterranean, Asia Minor, and the Americas to areas to the north of the Alps, as well as garden cultivars of different or new colours and/or with double petals.[2] This heavy emphasis on rare and curious flowers marks the album as a seventeenth-century florilegium. The flower depictions in the album were commissioned by and produced for Agneta Block (1629–1704), a renowned seventeenth-century rare plant collector in the Dutch Republic. Block is most celebrated for being the first European to cultivate a pineapple to fruition.[3] In addition to her rich rare flower and plant collection, Block is documented to have had a large assemblage of drawings, watercolours, and paintings of the plants by many artists, several of whom were well-known natural history watercolourists of the time, such as Maria Sibylla Merian (1647–1717).[4] The Rijksmuseum florilegium, while being the most intact volume, is only a small portion of Block's image collection, much of which is now dispersed as loose sheets across institutions in Europe and North America.[5]

1 Multiple contributors for Agneta Block, *Plusieurs especes de fleurs dessinees d'apres le Naturel*, Amsterdam, Rijksmuseum, RP-T-1948-119.

2 See Chapter 1 for discussion on the characteristics that made flowers rare and curious in the long seventeenth century.

3 For the biography of Block and an inventory of her collection, see Van de Graft, *Agnes Block: Vondels nicht en vriendin*. On account of Block's pineapple, see Levitt, '"A Noble Present of Fruit",' 106–119; Powell-Warren, 'Pineapple Lady,' 95–99.

4 The items are mentioned in a contemporary poem by Block's cousin, which gives an impression of her estate and collection, even though the constructed image though the poem may not fully reflect reality, see Blok, *Vijver-hof van Agneta Blok*. See also Powell[-Warren], 'Locating Early Modern Women's Participation in the Public Sphere of Botany,' 234–258.

5 Some surviving sheets including Herman Saftleven (attributed), *Studies of Five Flowers and Plants*, New York, The Morgan Library & Museum, 1984.10, https://www.themorgan.org/drawings/item/128399; Herman Saftleven (attributed), *Studies of Violets and Buttercups*,

© JESSIE WEI-HSUAN CHEN, 2025 | DOI:10.1163/9789004735149_007

In correspondence with the Bologna botanist Lelio Trionfetti (1647–1722), Block wrote about the reason behind her image collection:

> When I have some rare plants or flowers, I have them painted from life in true colours, so that when they die, they are still preserved on paper for me.[6]

This statement casts light on two points. First, Block's Rijksmuseum florilegium and other images functioned primarily, although not exclusively, as visual records of her living collection. Based on the rare and curious flowers and plants growing in her garden, Block's image collection is similar to the big flower book of the *Hortus Eystettensis* (1613). Compiled by the Nuremberg apothecary Basilius Besler (1561–1629), the big flower book worked, to an extent, as an elaborate pictorial catalogue of the rare plant collection in the garden of the Prince-Bishop Johann Konrad von Gemmingen (1561–1612).[7] Block's florilegium and image collection, likewise, catalogued her living rarities and curiosities through pictures. Second, being a well-known and skilful gardener of rare and curious flowers and plants could not prevent Block from losing living specimens. Having her living rarities and curiosities captured and depicted on paper was a means of 'keeping' and 'preserving' ephemeral living collections.

This chapter investigates how seventeenth-century flower books 'preserved' rare and curious flowers by comparing florilegium images to dried plants in early modern herbaria. It looks into what features were lost in the process of drying and pressing flowers and plants, as well as the artistic manipulations herbaria makers implemented in order to prevent the losses. This chapter then examines how florilegium images employed different modes of representation to capture and record living flowers, and further visually 'press' them onto papers and parchments. The last section delves into how the accumulative process of

---

1984.11, https://www.themorgan.org/drawings/item/128402; Guilliam de Heer, *Datura with Butterflies*, 2010.160, https://www.themorgan.org/drawings/item/303893. Herman Saftleven, *Study of a sticky nightshade (solanum sisymbriifolium)*, Stockholm, National Museum (Nationalmuseum), NMH 516/2016, https://collection.nationalmuseum.se/en/collection/item/183386/.

6 '*Quando io tengo alcuna pianta rara o fiore, le faccio depingere al vivo con li loro colori accio che quando moriscono mi si conservano tuttavia in carta.*' The letters were translated into Italian by Block's nephew Anthonio Block. Transcriptions of the letters can be found in Poelhekke and Oomen, 'Elf brieven van Agnes Block in de universiteitsbibliotheek te Bologna,' 2–28; quoted sentence on page 24. This quote is translated into English with the help of Carlotta Cappuro.

7 Besler, *Hortus Eystettensis*. See Chapter 1 for an introduction to this florilegium.

gathering (visually) pressed flowers preserved the living collections. By understanding the usefulness and limitations of dried plants, this chapter argues why images were a more suitable means to record and preserve the rare and curious features of flowers and decorative plants in the long seventeenth century.

While intaglio-printed florilegia continue to be part of the discussion, this chapter primarily focuses on manuscript flower books as examples. The contents of intaglio-printed and manuscript florilegia are the same, as both are image collections of rare and curious flowers of the seventeenth century. However, there are a few differences between the two mediums. First, while text is already a less prominent feature in printed flower books, painted ones usually contain virtually no text, with the exception of flower names and the occasional titlepages and/or frontispieces.[8] As a result, manuscript florilegia did not aim to excerpt and extract pictorial and textual knowledge, as did some engraved flower books.[9] Second, painted florilegia had the advantage of being coloured works from the beginning. Thus, the challenge of presenting colour information through the black-and-white medium of engraving and etching, as discussed in Chapter 4, was not a concern for manuscript flower books. Third, the production length and methods for printed and painted florilegia differ drastically from one another. This issue of production time will be especially pertinent to later discussions in this chapter.

## 2 Preserving Flowers on Paper and Parchment

Images in florilegia, generally speaking, capture the fleeting stage of bloom when the two visual characteristics of vibrant colours and double flowers are most prominent and recognisable for many flowers, thus preserving their rareness and curiousness through images. For example, Block's *Bloemenboek* includes seven varieties of anemones, and the variety '*sanguineo coxineo hexapryllo*' appears on leaf 164 (Fig. 25). The name of the flower is written on the top left corner of the page and the page number is on the opposite corner on the top right. The page shows two of the *sanguineo* variety at different stages of bloom. The one on the left is a budding flower that would soon be in full bloom, and the one on the right is blooming, perhaps close to the point of wilting. Both anemones display bright red petals. The one in full bloom also shows

8 The Nassau Florilegium is an exception, as the leaves often have short flower descriptions on the verso side of the depictions. See section 'Pressing Flowers for Rare and Curious Features' in this chapter for an introduction and more information on this florilegium.

9 See section 'Excerpting Flowers, Extracting Knowledge' in Chapter 3 for a discussion on this topic.

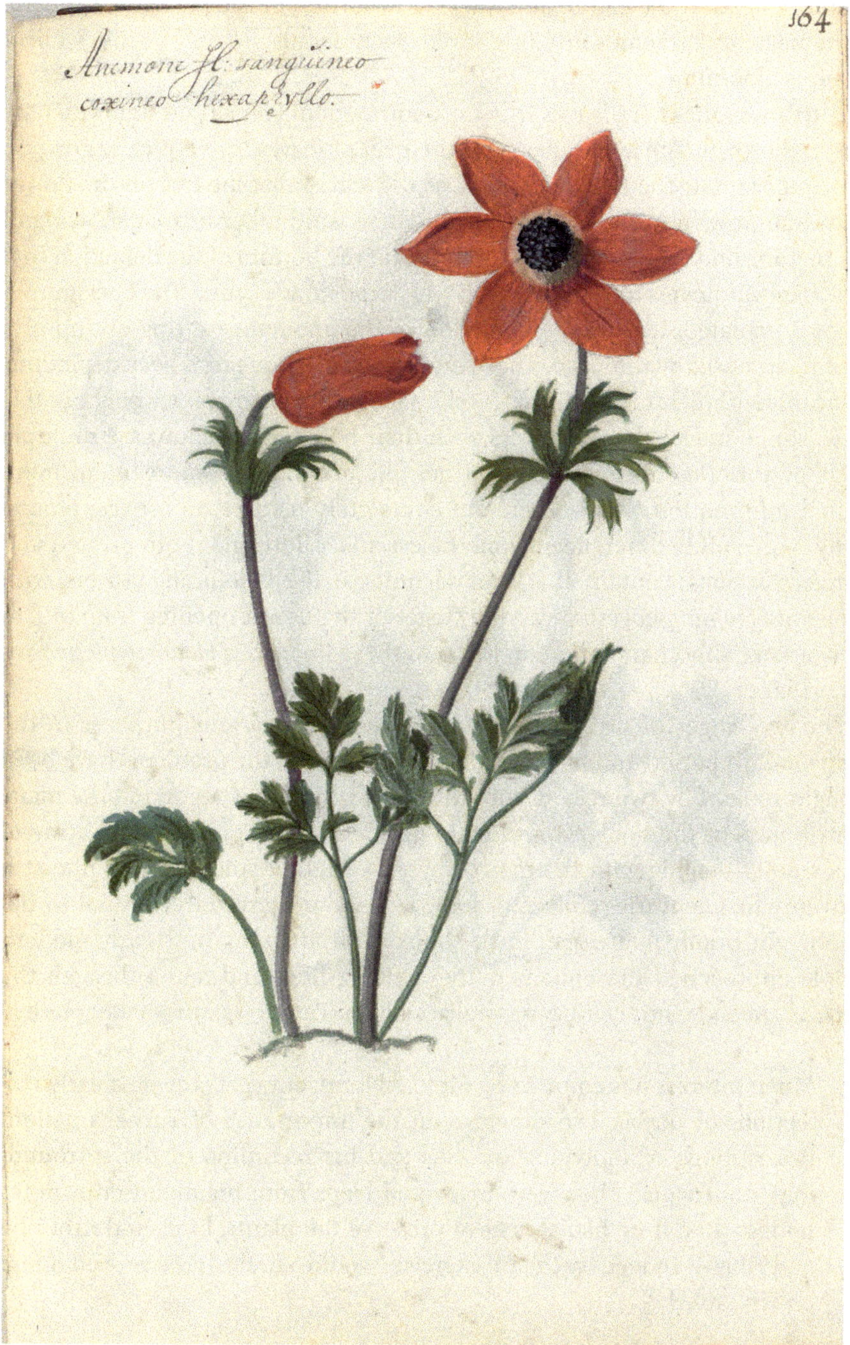


FIGURE 25 *Anemone sanguineo coxineo hexapryllo* from the *Bloemenboek* for Agneta Block (second half of the seventeenth century), Rijksmuseum, Amsterdam, RP-T-1948-119

the black head in the centre and the yellow portion of the petals that is close to the head and gradates into bright red; such details are only visible when a flower is blooming.

Both anemones have leaves with an assortment of greens and stems that are of a cool brown. The anemones have their set of leaves that typically grow on the same stem as the blooms, as well as the leaves that are not on the flower stem, but grow from the same tuber(s). There is no tuber or root showing in this picture, and there is an ambiguous line at the bottom of the depiction that most likely indicates the ground from which the flowers grew. The background is not articulated, leaving the depiction of the anemones sitting on top of a decontextualised white page. There are many variations of flower depictions in seventeenth-century florilegia. As shown in other examples throughout this book, some images depict flowers with their bulbs; some showcase multiple kinds or varieties of flowers within one picture plane; some contain more textual information while others are completely textless on a page; among many more subtle differences. Nonetheless, most florilegia (both printed and manuscript ones) contain images of decontextualised blooming flowers, with the word *blooming* covering varying stages of the flower opening from bud to over-bloom. This chapter hereon refers to these images as *bloom depictions* as a shorthand.

The challenges of preserving natural objects for various purposes in the early modern period and how people worked around the problems have been brought to light by two recent publications. Arthur MacGregor and the many contributors of the book *Naturalists in the Field* (2018) presented an array of case studies dealing with the many ways in which people from the fifteenth to twenty-first centuries collected, recorded, and preserved the natural world. MacGregor briefly addressed, in the book's introduction, how early modern people approached the ephemerality of living flora and fauna through the works of the sixteenth-century naturalist William Turner (1509/10–1568), stating,

> Much interest has come to be directed in recent years towards early collections of physical specimens, but the importance of Turner's patient descriptions of individual species and his recording of the surroundings from which they were recovered leaps from his manuscript notes no less than if he had sought to preserve the plants, birds and fishes he describes—indeed such an exercise would surely have proved more ephemeral.[10]

10 MacGregor, ed., *Naturalists in the Field*, 3.

Text, in the works of Turner, has the longevity that prepared specimens may not have, and thus the text not only records, but also preserves and conserves the information and understanding of the specimens without their physical presence.

More notably, the role of images in preserving and conserving natural or material objects is discussed by Morwenna Blewett and Lucy Wrapson. In their recent special issue of the journal *Notes and Records*, the two authors drew attention to several methods employed by early modern people to press and conserve objects that may not seem to be valid today.[11] Blewett and Wrapson discussed how the English antiquary, collector, and government official, Elias Ashmole (1617–1692), ordered watercolours or black-and-white images to be made after the rare birds, insects, fishes, that were prone to decay, as a way to accept the ephemeral nature of many specimens that were difficult to preserve and conserve. Blewett and Wrapson argued,

> This acceptance was attached to contingency planning in the form of generating visual documentation as a record, but which also served as a material proxy—a transposition for when the condition of an object was no longer representative or when it was totally destroyed.[12]

MacGregor and Blewett and Wrapson demonstrated that objects, text, and images were interconnected. Instead of each functioning as its own separate category, they formed a collective of multi-faceted tactile, textual, and visual information on flora and fauna for early modern people to understand the natural world.

Through this lens, the boundary between bloom depictions and living flowers becomes fluid, and the genre of the herbarium can offer insights into why Block and other collectors who commissioned to have florilegia made used images to conserve their rare plant collections instead of preserving the once living plants. Generally speaking, a herbarium is a collection of plants that are dried by being pressed flat and fixed onto sheets of paper with glue, string, and/or tape.[13] Emerging in the sixteenth century, the herbarium was more often called *hortus siccus* (dried garden) or *hortus hyemales* (winter garden) in the early modern period.[14] As one of its early modern names suggests,

---

11 Blewett and Wrapson, 'Introduction: Cabinet, Elaboratory, Gallery 1500–1800,' 229–235.

12 Blewett and Wrapson, 'Introduction: Cabinet, Elaboratory, Gallery 1500–1800,' 3.

13 The word *herbarium* was a term for herbal in the sixteenth century. More on the connection between herbals and dried plant collections in the sixteenth century, see Arber, *Herbals: Their Origin and Evolution*, 138–143; Ogilvie, *The Science of Describing*, 165–174, 313.

14 There have been many efforts to study early modern herbaria from multiple fronts, including historians, biologists, and conservators. The word *everlasting* for this book title

the herbarium was a tool for dealing with the ephemerality of plants so that naturalists and their students could have something to study during wintertime, when many of the living plants were not available. The Italian physician and botanist Luca Ghini (1490–1556) is the earliest recorded individual to be known for systematically making herbaria and teaching his students to do so. By the seventeenth century, the use of dried plants for studying had become common, or at least its usefulness recognised, as accounted in the *Isagoges in rem herbariam libri duo* (1606, 1633, 1667).[15] Written by the Flemish physician Adriaan van de Spiegel (1578–1625), the title is one of the earliest printed books describing the genre. The small book is mostly a textbook to show medical students ways of studying plants, both their morphology and medicinal uses. Van de Spiegel described two methods for recording and preserving the morphological information of plants in order to study them on bad weather days and during winter. The first method shows readers how to make nature prints, and the second instructs how to prepare specimens for collecting and compiling herbarium sheets.[16]

Early modern herbaria, as described in Van de Spiegel's book, were made to primarily provide continuous access for students to familiarise themselves with the morphology of the plants without the constraint of plants' ephemerality.[17] The author's how-to text instructs his readers on what to look for in a specimen when selecting and collecting plants to dry. In addition to the practicalities and processes of pressing and glueing, Van de Spiegel wrote,

arrived independently, but the selected studies were useful for better understanding the usage production, and material culture of early modern herbaria during this research, Carrión, 'Planted Knowledge,' 47–67; Findlen and Toledano, 'The Materials of Natural History,' 151–169; Jarvis, '"The Most Common Grass, Rush, Moss, Fern, Thistles, Thorns or Vilest Weeds You Can Find",' 303–328; Stefanaki, et al., 'Breaking the Silence of the 500-Year-Old Smiling Garden of Everlasting Flowers'; and Thijsse, '"Everlasting Gardens": Origin, Spread and Purpose of the First Herbaria,' 72–107.

15 I consulted the 1633 edition.

16 Van de Spiegel, *Isagoges in rem herbariam*, 130–135. See the book chapter by Dominic Olariu that discussed some issues and benefits of nature printing as an image-making technique for plant illustrations, 'Herbs under Pressure,' 9–32.

17 Herbaria have developed conventions for collecting and presenting dried specimens over the centuries. However, early modern herbaria were much more fluid in their formats and presentations of dried plants. Two recent books are a good starting point to see the changes throughout the ages through selected herbaria, which include several examples demonstrating that early herbaria did not have fixed rules on how to do things, see Thiers, *Herbarium: The Quest to Preserve & Classify the World's Plants*; Flannery, *In the Herbarium*.

> If it can be found, a plant you want, or its green branch with flower, leaves, and seed, that is without defect, and it shall be placed between two sheets of white paper.[18]

The instruction advises picking a specimen during its flowering time, and, if possible, closer to the end of a plant's flowering time to potentially include seeds. This way, one could preserve morphological information on both the leaves and flowers, and sometimes seeds, for studying during the wintertime. Picking and drying the particular plant section of flowers and leaves at the stage when it was blooming keeps similar morphological information as a bloom depiction in florilegia. In other words, while the purposes of producing a florilegium image and a herbarium specimen may have been different, the visual plant information the two genres preserved and presented are comparable. It is the forms of the plants, and not the medicinal and utilitarian functions, that herbarium users were trying to extract by studying the dried plants.

Furthermore, seventeenth-century authors sometimes grouped dried and pictured flowers and plants together, blurring the boundaries of the florilegium and the herbarium genres. An example is *De florum cultura* (1633) by Giovanni Baptista Ferrari (1584–1655), whose illustrated plates were excerpted in the *Florilegium renovatum et auctum* (1641).[19] In book four on the uses of flowers, Ferrari included a chapter on dried and imitation flowers (*Florum siccitas vivax, & verisimile mendacium*). The first half of the chapter mostly provides examples of ways of keeping and preserving living specimens, including sand-drying flowers for bouquet arrangements and pressing flowers dry, the latter of which referenced Van den Spiegel's *hortus hyemales*.[20] The second half of the chapter turned to several ways of making flower images, including nature printing and drawing flowers on paper and parchment. In particular, Ferrari wrote about the *Hortus Eystettensis,* describing how the flowers bloom more enduringly on paper than in soil with their images engraved to their natural size.[21] Dried flowers and images of flowers were not the same thing in Ferrari's chapter, but the common thread was that they both keep flowers visually blooming for a much longer time than their lifespan in nature.

---

18 '*Planta quam vultis, aut ejus ramus viridis cum flore, foliis, atque semine, si haberi potest, perfectaque sit Planta, inter chartae candidae duo folia ponantur.*' in Van de Spiegel, *Isagoges in rem herbariam*, 133.

19 De Bry, *Florilegium novum*. See Chapter 3 for an introduction to this florilegium.

20 Ferrari, *De florum cultura*, 433–435.

21 Ferrari, *De florum cultura*, 439.

## 3 Losing Rareness and Curiousness

Despite being able to keep the flowers in their blooming moments, the many petals of double flowers and the growing varieties of flower colours were difficult to preserve through pressing blooms. While there is no surviving *hortus siccus* with only rare and curious flowers, many herbaria have a florilegium component.[22] The majority of the specimens in most early modern herbaria are common utilitarian and medicinal plants. Also making frequent appearances are the dried flowers that were rare and curious, such as variations of anemones, irises, hyacinths, narcissi, and so on, that commonly graced the pages of seventeenth-century flower books.[23] These flower pressings show that early modern naturalists were also interested in the same flowers as the *liefhebbers*, but they preserved the flowering plants through different means. The pressed specimens in herbaria demonstrate that dried plants can preserve many aspects of a flower that an image cannot, most significantly the textures of different plant parts. However, the vibrant colours and full forms are often lost, or at least obscured, through the process of being pressed and dried.

Many pages in a herbarium associated with the Dutch botanist and chemist Herman Boerhaave (1668–1738), for example, demonstrate such a loss and obscurity.[24] Boerhaave is known to have collected and pressed many plants for his herbaria, potentially multiple sets with different contents.[25] A four-volume set, now preserved at the Natural History Museum in London as part of the vast collections of Hans Sloane (1660–1753), used to be attributed to Boerhaave. The attribution is now questioned, as the botanist and chemist would have been too young to produce such extensive work.[26] This chapter addresses the

---

22 Sixteenth-century herbals tend to be encyclopaedic in the plants they include. Crops, (fruit) trees, lichens, mushrooms, and so on, compose the majority of plants in these books. Many herbaria are similar to the *Libri picturati A16–30*, a watercolour collection discussed in Chapter 1 that has a scope of plant depictions comparable to sixteenth-century herbals, including having a florilegium component. See reproduced pictures and studies in De Koning et al., eds., *Drawn after Nature*.

23 Lorraine Daston also pointed out that flowers are among some of the more difficult ones to preserve as dried plants, and images were (and are) used as the type specimens of some plants, see 'Type Specimens and Scientific Memory,' 160.

24 Herman Boerhaave (associated with), [*Hortus siccus ex*] *Hermanni Boerhaave* [*Collectione, a*] *D. Hans Sloane* [*comparatus*] (4 volumes), London, The Natural History Museum, H.S.319–H.S.322.

25 Thijsse and Wesseling, 'Herman Boerhaave: A Search for His Herbarium Collections,' 170–181. Boerhaave is more often studied for his role as a chemist in the history of science, but some discussions on his experimentations with plants can be found in Klein, 'Experimental History and Herman Boerhaave's Chemistry of Plants,' 533–367.

26 Thijsse and Wesseling, 'Herman Boerhaave: A Search for His Herbarium Collections,' 172.

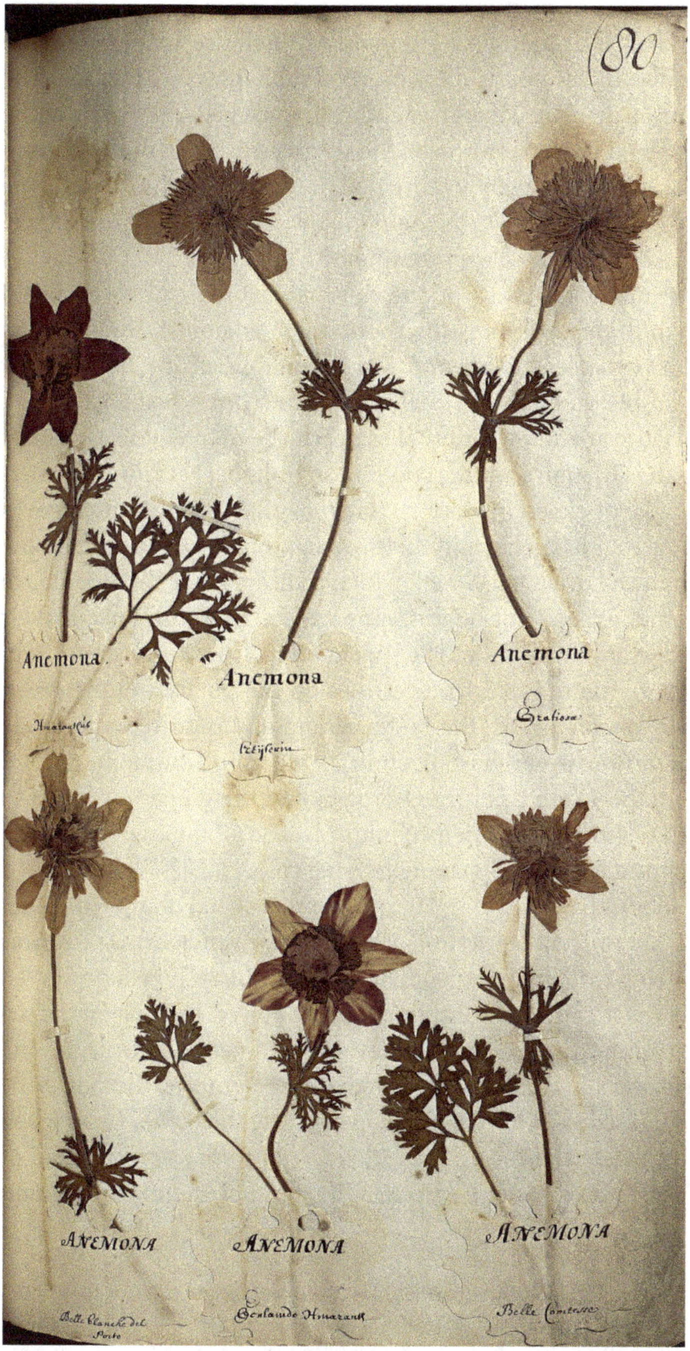


FIGURE 26 Poppy anemones from the [*Hortus siccus ex*] *Hermanni Boerhaave* (second half of the seventeenth century), The Natural History Museum, London, HS320 f. 080

collection as Boerhaave's herbarium, since who the precise maker was is not the central question here. Boerhaave's herbarium includes many anemone varieties, and folio 80 in volume II displays six double flowers (Fig. 26). The majority of the double petals are intact, as shown in the cluster of petals in the centres of the preserved flowers. However, the volumetric effect of the blooming flowers is lost, or at least much less defined. By flattening the flowers, the anemones lost the fullness achieved by their double petals, making this aspect of their rareness and curiousness less prominent.

Similarly, even though the hues of the flowers' colours can be retained to a degree through drying and pressing them, there is almost always a loss in saturation.[27] For instance, the anemone in the middle of the bottom row on Page 80 is a double variety with red and white striped petals (Fig. 27). A watercolour of a close anemone variety (Fig. 27) can be found in the Gottorfer Codex, painted by the German painter Hans Simon Holtzbecker (1610/20–1671) for Frederick III, Duke of Holstein-Gottorp (1597–1659).[28] The depicted leaves in the watercolour are a bit thicker and there are more petals in the outer ring of the flower than in the dried specimen, which indicates that the Boerhaave and Gottorfer anemones may not be the same variety, but both have similar petals with red and white variegation. The specimen in Boerhaave's herbarium still has its colours, as the red and white stripes on the petals and the green leaves are visible. However, unlike the highly saturated bright red and white and lush green the flower would have been in its blooming moment, the red has become muted, the white has turned brown, and the green has faded to brown in the dried specimen. The dried anemone lost its vibrancy, diminishing its level of rareness and curiousness through being colourful.

A specimen of the *Iris Susiana flore* (Fig. 28) on Page 59 in Volume I of Boerhaave's herbarium not only displays similar losses of rare and curious features in the process of being dried and pressed, but further shows how a preserved specimen can look drastically different from a living flower. The Susian iris is a variety that made common appearances in seventeenth-century florilegia. The specimen in Boerhaave's herbarium preserved all the petals of the flower by pressing them flat individually, and then reconstructing the flower head when gluing them to the sheet. The result is an iris as seen from above instead of from the side, which is possible to see in real life, but not the most common

27 The challenge of retaining colours was partially discussed by Pugliano, 'Ulisse Aldrovandi's Color Sensibility,' 70–108.

28 Hans Simon Holtzbecker, *Gottorfer Codex*, Copenhagen, National Gallery of Denmark (Statens Museum for Kunst), kksgb2950, https://open.smk.dk/en/artwork/image/KKSgb2950. See Chapter 1 for an introduction to this florilegium.

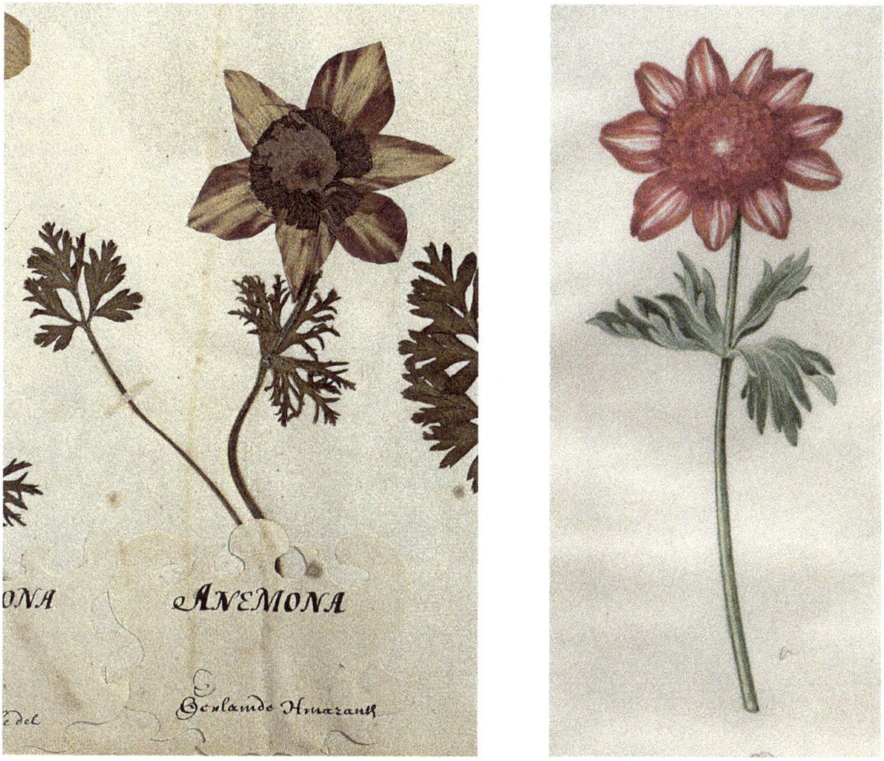


FIGURE 27 Pressed poppy anemone from the [*Hortus siccus ex*] *Hermanni Boerhaave* (second half of the seventeenth century) (left) and watercolour poppy anemone from the Gottorfer Codex (1649–1659) (right), The Natural History Museum, London, HS320 f. 080 / SMK, National Gallery of Denmark, Copenhagen, KKSgb2950/28. Cropped and edited by the author

nor intuitive view when encountering the flower in a garden. The volume and structure of the *Iris Susiana* are largely lost in the pressed specimen, making the flower more difficult to recognise. The Susian iris was also known to have impressive colours of deep purple that was almost black, in addition to having intricate patterns on the petals.[29] While the patterns are preserved, as the dotted lines can still be seen on the petals, the deep purples have faded to several kinds of browns on the specimen. The colours of these pressed flowers would have been much more vivid in the seventeenth century than they are

29 For example, the dark colour was commented by Crispijn van de Passe II in the *Hortus floridus*, 3 (summer garden). See Chapter 1 for an introduction to this florilegium.

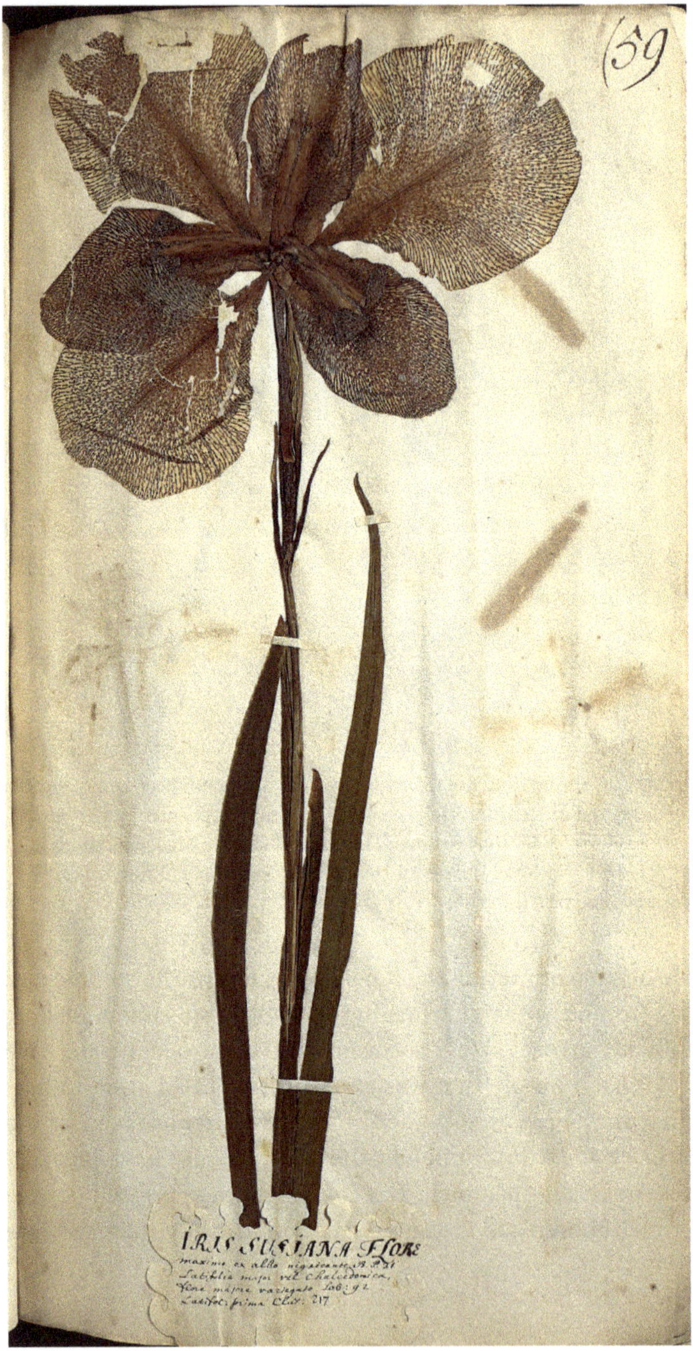


FIGURE 28 *Iris Susiana flore* from the [*Hortus siccus ex*] *Hermanni Boerhaave* (second half of the seventeenth century), The Natural History Museum, London, HS319 f. 059

today.[30] Nonetheless, there is always a change in the colours of the flowers, most often becoming less vibrant and more muted, after they have been dried and pressed.

Among the surviving dried plant collections from the early modern period, the herbarium of Felix Platter (1536–1614) is an insightful example of how the makers of early modern *horti sicci* employed creative methods to work around the losses of certain plant features.[31] In addition to the dried plants, the Swiss physician Platter is also known for his large curiosity collections. As it stands, Platter's herbarium is not complete; only eight volumes of pressed specimens out of the supposedly eighteen have survived and are now preserved at the Burgerbibliothek in Berne.[32] The herbarium is well-known for Platter's artistic manipulation of the pressed plants. An exemplary instance is that, for a specimen of a campanula, Platter used the petals of a dried delphinium.[33] Both flowers have a blue-purple colour when they are fresh. Campanula petals easily turn brown when they wilt, but delphinium petals are comparatively more preservable, usually retaining their blue colours for a much longer time, albeit more muted than when fresh. A campanula has several small blooms on its stem. Platter preserved one of the small blooms with a real campanula flower, which is visibly brown, and pasted delphinium petals to form the other buds and blooms on the stem. By constructing a campanula bloom out of another kind of flower, this 'specimen' does not preserve the flower, but the 'image' of the campanula.

A significant feature of Platter's herbarium, moreover, is that the physician usually paired a dried specimen with an image of the plant to supply visual information that is lost in the pressed plant.[34] For instance, the recto side of folio 56 in volume 2 (Fig. 29) of Platter's herbarium shows a specimen of

---

30 Many historical adhesives are comparatively less corrosive than some modern alternatives, but the common method of glueing specimens can contribute to their decays and damages, see Grenda-Kurmanow, 'Adhesives Used in Herbaria,' 1–15.

31 Felix Platter, Platter-Herbarium, Bern, Burgerbibliothek of Berne (Burgerbibliothek Bern), ES 70 (ES 70.1–ES 70.9), https://www.burgerbib.ch/de/bestaende/privatarchive/einzelstuecke/platter-herbarium.

32 There is an additional volume with only cutouts of woodcuts from sixteenth-century herbals and botanical treatises. For more descriptions and studies on this collection, see Dauwalder, 'Felix Platter's Herbarium: The Preservation of a Historical "Bound Herbarium",' 26–32; Dauwalder and Lienhard, *Das Herbarium des Felix Platter*.

33 See especially study by Davina Benkert, who first wrote about this example of the campanula among other kinds artistic manipulation by Platter, 'The 'Hortus Siccus' as a Focal Point,' 211–239.

34 Benkert, 'The 'Hortus Siccus' as a Focal Point,' 230.

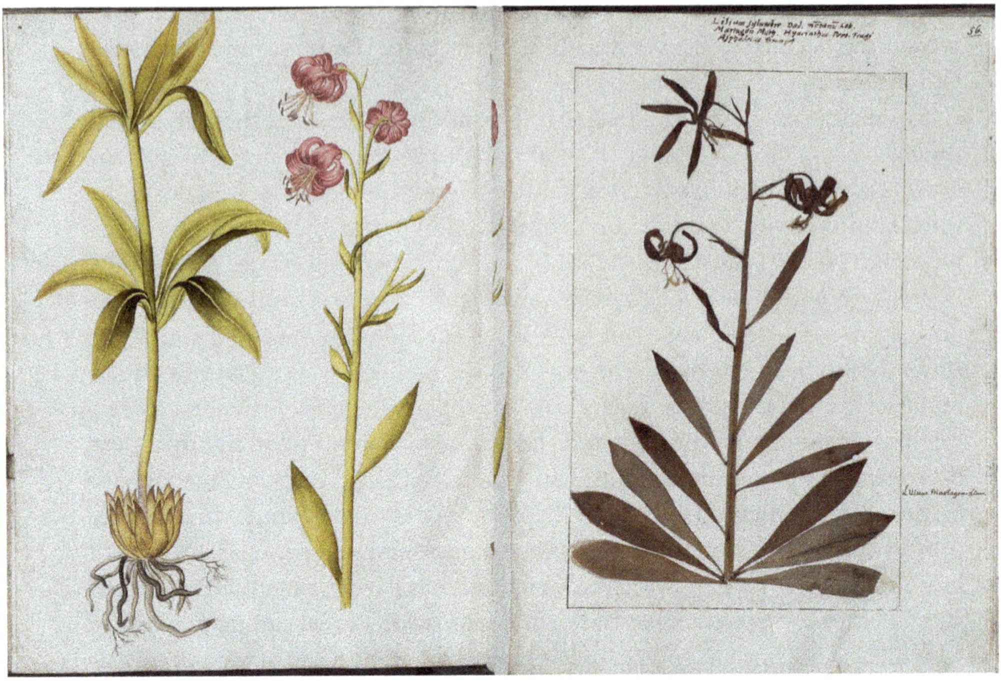

FIGURE 29 Watercolour and pressed martagon lilies in the Platter-Herbarium (second half of the sixteenth century), Burgerbibliothek of Berne, Bern, ES 70.2

the martagon lily.[35] Platter pressed the stem with the blooms and the leaves separately, and then reconstructed the dried specimens into one piece. On the opposing page on the verso side of folio 55 is a watercolour of the flower. The watercolour depicts the martagon lily in bloom, in the same stage of the flower's life as the dried specimen. The bulb and roots that are not preserved through the pressed flower, as well as the vibrant magenta colour of the petals, are presented in the picture. In this spread, Platter utilised the watercolour—a bloom depiction—to visually preserve the plant parts and visual information that the dried martagon lily could not retain. Many of the dried plants in Platter's herbarium are accompanied by cutouts of woodcuts (both coloured and uncoloured) from printed herbals. Decorative flowers, on the other hand, are most often paired with watercolours that are stylistically similar to bloom depictions in florilegia. Platter's herbarium demonstrates how images could retain the visual information on forms and colours that dried specimens lost through the process of being pressed. In the same manner, bloom depictions

35 The volume ES 70.2 of the Platter-Herbarium contains mostly garden flowers, many of which are the rare and curious ones discussed in this book.

are able to prevent such losses and preserve the forms and colours of rare and curious flowers.

## 4 Capturing Flowers in the Moment(s)

The making of a florilegium, especially a manuscript one, is the making of a series of bloom depictions that preserve rare and curious flowers in their most volumetric and colourful moment. As previously mentioned, bloom depictions include budding to over-blooming flowers. While it is a relatively short period of a flower's lifetime, it still covers multiple steps of transformation within this period, and depicting and visualising these stages within one picture plane is not a straightforward act. The final images, either in print or in watercolour, are a result of selecting and prioritising certain visual information from these states of growth. How to capture and visually represent the various stages of a plant's lifetime in one image, or occasionally a series of images, was (and still is) an issue and challenge for image makers of plants, not only for makers of bloom depictions. An image maker may not have been fully aware that they were making such decisions, as oftentimes they were made intuitively based on the makers' training and the wider visual culture around them. Nonetheless, which moment and what to depict from the blooming stage of a flower's life were deliberate choices, and they subtly affected what bloom depictions precisely preserved for each image in a florilegium. The previous chapters have shown how printed florilegia excerpted and extracted visual and textual information from several sixteenth-century herbals, most significantly from the works of the physician-botanists Rembert Dodoens (1517–1585), Carolus Clusius (1526–1609), and Matthias Lobelius (1538–1616) in the southern Low Countries.[36] The visual strategies for how to represent a plant pictorially that have been more explicitly discussed in herbal illustrations can assist in better understanding the different kinds of bloom depictions in florilegia.

The contested strategies—specific plant portrait and generic plant image—are the two important modes of representation that started the use of copious illustrations by physician-botanists in the sixteenth century, subsequently excerpted into florilegia.[37] The *Herbarum Vivae Eicones* (1530) by Otto Brunfels

36 See sections 'Excerpting Flowers, Extracting Knowledge' in Chapter 3 and 'Describing Flowers and Their Colours' in Chapter 4 for a discussion on how compilers adapted visual and textual information in herbals for their florilegia.

37 For a clear layout of these two modes and their application in natural historical images, see Egmond, *Eye for Detail*, 126–144.

(1488–1534) and *De historia stirpium commentarii insignes* (1542) by Leonhard Fuchs (1501–1566) both employed well-known image makers to create the illustrations from living flowers and plants, but the images display rather contrasting decision making over depicting and visualising what the image makers observed.[38] The woodcuts in Brunfels' herbal were reproduced from the drawings by the German painter and printmaker Hans Weiditz (1500–*c.*1536).[39] The Brunfels/Weiditz illustrations (both the drawings and the subsequent woodcuts) are most significant in showing many details that suggest the image maker closely drew what he saw, based on the specific specimens that were in front of him. For example, one of the woodcuts depicts a lily-of-the-valley (*Lilium convallis, vel sylvestre Lilium; Meyenblümlin*) (Fig. 30).[40] Weiditz presented the flower in profile view, including the roots, two leaves, and a stem with five small white bell-shaped blooms. The bloom on the very top is budding, with the rest in full bloom; half of the blooms are positioned in front of the leaf on the right. This overlapping of plant parts is close to the common experience of how one usually sees a living specimen of the lily-of-the-valley. Furthermore, the detail that is most indicative of the woodcut depicting a specific specimen, rendering this illustration a plant portrait, is that the breakage at the very tip of the root on the left was fully captured by Weiditz. This portrayal shows the lily-of-the-valley with its imperfection and visually freezes the flower in the very moment when Weiditz drew the specimen.

Fuchs' herbal (Fig. 31) presents generalised images that were based on close observation of living specimens, but did not capture them as directly seen by the image makers. German painters Albrecht Meyer (active 1540) and Heinrich Füllmauer (active 1540) made the drawings that were then reproduced into woodcuts. The lily-of-the-valley (*Ephemerum non letale; Mayenblümle*) by Meyer or Füllmauer has many of the same features as the one drawn by Weiditz, showing the flower with roots pointing to the left, two leaves, and several blooms.[41] The angle of the leaves, however, is artificially turned towards the reader, making it more of a frontal view instead of a profile view. Even though the stem of the blooms is also positioned in between the leaves as in Weiditz's drawing, the blooms do not overlap with the leaves, which makes the string of blooms more easily visible. This frontal showing of both sides of the leaves was also practised for pressing plants in herbaria, and the pictorial

38 On the use of images of these herbals, especially on Fuchs' title, see Kusukawa, *Picturing the Book of Nature*.

39 Close to 80 of these drawings found their way to the pages of the Platter-Herbarium, Dauwalder and Lienhard, *Das Herbarium des Felix Platter*, 46–47.

40 Brunfels, *Herbarum vivae eicones*, 211.

41 Fuchs, *De historia stirpium commentarii insignes*, 240.

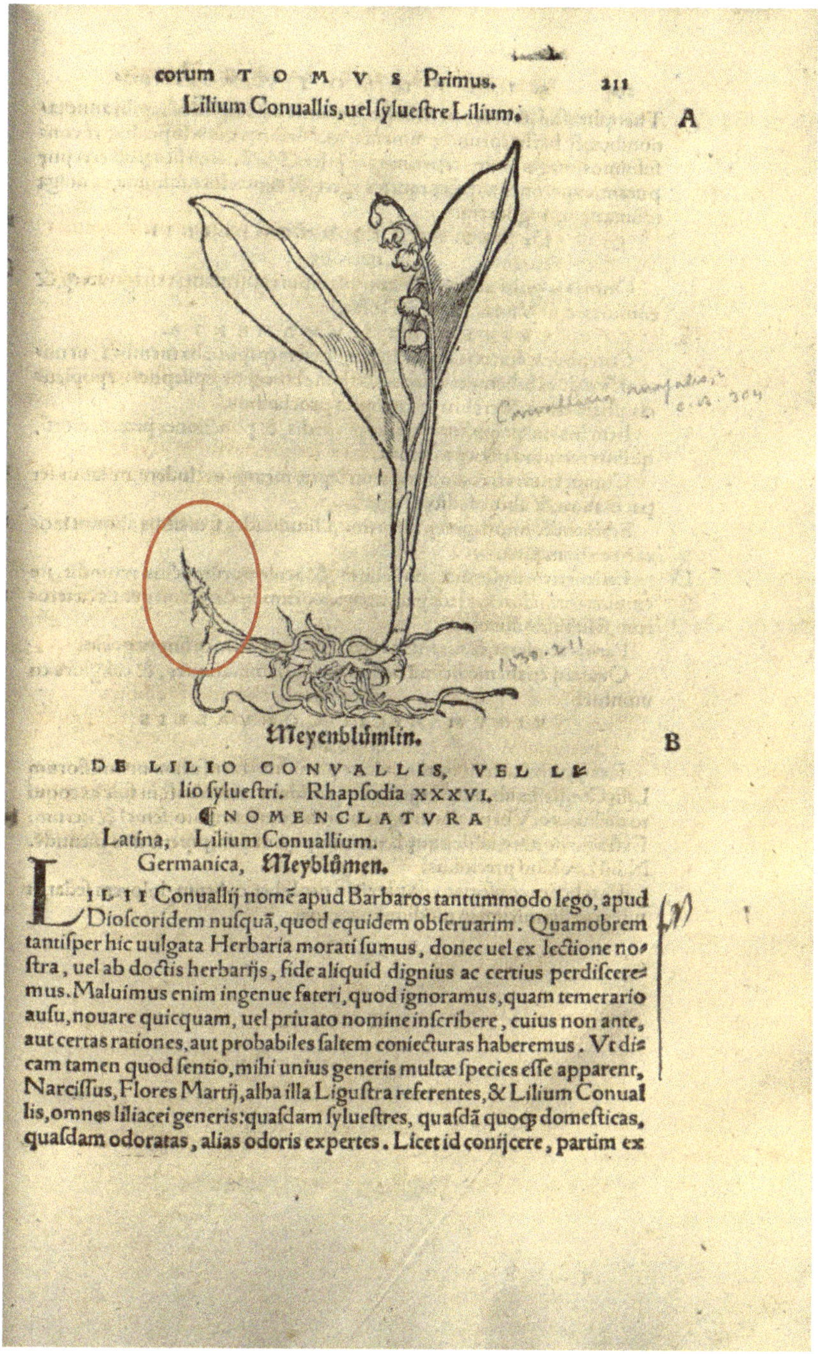

corum TOMVS Primus. 211

Lilium Conuallis, uel ſylueſtre Lilium.

A

Meyenblůmlin.

B

DE LILIO CONVALLIS, VEL LE lio ſylueſtri. Rhapſodia XXXVI.

NOMENCLATVRA

Latina, Lilium Conuallium.

Germanica, Meyblůmen.

LILII Conuallij nomẽ apud Barbaros tantummodo lego, apud Dioſcoridem nuſquã, quod equidem obſeruarim. Quamobrem tantiſper hic uulgata Herbaria morati ſumus, donec uel ex lectione noſtra, uel ab doctis herbarijs, fide aliquid dignius ac certius perdiſceremus. Maluimus enim ingenue fateri, quod ignoramus, quam temerario auſu, nouare quicquam, uel priuato nomine inſcribere, cuius non ante, aut certas rationes, aut probabiles ſaltem coniecturas haberemus. Vt dicam tamen quod ſentio, mihi unius generis multæ ſpecies eſſe apparent, Narciſſus, Flores Martij, alba illa Liguſtra referentes, & Lilium Conuallis, omnes liliacei generis: quaſdam ſylueſtres, quaſdã quoq; domeſticas, quaſdam odoratas, alias odoris expertes. Licet id conijcere, partim ex

FIGURE 30 Lily-of-the-valley woodcut from the *Herbarum vivae eicones* (1530), Missouri Botanical Garden, Saint Louis, MO, QK41.B75 1536–37 [#596]. Edited by the author

representation became a convention for herbal illustrations. This correlation often leads to the conclusion that early modern image makers drew from dried plants in herbaria. However, the process of drawing a plant can be quite complex, often involving a combination of living and dried specimens.[42] There is no detail that suggests the lily-of-the-valley by Meyer or Füllmauer is based on a specific living specimen, as the image is a synthesised depiction of the flower. Most importantly, the progression of time is presented in this woodcut, with the small white bell-shaped blooms representing several stages of the flower. The three on the very top show the budding stages as the flower gradually opens. The section in the middle depicts the flower in full bloom, with four of them in profile view and one that shows the six petals as when one was looking up to the centre of the bloom. The last two at the bottom are the fruit of the lily-of-the-valley. The close-to-full life cycle of the flower from bud to fruit, which usually does not exist at the same time in nature, is visualised in this woodcut. This *completeness* of a plant, with all the major plant parts and stages of growth visualised, is a signature mode of representation in Fuchs' herbal.[43]

Both modes of plant portrait and generic/complete image are present in seventeenth-century flower books. For example, a florilegium attributed to the Dutch glass painter Pieter Kouwenhoorn (1599–c. 1654) contains mostly portraits with observable details that could realistically come from the painter recording the specimens faithfully to what he saw.[44] Most watercolours of flamed tulips are considerably portraits or portrait-like (Fig. 4). The distinctive striped patterns on each and every living specimen were what made them rare and curious.[45] Capturing these blooming flowers through plant portraits can better preserve their rareness and curiousness which are largely connected to the specificity of each specimen. Tulip portraits also often include details such as wormholes, blemishes, and/or other flaws on the flowers, adding to the naturalistic features that could be observed from a specific specimen as the model for the image. It should be noted, however, that an image with blemishes, wormholes, and/or other similar details does not automatically make it a plant

42 See Kusukawa, *Picturing the Book of Nature*, 147–152.

43 The concept of a 'complete' picture of a plant came from Kusukawa, *Picturing the Book of Nature*, 114–119.

44 Pieter Kouwenhoorn, *Verzameling van Bloemen naar de Natuur geteekend door Pieter van Kouwenhoorn*, London, The Royal Horticultural Society Lindley Library, 38968–1001. I only had a brief moment to look at the volume with the curator Charlotte Brooks. Not much has been written about this this florilegium; see most updated notes in Segal and Alen, *Dutch and Flemish Flower Pieces*, vol. 2, 1027 (note no. 62). On Kouwenhoorn, see Van Ruyven-Zeman, 'Pieter Kouwenhorn,' 162–179.

45 See Anne Goldgar's book for a discussion on tulip portraits and why made them, *Tulipmania*, 98–112.

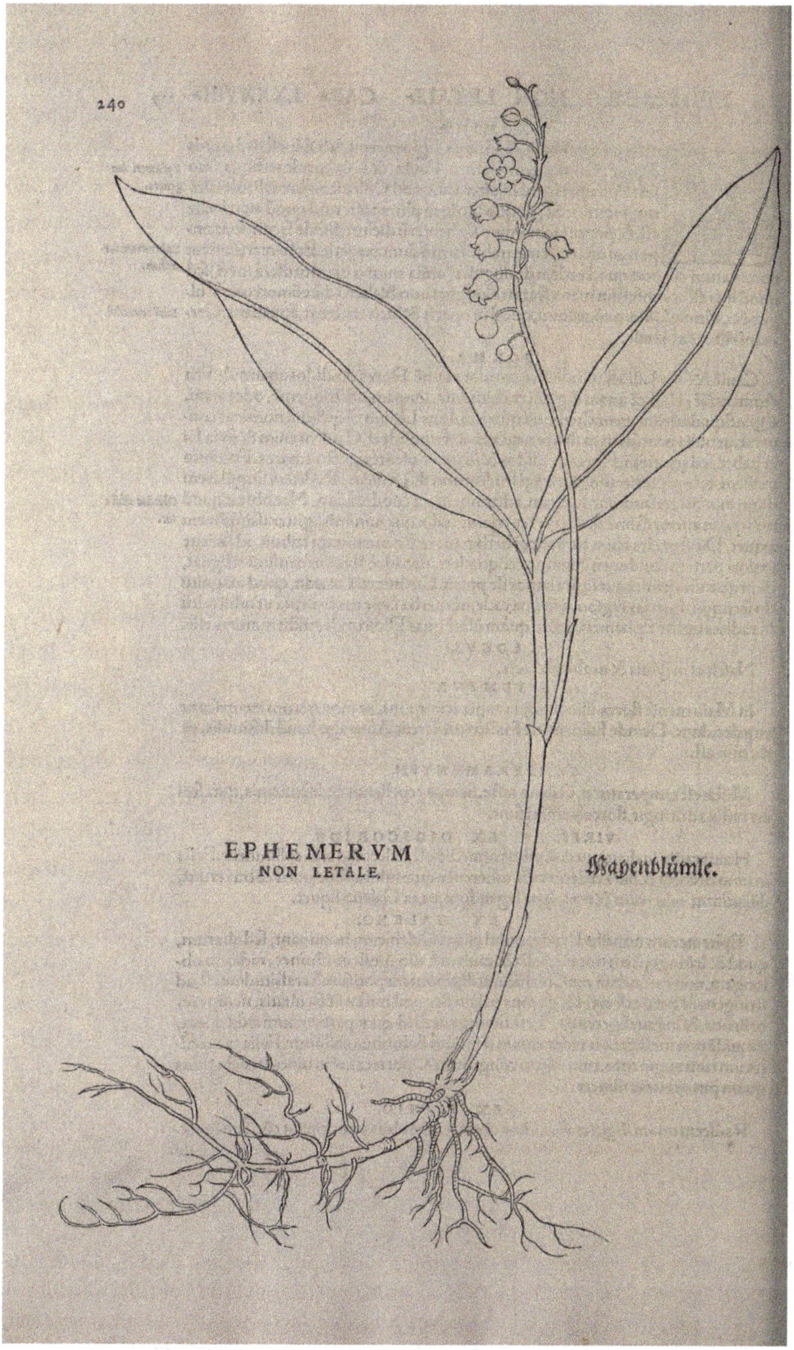


FIGURE 31 Lily-of-the-valley woodcut from *De historia stirpium commentarii insignes* (1542), Smithsonian Libraries and Archives, Washington D.C., QK41.F7 1542

portrait. A skilful image maker could easily depict generalised flaws. The specificity needs to be assessed from a multitude of details of how a flower is drawn.

Conversely, generic images of flowers and decorative plants that combine several stages of a flower's lifespan occasionally appear on plates and leaves in florilegia. A plate in Besler's *Hortus Eystettensis*, for example, depicts two kinds of lily-of-the-valley and a paradise lily (Fig. 32).[46] The lily-of-the-valleys have two branches, one with many small blooms, and one with the (red) round berries, and the roots of the flowers are present. Similar to Fuchs' illustrations of the 'complete' lily-of-the-valley, the two varieties on Besler's plate provide general images of the flowers that show different stages of their growth and more plant parts, but no details suggesting they are portraits of specific specimens. The mode of generic/complete image is only occasionally implemented, with lily-of-the-valley one the most noticeable cases among florilegia.[47] This suggests that the fruit was one of the rare and curious features of this flower, and a complete image was suitable for preserving the features that occurred at different moments of the flower's life.

A major mode of representation among bloom depictions, however, is a mixture of capturing the flower in the specific moment of the bloom with a generalised depiction of its features. Already in the sixteenth century, a large number of herbal illustrations do not present a strong opposition of specificity and generalisation as in Brunfels' and Fuchs' works. Even within the herbals of Brunfels and Fuchs, not every woodcut strictly follows the physician-botanists' ideals of specificity versus generalisation/completeness. Chapter 3 particularly discussed the woodcuts of two kinds of snowdrops (Fig. 17) that illustrated the descriptions for the herbals of Dodoens, Clusius, and Lobelius.[48] Multiple drafts-people contributed to the drawings and/or watercolours that were then reproduced into woodcuts for the three physician-botanists, including Pieter van der Borcht (1535/40–1608) and Jacques van den Corenhuyse (c. 1530 –c. 1584).[49]

46 Besler, *Hortus Eystettensis*, nonus ordo. fol. 6 (spring garden); Littger and Dressendörfer, *Florilegium: The Book of Plants*, plate 131.

47 Such depiction of the lily-of-the-valley is also found in also found in Sweerts, *Florilegium Emanuelis Sweerti Septimonti Batavi*, plate 7 (*Pars Secunda*); it also appears on leaf 78 in the *Hortus amoenissimus* by Franciscus de Geest that is discussed later in this chapter.

48 See section 'Copying Flowers' in Chapter 3.

49 Van der Borcht was employed by the printer-publisher Christophe Plantin, and a large number of the woodcuts in the herbals of Dodoens, Clausius, and Lobelius were drawn by him, see Voet, *The Golden Compasses*, vol. 2, 199, 211. Jacques van den Corenhuyse may have been the watercolourist for many of the sheets in the *Libri picturati*, thus the woodcuts made after the watercolours might be attributed to him, see De Koning et al., eds., *Drawn after Nature*, 18.

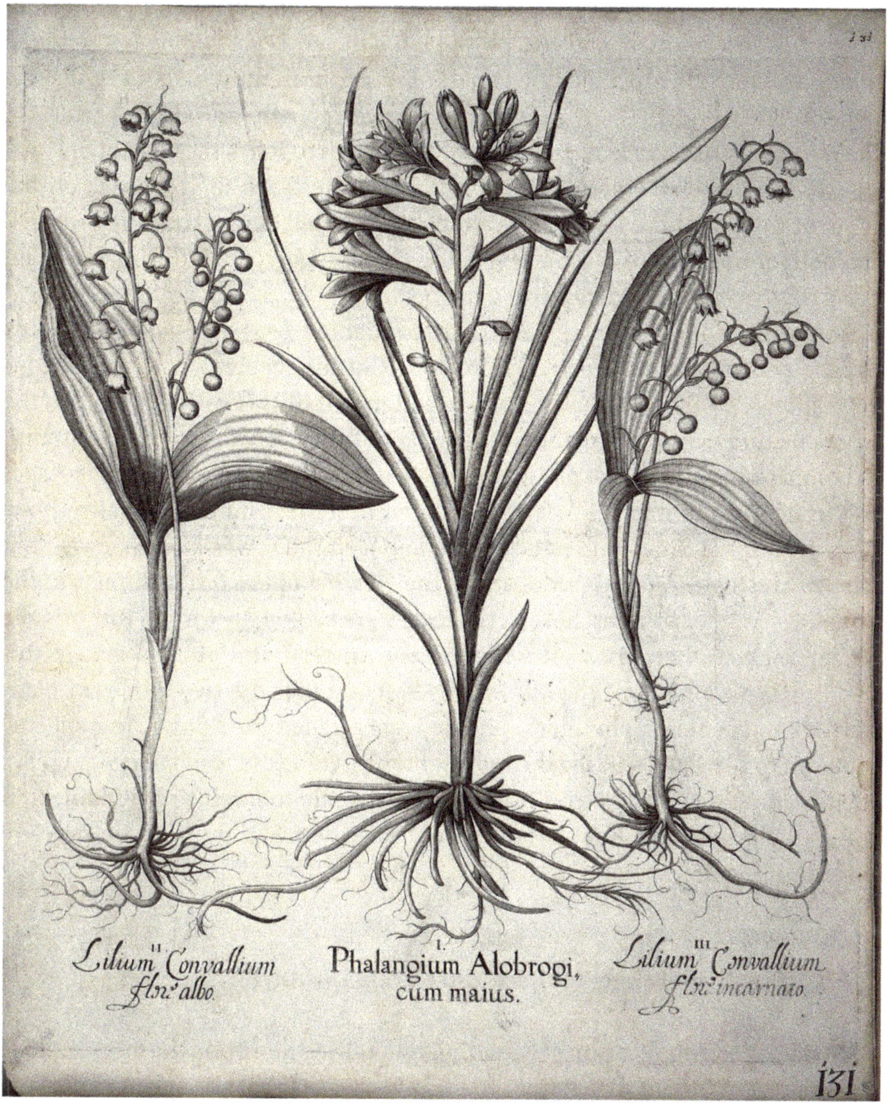


FIGURE 32 Lily-of-the-valley plate from the *Hortus Eystettensis* (1613), this digitised impression is from a copy of the 1640 edition, Missouri Botanical Garden, Saint Louis, MO, QK41.B5 1640 [#1054]

The two snowdrops are printed side by side (Fig. 17) in Clusius' *Rariorum plantarum historia* (1601).[50] The one on the left, labeled *Leucoium bulbos. praecox minus* and most likely drawn by Van der Borcht, presents a high level of specificity.[51] The woodcut shows a blooming snowdrop with an offshoot that is in the middle of growing leaves. The tip of the sprout touches the edge of one leaf on the main flower, creating a small tangent. Similar to the breakage on the roots of the lily-of-the-valley by Weiditz, these small details and the captured moment of the flower's life are so specific, that it suggests that the image maker drew this image based on a particular specimen.

In contrast, the snowdrop on the right, named *Leucoium bulbos. praecox Byzant* and modelled after a watercolour Clusius received from one of his correspondents, shows a more generalised image of this variety.[52] Unlike the woodcuts in Fuchs' herbal, the image maker of this snowdrop did not attempt to compress time into one picture plane. Instead of showing the various stages of the flower, the image is set at the state of bloom. The image maker, however, positioned the bloom and the bud between the leaves in such a way that each of the features does not obstruct the viewing of the other. Similar to the lily-of-the-valley in Fuchs' herbal, the leaves are artificially turned towards the reader, making them more of a frontal view instead of a profile view. In this manner, the woodcut of the *Byzant* variety combines the two modes of plant portrait and generic plant image and delivers a generalised flower in a specific moment of bloom. Recalling the red anemone in Block's *Bloemenboek* (Fig. 25), the bloom depiction likewise presents the specific moments of budding and (over-)blooming with flowers that do not have details that proclaim they are portraits of actual specimens.

## 5 Pressing Flowers for Rare and Curious Features

The widespread application of depicting generalised flowers in the more specific moment(s) of bloom throughout the genre suggests that the specificity of the living specimens was not the priority. Presumably, based on Blocks' statement to Trionfetti, the red anemone could have been one of the rare flowers the collector had growing in her garden at Vijverhof in Utrecht. One of the artists and

50 Clusius, *Rariorum plantarum historia*, 169.

51 The woodcut first appeared in Dodoens, *Florum*, 192.

52 This snowdrop is modelled after a watercolour Clusius received through one of his correspondences. More on the watercolours and drawings Clusius received, see Van Zanen, *Planten op papier*, 216–228.

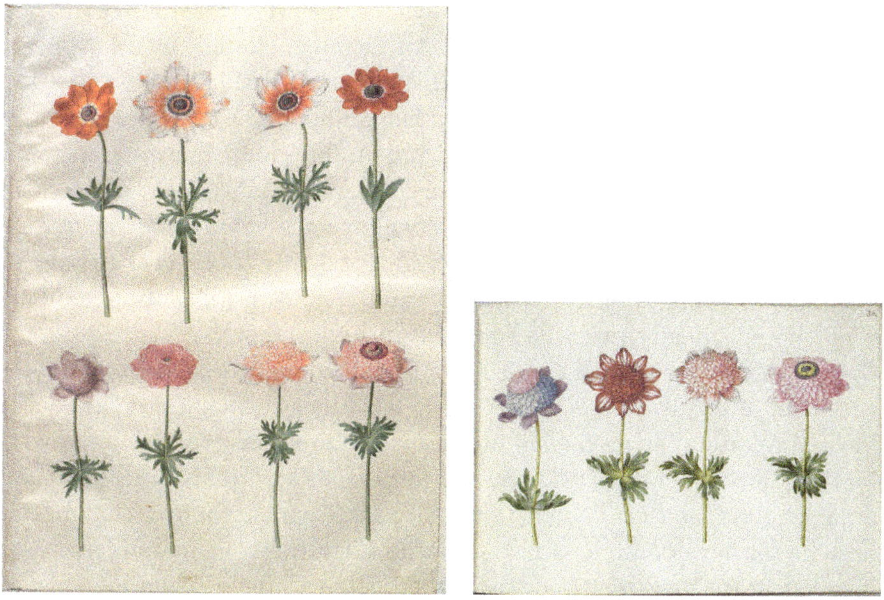

FIGURE 33 Poppy anemones from the Gottorfer Codex (1649–1659) (left) and the Eberhard Anckelmann Florilegium (1660) (right), SMK, National Gallery of Denmark, Copenhagen, KKSgb2949/35 / The British Museum, London, 1888,1211.1.1-74

image makers with whom she collaborated may have captured the likeness of the anemones on site, yet the depictions show little specificity of a living specimen. Similarly, aside from the Gottorfer Codex, Holtzbecker painted florilegia for multiple patrons, and many of the flowers were drawn and painted in the same way for different collections. For example, the Gottorfer Codex includes more than 70 variations of single and double anemones in many colours, and Leaf 36 from the third volume depicts eight red and pink varieties. One can find the same pink double anemone depictions on the bottom right (Fig. 33) in another florilegium, the Eberhard Anckelmann Florilegium (Fig. 33), now preserved at the British Museum in London.[53] Eberhard Anckelmann (1599–1664) was a wealthy merchant in Hamburg, and his florilegium contains 72 painted leaves, each with multiple (usually four) flowers per page.[54] The same

53 Hans Simon Holtzbecker, Eberhard Anckelmann Florilegium, London, The British Museum, 1888,1211.1.1-74, https://www.britishmuseum.org/collection/object/P_1888-1211-1-1-74. This flower book is usually discussed in conjunction with the Caspar Anckelmann Florilegium, which belonged to Eberhard's son Caspar, a merchant and town councillor who had a more famous garden in Hamburg, see De la Fuente Pedersen and Kolind Poulsen, eds., *Flowers and World Views*, 109–111; Schattmann, 'Von der Leidenschaft für Blumen,' 117–125.

54 There are and additional 25 unpainted leaves bound in the manuscript.

depiction of the red double anemone with stripes (Fig. 33), the second one to the left on the page, in the Eberhard Anckelmann Florilegium also appears in the Gottorfer Codex on the previous leaf of number 35 in volume 4 (Fig. 27). The anemones in the two florilegia are not carbon copies of one another, as the leaves have slightly different shapes, but the flowers are drawn and painted in the same manner, with no details suggesting the individuality of the flowers as plant portraits. There are many such instances of Holtzbecker reusing a depiction with minor alterations for different flower books, despite most of his patrons having extensive rare flower and plant collections for potentially supplying specimens from which to make plant portraits.[55]

Florilegium image makers used bloom depictions to visually represent a spectrum of subtle modes between generals and specifics. This spectrum is highly important, as it enabled image makers to visually 'press' the flowers for their rare and curious features. A closer look at *how* Holtzbecker drew and arranged the anemones demonstrates the ways in which bloom depictions can push for the flowers' rareness and curiousness. The sheets containing anemone varieties in the Gottorfer Codex and the Eberhard Anckelmann Florilegium (Fig. 33) have comparable arrangements of flowers to the anemone page in the Boerhaave's herbarium (Fig. 26). Holtzbecker arranged the depictions of eight anemone varieties in two registers of four on leaf 36 of volume 3. The eight bloom depictions are placed in front of an empty background, apart from one another with a little distance in between. This arrangement presents the anemones in a specimen-like manner in early modern herbaria. Moreover, the anemones are drawn in such a way that they are visually unfolded, as they would have been if they were living specimens that were being prepared for pressing. The flower head is turned towards the viewers. This presents a top view of the flower, usually achieved when viewers look downward at the flowers from above. The stems, however, are in profile view, which is the perspective when viewers look at the flowers from the side. Likewise, Holtzbecker artificially arranged the leaves by raising the leaves behind the stems and lowering the leaves in the front. The way that Holtzbecker manipulated the perspectives of the anemones is close to how a flower is often unfolded and prepared before being pressed to dry. The top right anemone on folio 80 in volume 11 of the Boerhaave's herbarium (Fig. 26) shows a specimen that was prepared in this manner, but in the dried form.

Such visual unfolding is also noticeable when image makers compose bloom depictions to create a scene, such as showing a picture of flowers growing in a

55 De la Fuente Pedersen and Kolind Poulsen, eds., *Flowers and World Views*, 157–161.

garden. Similar to pressed flowers in herbaria, bloom depictions often lose more information than they preserve. The environment and ecology of the flowers, for instance, are almost entirely lost, with the bloom being decontextualised.[56] Sometimes, flowers are placed on or in front of generic grounds.[57] For example, the sheets in the Nassau Florilegium often place the flowers in front of a ground (Fig. 34). The Nassau Florilegium is a multi-volume collection, now housed as three separate sets in Paris, London, and Frankfurt.[58] Painted by Johann Jakob Walther (or Johann Walter) (1650–1717) between 1650 and 1670, the large image collection was commissioned by Count John of Nassau-Idstein (1603–1676) based on the living rare flower collections at the garden of Idstein. Walther was documented to have visited the site eight times to study and draw the flowers in the garden.[59] The example sheet (Fig. 34) shows several species and their varieties, including one fritillary, five narcissi, and four anemones. The flowers are depicted in profile view, intersecting one another. It should be noted that the ground might not have been part of the sheet's original arrangement, as it is painted with and in a slightly different material and technique. Even if the ground was part of the intended composition, it suggests a generic garden ground. The visual information is so minimal that the ground does not contextualise the flowers' (humanmade) habitat much more than a plain white background.[60] Despite being more dynamically composed, the flowers

56 A famous set of illustrations that took a drastically different approach is a manuscript by Gherardo Cibo, who placed the plants in front of landscapes that contain naturalists doing fieldworks in the background, *Pietro Andrea Mattioli, Discorsi*, London, British Library, Add. Ms. 22332; Tongiorgi Tomasi, 'Gherardo Cibo,' 199–216. The works by Maria Sibylla Merian, on the other hand, are a prominent example of visualising small-scale ecologies though a combination of flowers, plants, and insects, see Etheridge, *The Flowering of Ecology*.

57 The *Hortus floridus*, which has been extensively discussed in the previous chapters, is most known for this practice. Many plates show the flowers planted in the ground or positioned in front of one.

58 It is not entirely clear if they were ever part of a big florilegium or already as multiple smaller ones with similar contents, *Le Florilège de Nassau-Idstein* (54 watercolours of flowers on parchment), Paris, The National Library of France (Bibliothèque nationale de France), Manuscript. Rés. Ja-25-Fol; Nassau Florilegium (133 flower studies plus several views of the garden), London, Victoria and Albert Museum, 9174 & 9175, https://collections.vam.ac.uk/item/O101038/simulacrum-scenographicum-celeberrimi-horti-itzsteinensis-album-walther-johann-jakob/; Nassau Florilegium (130 watercolours of flowers on parchment), Frankfurt, Städel Museum, Bibl. 1596.

59 See the most recent essay on the historical and production background of the Nassau Florilegium by Graser and Sonnabend, 'Johann Walter der Altere und das Florilegium des Grafen Johannes von Nassau-ldstein,' 101–115.

60 The *Hortus floridus*, in addition to generic garden grounds, sometimes displays flowers in front of landscapes, such as hills or buildings in the distance. However, these landscapes

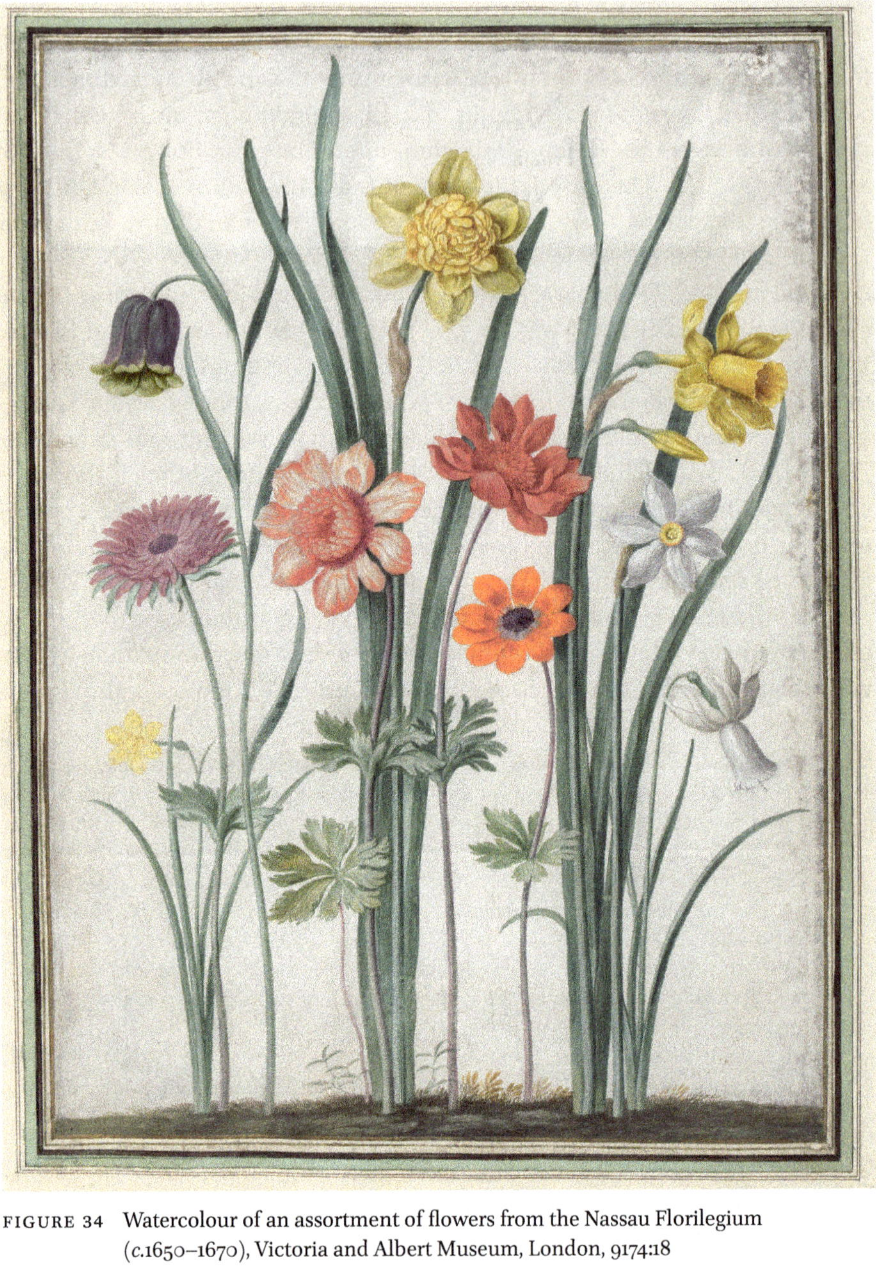

FIGURE 34 Watercolour of an assortment of flowers from the Nassau Florilegium (*c*.1650–1670), Victoria and Albert Museum, London, 9174:18

are not depicted in a consistent perspective, as one would see the living flowers in a garden. Rather, the flowers are artificially unfolded and positioned in a way such that each of the flowers can easily be disentangled from the others for viewers to look at or study separately.

Visually unfolding the flowers brings clarity and visual emphasis on what *is* preserved, as it reduces flower parts from obstructing one another and hence allows petals to be shown in their full form and colour(s). By unfolding the flowers in the bloom depictions, Holtzbecker and Walther not only visually pressed the flowers without flattening them to preserve their rare and curious features, but further *pressed for* the rareness and curiousness to be more prominent. Pressing for rare and curious features explains, to an extent, why plant portrait was not a necessary nor default mode for the genre of the florilegium, even for the flower books that had close connections to specific gardens and rare flower collections. Visually pressing the flowers was already an artificial way to manipulate and enhance the specimens if there were living models for the image makers to draw from life when creating a bloom depiction. Removing the blemishes and flaws and generalising the features from the observed specimens further idealised the flowers, showing them in their prime moment of rareness and curiousness.[61] Depictions with generalised features of a blooming flower, in this regard, better preserve the most desirable qualities of rare and curious flowers.

## 6 Keeping the Collection as a Whole

For flower books that had close ties to a garden, the very process of producing the series of bloom depictions in watercolour contributed to the preservation of the living collections as a whole. Manuscript florilegia usually took a long time to compile and produce, oftentimes taking several years and sometimes decades. The production of these flower collections on paper and parchment was an accumulative process, preparing and making one sheet or page at a time. Most sheets in manuscript florilegia are not dated, but sporadic dates that appear in a few flower books offer clues as to how long it might have taken

---

do not offer much visual information that contextualises the flowers' (natural) habitat either.

61 It is common to see claims that florilegium images present idealised flowers in secondary literature. However, as shown in this chapter, there are multiple modes for bloom depictions and the use of a mode depends on what was considered useful by each image maker to present certain visual characteristics of the flowers.

for these collections to be compiled. Block's *Bloemenboek*, for example, was possibly compiled over several decades of the seventeenth century. Leaf 27, in particular, has three bloom depictions, two of which are dogtooth violets in pink and white, and one the purple mallow (Fig. 35). The image of the mallow, together with four butterflies, is pasted onto the blank space of the paper below the dogtooth violets. Two of the butterflies and some parts of the mallow are cropped from the original piece of paper, and the missing parts are repainted on the pasted paper outside the cut picture plane to complete the image. The name Octavius M. de J. is attributed as the image maker for the mallow on the original piece of paper, along with the year 1637. A portion of the watercolours in the *Bloemenboek*, however, was most likely not produced until the 1680s, as the depicted rare flowers, such as varieties of the hibiscus, only began to be cultivated in the Dutch Republic around this time.[62]

Similarly, the *Hortus amoenissimus* by Franciscus de Geest is most likely a project that lasted many years of the compiler and image maker's lifetime.[63] The frontispiece of the *Hortus amoenissimus* depicts De Geest kneeling in front of and presenting a florilegium to the personification of Flora; an angel plays a trumpet above the two figures, holding a piece of paper on which is written '*Florilegium auctore Francisco de Geest A. 1668.*' Within the volume, leaf 64 depicts a blooming composite martagon lily, a flower that is made up of many individual martagon lilies. It is written on the page that this composite flower was witnessed in 1658 in the garden of a Petrus van Nys.[64] In many cases, manuscript flower books were collections of loose sheets that grew in number and were only bound some years after the death of the compiler, who was either a collector who commissioned the images to be made after their living rarities or an image maker who drew and painted the flowers by visiting local curiosity gardens.[65]

While seventeenth-century herbaria tended to be produced in manuscript format from the beginning instead of as collections of loose sheets, the

62 Kistemaker, Kopaneva, and Overbeek, eds., *Peter de Grote en Holland*, 245 (catalogue number 219).

63 Franciscus de Geest, *Hortus amoenissimus*, Rome, National Central Library of Rome (Biblioteca Nazionale Centrale di Roma), MS Varia 291, http://digitale.bnc.roma.sbn.it/tecadigitale/manoscrittoantico/BNCR_Ms_VARIA_0291/BNCR_Ms_VARIA_0291/201. See Chapter 1 for an introduction to this florilegium.

64 '*Wonderlijcken Plant. gewossen inden Jaere 1658. inden Hoff van den Edelen Heere myn Heere. Petrus van Ny*[*s*].' in De Geest, *Hortus amoenissimus*, 64.

65 See, for example, the posthumous binding that caused confusion for the dating and attribution for the Gottorfer Codex in De la Fuente Pedersen and Kolind Poulsen, eds., *Flowers and World Views*, 103–104, 114–118.

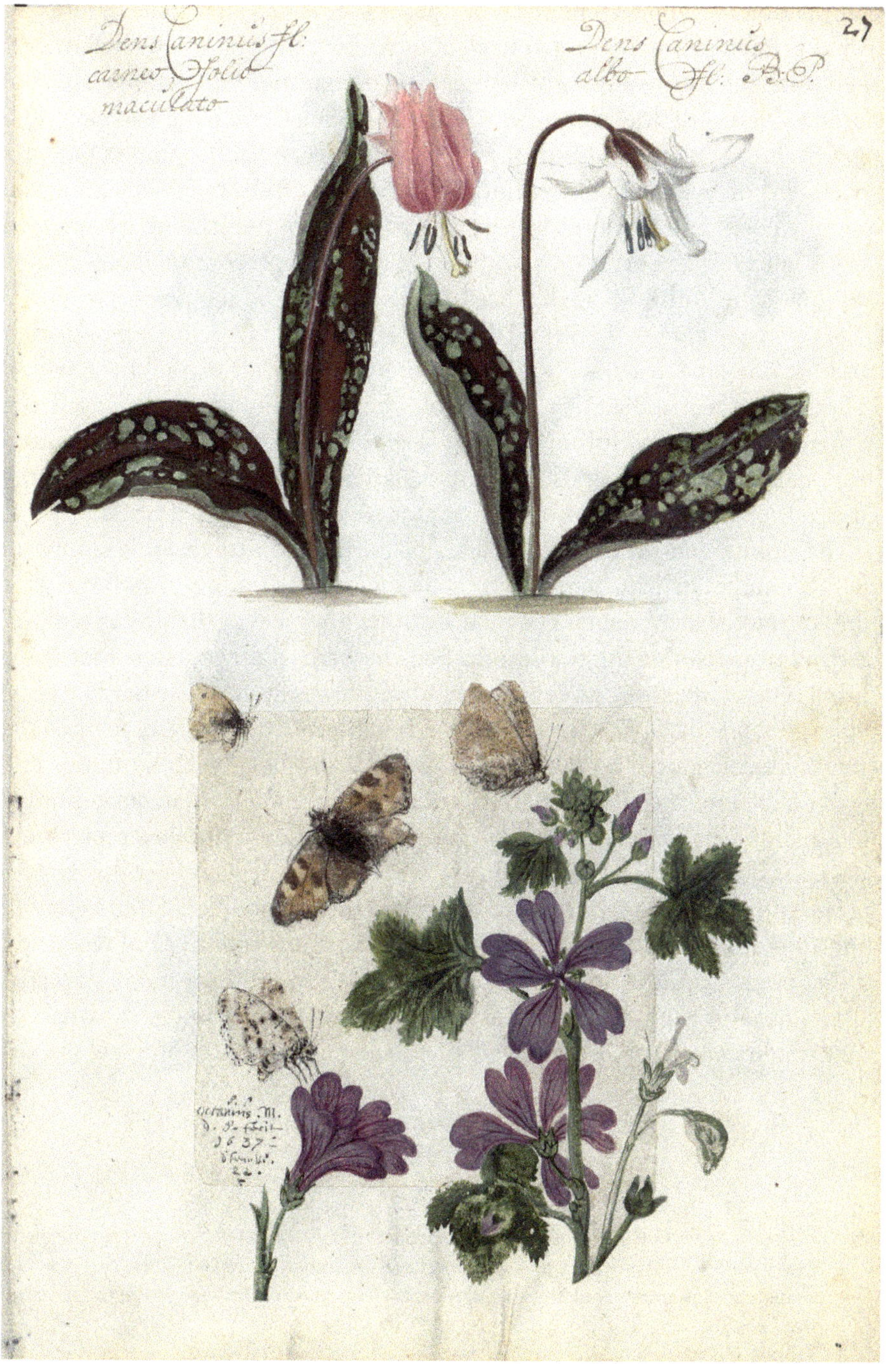


FIGURE 35 Dogtooth violets and mallow from the *Bloemenboek* for Agneta Block (second half of the seventeenth century), Rijksmuseum, Amsterdam, RP-T-1948-119

compilation of plant specimens was a process covering a prolonged period.[66] The specimens were not prepared all at once, but rather in small batches over the years. The four volumes of Boerhaave's herbarium, for example, are supposedly filled with dried specimens from the Leiden Garden between 1685 and 1693, including many rare and curious flowers.[67] Folio 64 in volume 1 includes specimens of five kinds of hyacinths (*Hyacinthus*). The specimens are glued onto the paper and reinforced with glued paper strips (which might have been added later for conservation purposes). Written on the ornate labels, which were made by cutting a small piece of folded paper, are the plant names and page references to the works of Dodoens, Clusius, and Lobelius, among other authors. The label for the *Hyacinthus* variety on the left of the top row also includes the year 1688. Such records of when a specimen might have been gathered are scattered throughout the four-volume herbarium, and the many years between 1685 and 1693 that are inscribed on the labels show that the compilation of the dried flowers and plants was not a speedy process.

The production process of manuscript florilegia, in this way, is similar to the accumulation of a herbarium, collecting and pressing one flower at a time through watercolours. The most significant aspect of the prolonged production processes of the florilegium and the herbarium genres is that going through the collections of depicted or dried flowers was a constantly changing experience as the collections grew. The viewers in the past, such as the compilers/collectors and their garden visitors, did not see these flower and plant collections on paper or parchment in their 'completion' as viewers today. Rather, florilegia and herbaria were growing collections with new images and specimens continuously added for years. Printed florilegia reflect this growth to an extent, but readers did not often see the printed collections grow in real-time. As discussed in the previous chapter, many reprints and reeditions of the intaglio-printed florilegia would grow in terms of the number of plates, sometimes doubling their initial sizes in later editions.[68] However, owners of the printed copies usually received all or most of the plates at once, instead

66 Herbaria moved in an opposite direction from florilegia, with collections from the eighteenth century and onward utilising loose sheets instead of bound volumes. This way, naturalists and botanists could order and reorder them more easily, see Fleischer, 'Leaves on the Loose,' 117–135.

67 Thijsse and Wesseling, 'Herman Boerhaave: A Search for His Herbarium Collections,' 172.

68 See especially the growth of the flower book by Johann Theodor de Bry in section 'Modifying and Reprinting Seventeenth-Century Florilegia' in Chapter 4.

of purchasing one plate at a time, even though the production time leading to the expanded republications could have also taken many years to decades.[69]

The process of adding pages, images, and specimens was a useful method for preserving the *whole* of the living collections. Rare flower and plant collections seldom reached a state of completion, as new species and varieties were continuously introduced and/or cultivated through breeding. Despite the gardeners' aid, care, and support in helping flowers survive, grow, and thrive, the flowers and plants remained an ephemeral matter, as they were easily affected by climates and environments. Living collections did not always expand; rather, they sometimes diminished in size and scope through losing living specimens. By having the pictorial and dried gardens grow with the living gardens, rare and curious flowers could be preserved through images and/or specimens upon their entry into a living collection, or shortly afterward. Whether or not the rare flowers and plants lived, and a living collection grew or shrank, the pictorial and dried gardens could preserve all that had once been part of the collection. The accumulative process of the compilation of the florilegium and the herbarium thus preserved the flowers that could no longer be spotted among the living flora outdoors.

This layer of visually preserving ephemeral flowers also connects florilegia to the pictorial genre of the flower still life.[70] Seventeenth-century flower still lifes, which are usually oil paintings on panels or canvases, feature many of the same rare and curious flowers as florilegia.[71] Single and double tulips, crown imperials, sunflowers, anemones, and so on, are assembled into bouquets. Most often, the painted bouquets are impossible to achieve in real life, as the depicted flowers bloom at different times of the year. Seasons are prominently present within the pictorial boundaries of florilegia and flower still lifes.[72] However, the ways the two genres engage with seasonality and deliver

69 The *Hortus Eystettensis* supposedly took more than a decade to make, but the viewers of the book saw all the plates at once. This florilegium also preserves the end point of the rare flower and plant collection at the garden of Eichstätt, instead of growing with the living collection as it was developing in the sixteenth and seventeenth century.

70 Frank Fehrenbach discussed some issues along the line of preserving cut flowers through still lifes, despite not using the word preservation, see 'Cut Flowers,' 583–614.

71 Sam Segal was a main figure in identifying a lot of the species and varieties used in flower still lifes and florilegia. For the most updated and comprehensive information on the genre, including flower identification and iconography, see Segal and Alen, *Dutch and Flemish Flower Pieces*. Additionally, Paul Taylor discussed some aspects of the commercial and cultural backgrounds that produced flower still lifes, see Taylor, *Dutch Flower Painting*.

72 See section 'Putting Flowers in Seasonal Order' in Chapter 3 on discussions of seasons in florilegia.

visual information differ. Whereas viewers of florilegia need to engage with the action of flipping through the pages in order to encounter flowers from different seasons, viewers of flower still lifes see blooming flowers from different seasons on display at once. While viewers can look at or study each flower separately, if they so desire, the arrangements of flowers in still lifes are designed for the flowers to work together to present dynamic compositions as a whole. Furthermore, the painting and viewing processes were not accumulative, even though flower still lifes could take many years for the image makers to produce. The viewers only saw the state of completion, and the bouquets no longer grew once the paintings were finished and sent to the owners.

Despite these differences, the two genres offered early modern viewers ways to access natural objects that easily perished and decayed through images.[73] The two genres could also evoke different kinds of conversations on the theme of ephemerality. The additional objects in flower still lifes that are often associated with the inevitability of death, such as a timepiece of a skull to represent the allegory of *vanitas*, provided ample discussion points on symbolic views of the world and the fleeting nature of life.[74] The bloom depictions in florilegia embodied empirical experiences of the skills, knowledge, and challenges to grow and cultivate rare and curious flowers that were often not easy to keep alive. By visually preserving the flowers on canvases, panels, parchments, or papers, early modern image makers counteracted the flowers' ephemerality by making everlasting pictures.

## 7 Conclusion

The loss of certain visual information associated with the genre of the herbarium offers some insight into why manuscript florilegia became prevalent in the second half of the seventeenth century. The rareness and curiousness of many flowers show most prominently when they are in bloom. The forms of the flowers are fullest with double petals, and their colours most vibrant. While pressing the once-living flowers can preserve the very specimens, it is challenging to prevent the flowers from losing volume and colour in the drying process, hence they lose the crucial characteristics of their rareness and curiousness, at

73 Marlise Rijks made a comparable argument of using paintings to preserve aquatic creatures that were difficult to keep as specimens. See Rijks, 'Scales, Skins, and Carapaces in Antwerp Collections,' 295–320.

74 On *vanitas* and other allegories, see Chapter 2 'On the Symbolism of Flowers and Animals in Still Life Paintings' in Segal and Alen, *Dutch and Flemish Flower Pieces*, 25–86.

least to a degree. The lack of usage by the *liefhebbers* does not mean that they were not interested in dried plants. For example, Chapter 1 briefly mentioned the eighteenth-century florilegium of Mary Somerset, the Duchess of Beaufort (1630–1715).[75] Somerset also owned an extensive herbarium, with hundreds of dried specimens from her garden filling twelve volumes of albums.[76] Many seventeenth-century florilegium owners who were plant collectors, such as Agneta Block, could have pressed plants, but they did not systematically do so and/or the pressings are now lost or decayed.

At any rate, images ensured the rare and curious features to be everlasting. Prints can visually preserve the forms, and watercolours can further preserve the colours.[77] By commissioning image makers to produce a 'herbarium' with watercolours instead of dried specimens of rare and curious flowers, Block and her contemporary rare flower and plant collectors used images to preserve their living collections in their prime. For florilegium compilers and image makers who did not own living rarities and curiosities, pressing flowers through depictions provided a way to assemble a collection of floral rarity and curiosity on paper and parchment. As images were constructed, there was a further opportunity to not only preserve the flowers' rareness and curiousness, but to enhance such features. As a topic for further discussion in the next chapter, bloom depictions often accentuate the rareness and curiousness, making the features more prominent than in real life. For people who were primarily interested in the forms and colours of decorative flowers, images—particularly hand-coloured prints or watercolours—were thus more suitable for preserving these rare and curious qualities.

75 See section 'Conclusion' in Chapter 1 on the Badminton Florilegium (1703–1705).

76 See description and photos of this herbarium in Pickering, 'Mary Somerset, Duchess of Beaufort,' 60–65.

77 In a way, the colour information florilegium compilers provided through the forms of flower descriptions and colouring instructions, as discussed in Chapter 3, are means to revive this aspect of the flowers' rareness and curiousness.

CHAPTER 6

# Making Colourful Images of Flowers in Life-Size

## 1 Introduction

Seventeenth-century florilegia, particularly the intaglio-printed ones, often adopted a double-title approach, starting with a short (and usually better-known) primary title and followed by a descriptive title that provides more information on what readers will find between the pages.[1] Aside from announcing the depicted subjects—flowers and decorative plants that were rare and curious to the period—the long titles also describe the major mode of representation; that florilegium images were made *ad vivum*. For example, the full Latin title for Johann Theodor de Bry's (1561–1623) flower book is printed as,

> *Florilegium novum hoc est: variorum maximeque rariorum florum ac plantarum singularium una cum suis radicibus & cepis, eicones diligenter in aere sculptae &* ad vivum-*ut plurimum expressae.*[2]

Similarly, the full title of the florilegium compiled by Basilius Besler (1561–1629) is,

> *Hortus Eystettensis, sive, diligens et accurata omnium plantarum, florum, stirpium ex variis orbis terrae partibus singulari studio collectarum, quae in celeberrimis viridiariis arcem episcopalem ibidem cingentibus hoc tempore conspiciuntur, delineatio et* ad vivum *repraesentatio.*[3]

---

1 This was a common practice for early modern books. Most have long descriptive titles written in a similar structure as the provided florilegium examples.

2 'This is a new florilegium: of various and extremely rare flowers and unique plants, together with their roots and bulbs, icons carefully carved in copper and represented as ad vivum as possible.' on the frontispiece of De Bry, *Florilegium novum*. See Chapter 3 for an introduction to this florilegium. Quote translated with the help of Ada Migliazza.

3 'The Garden of Eichstätt, or, diligent and accurate drawing and ad vivum representation of all the plants, flowers, and species collected by a singular study from various parts of the world, which can be seen in the celebrated pleasure garden surrounding the episcopal castle at this time.' on the frontispiece of Besler, *Hortus Eystettensis*. See Chapter 1 for an introduction to this florilegium. Quote translated with the help of Ada Migliazza.

 | DOI:10.1163/9789004735149_008

Time and again, across the long titles and prefaces of seventeenth-century florilegia appears the term *ad vivum* or similar iterations in vernacular languages, such as the *nach dem Leben* in the German title, which is engraved together with the Latin one, on the frontispiece of the *Florilegium novum*.[4]

The term *ad vivum* and its iterations is not an isolating concept for florilegium images. Rather, *ad vivum* as a mode of representation was central to images across many sectors in the early modern period and has been a heavily discussed and debated topic among historians of art and science.[5] The 2019 anthology, *Ad Vivum?,* edited by scholars Thomas Balfe, Joanna Woodall, and Claus Zittel, in particular, presents a comprehensive critical reflection through the various studies of which the volume consists. As Balfe and Woodall unpacked and defined in the book's introduction, 'Simplifying to the maximum, *ad vivum* terminology normally advertised the claim of an image to be a faithful likeness or a bearer of reliable data.'[6] *Ad vivum* images are highly associated with first-hand witnessing and observation, and the term is used to assert authenticity or trustworthiness on some occasions, but the images did not have to be made from living models.[7] The term is elusive and comes with a spectrum of meanings, but 'from life,' 'lifelike,' and 'lively' are the three possibilities that run most prevalent in early modern *ad vivum* images.[8]

Balfe and Woodall further emphasised the *life* in lifelikeness by arguing that they coexist with a variety of visual effects in the early modern period, listing 'morphological accuracy (paradigmatically in a nature casting), credible colour and effects of light on different surfaces, convincing three-dimensionality,

---

4 '*New Blumbuch Darinnen allerhand schöne Blumen undfrembde Gewächs / mit ihren Wurtzeln und Zwiebeln mehrer theils dem Leben nach in Kupffer fleissig gestochen zu sehen seind.*' on the frontispiece of De Bry, *Florilegium novum*.

5 Claudia Swan's article is important in decoding the term by bringing natural history into discussion with fine art pieces. Her assessment of florilegium images, however, is positioned in 'scientific' images and hence undervalues them, '*Ad vivum, naer het leven, from the life*: Defining a Mode of Representation,' 353–372. Additionally, Boudewijn Bakker looked into bibliographical evidence in the late Middle Ages to understand the use and development of the word on a longer timeframe, '*Au Vif—Naar 't Leven—Ad Vivum*: The Medieval Origin of a Humanist Concept,' 37–52. Robert Felfe examined various pictorial genres and how image makers depicted things *ad vivum* in the early modern period in '"Naer het leven": Eine sprachliche Formel zwischen bildgenerierenden Übertragungsvorgängen und ästhetischer Vermittlung,' 165–195.

6 Balfe, Woodall, and Zittel, eds., *Ad Vivum*?, 2.

7 Balfe, Woodall, and Zittel, eds., *Ad Vivum*?, 3.

8 Balfe, Woodall, and Zittel, eds., *Ad Vivum*?, 4, 10–11.

various kinds of animation, and so on.'[9] Historian Florike Egmond also addressed the issue of *ad vivum* representations through visual effects, most prominently in her book *Eye for Detail* (2017) and a follow-up essay 'The *ad vivum* Conundrum' (2020).[10] Egmond discussed a number of visual techniques for depicting things in a lifelike or lively way, using sixteenth-century image collections of flora and fauna, including showing animals in movement to capture their liveliness.[11] In her essay, Egmond raised many questions about the interplay between *from life*, *lifelike*, and *lively* through a fish watercolour painted by the Italian painter Jacopo Ligozzi (1547–1627). With its naturalistic form and detailed and vivid colours, the fish is unidentifiable to modern fish experts. There is a possibility that Ligozzi depicted a living specimen of a now-extinct species. More to consider, however, is the assessment from a contemporary of Ligozzi that the fish depiction was 'better than life, and certainly miraculously.'[12] *Ad vivum* images, in this sense, also have the ability and freedom to perfect or idealise depictions that were observed *from life*.

It is these aspects of visual effects and techniques employed by seventeenth-century florilegia and how they achieved *ad vivum* representations of rare and curious flowers that are of interest in this chapter. The interpretation of lifelike(ness) as sometimes going into the area of *better than life* is the primary definition of *ad vivum* and the more interesting concept to explore in this chapter. As discussed in Chapters 3 and 5, florilegium images were not always made from life. Copied and excerpted flowers did not pose any problem for their inclusion in a collection, and manuscript florilegia by the same image-maker but belonging to different patrons regularly featured bloom depictions with the same composition.[13] The line between lively and lifelike in florilegium images is somewhat blurrier than for images of fauna, as flowers grow, but do not move, at least not to the extent of insects and animals. Lifelike depictions in florilegia, however, require the flowers to have *life* and to look *alive*, as their rareness and curiousness in the forms of volumetric petal counts and a large variety of colours are most prominent when they are living and blooming.[14] Lifelike depictions, thus, largely return to Balfe and Woodall's

9 Balfe, Woodall, and Zittel, eds., *Ad Vivum?*, 5.

10 Egmond, *Eye for Detail*; 'The *ad vivum* Conundrum,' 33–62.

11 Egmond, 'The *ad vivum* Conundrum,' 49.

12 Egmond, 'The *ad vivum* Conundrum,' 51–53.

13 See section 'Pressing Flowers for Rare and Curious Features' in Chapter 5 for a discussion on this point.

14 See section 'Losing Rareness and Curiousness' in Chapter 5 for a discussion on this point.

simple definition of an image as a 'faithful likeness or a bearer of reliable data depictions.'

The understanding of faithfulness and reliability needs to be clarified. The two words can easily lead to another word, *accuracy*, which also often appears in florilegium titles, such as in the phrasing '*diligens et accurata omnium plantarum*' (diligent and accurate [drawing] of all plants) in the descriptive title of the *Hortus Eystettensis*. Faithful, reliable, and accurate flower depictions are not about what is considered 'botanical accuracy' by today's standards, or even historical ones set by naturalists or botanists. *Accuracy* in florilegia refers to identifiable and recognisable depictions of flower diversity and variety. While certain details may be obscured or synthesised, they are based on a certain level of observation. When viewers saw a bloom depiction, they could recognise the flower as a tulip, a hyacinth, an anemone, and so on, that had real-life references in the gardens, instead of something that came out of lore or fantasy.[15]

To analyse the visual effects and techniques seventeenth-century florilegia used to present *ad vivum* depictions of rare and curious flowers, this chapter starts with a discussion on the basic processes of making bloom depictions through the stages of drawing and colouring. The chapter then looks into how the materials and techniques of intaglio-printing and watercolour specifically affected and/or supported the image makers in two aspects of depicting flowers: drawing them in life-size and giving them an array of colour variations. Similar to the compilers of florilegia, the people who made flower depictions for flower books plied many trades, including printmakers, embroidery designers, illumination or miniature painters, and amateur practitioners. This chapter hence uses the terms image maker and colourist to encompass the range of people and professions that participated in the different stages of the making of seventeenth-century florilegia. Through examining material practices and artistic techniques used for making images in florilegia, this chapter aims to illuminate how the genre achieved lifelikeness, particularly through representing rare and curious flowers in life-size and in colour.

15 One exception is that some orchid petals transformed into birds, insects, and little dancing people in some manuscripts by the German painter Hans Simon Holtzbecker. However, this transformation took morphological characteristics that were already in the orchids and pushed them one step further, instead of inventing non-existing flowers. More on Holtzbecker's orchids, see De la Fuente Pedersen and Kolind Poulsen, eds., *Flowers and World Views*, 144–149.

## 2 Picturing Rare and Curious Flowers

The main artistic techniques for the production of seventeenth-century florilegia, as frequently mentioned throughout this book, were intaglio printing and watercolour. Most of the printed florilegia are engraved collections, such as the plates in the *Hortus Eystettensis* (Fig. 36), but *Le jardin du roy* ... (1608) by Pierre Vallet (*c.*1575–1657) is etched. Metal engraving and etching are both intaglio-printing techniques, and the mark-making and printing of the plates would usually have been performed exclusively by trained professionals.[16] Even though certain mark-making techniques are distinctive and unique for engraving or etching, both techniques support fine linework and volumetric shading that were applied to render lifelike flowers.

To make a printed bloom depiction, a simplified making process includes the following steps. First of all, draw a flower on a plate. This could be done either by drawing directly with a stylus on a prepared plate or, more often, by transferring a composed flower onto a plate from an original drawing or an existing print to be copied. Second, make marks on the drawn plate. To engrave, the image maker carved the lines by pushing a burin into the drawing to produce grooves on the plate. To etch, the image maker placed the plate into an acid bath. The acid corroded the drawn lines into the plate to create the grooves, while the undrawn part of the plate was protected by the coating on the prepared plate. Third, print the engraved or etched plate using a high-pressure rolling press. The printing ink was applied onto the plate and pushed into the grooves; the plate was then wiped clean and placed onto the press; the plate went under the roller, with a piece of paper in between, and the inked lines in the groves were transferred onto the paper in the process.[17] A printed flower in black and white was then created. In some cases, a fourth step was applied when a printed image was hand-coloured with watercolour. This could have been done by a professional colourist or an amateur practitioner, such as the owner of a flower book.[18]

16 For detailed survey to the long history intaglio printmaking techniques, see Stijnman, *Engraving and Etching, 1400–2000*; Griffiths, *The Print Before Photography*, 28–48.

17 General step-by-step demonstrations of the two intaglio-printmaking techniques can be seen on the webpage 'What Is Printmaking?,' The Metropolitan Museum of Art, https://www.metmuseum.org/about-the-met/collection-areas/drawings-and-prints/materials-and-techniques/printmaking.

18 On professional hand-colouring practices in the early modern period, see especially Goedings, 'Ingekleurde prenten en boeken,' 163–177; Goedings, *'Afsetters en meester-afsetters'*; Dackerman, ed., *Painted Prints*; Eiermann, ed., *Prachtvoll illuminirt*. See some

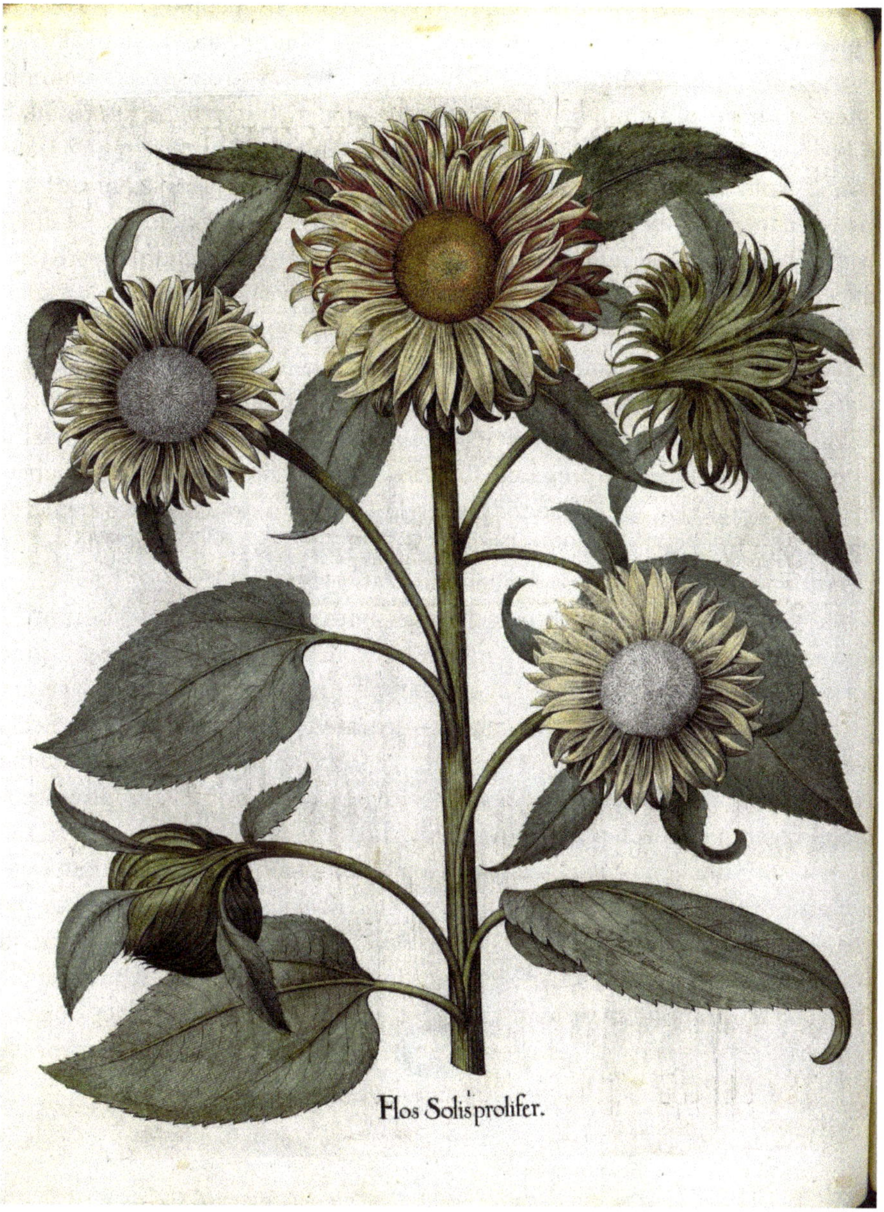


FIGURE 36 *Flos Solis prolifer* from the *Hortus Eystettensis* (1613), Biblioteca Real Jardín Botánico, Madrid, Spain, Af BES, https://bibdigital.rjb.csic.es/idviewer/10913/140

Watercolour, on the other hand, was practised by people from a wider range of professions and with various skill levels. Many of the surviving flower books were created by makers with demonstrative professional training, often in the art of painting. All of the collections that have been discussed in this book thus far, such as the Gottorfer Codex by the German painter Hans Simon Holtzbecker (1610/20–1671) (Fig. 37) for the Duke of Holstein-Gottorp (1597–1659), were produced by professional (watercolour) painters.[19] Albeit with a lower survival rate, there are also manuscript florilegia that were produced by amateur practitioners. Regardless of their skill levels, professional and amateur image makers of manuscript florilegia engaged with similar processes to make a watercolour bloom depiction. Generally speaking, making a watercolour has fewer essential steps than making an intaglio print. One, draw a flower. This could be done with a piece of chalk, pencil, or brush, and paint directly onto a piece of prepared support, either a sheet of paper or parchment. Alternatively, an image maker could transfer a drawing onto paper or parchment from an existing drawing or print. Two, colour the drawn flower by painting with watercolour, and a painted flower is made.

Despite being greatly different technically, both mediums of intaglio-printing and watercolour embody certain monetary value and status that were fitting for depicting flowers collected mainly as rarities and curiosities. Whether printed with engraved or etched plates or painted with watercolour, florilegia were luxurious books in the seventeenth century, and still are today. Intaglio printing was, in general, a more expensive method of printing pictures, as it took longer to ink each plate than a woodblock for printing woodcuts and the number of prints one could get from a plate was significantly lower than from a block. Additionally, intaglio-printing gained social status at the end of the sixteenth century and was perceived to be a more prestigious medium than woodcuts for making images.[20] Printed florilegia in black and white usually included a minimum of around 70 plates, and this made them pricy objects. Likewise, good-quality paints were expensive for making watercolours or for

discussions of how an amateur colourist coloured a copy of De Bry's *Florilegium novum* by Eriksen and Wen, 'Colouring Flowers,' 21–43.

19 Hans Simon Holtzbecker, *Gottorfer Codex*, Copenhagen, National Gallery of Denmark (Statens Museum for Kunst), kksgb2950, https://open.smk.dk/en/artwork/image/KKSgb2950. See Chapter 1 for an introduction to this florilegium.

20 Stijnman, *Engraving and Etching, 1400–2000*, 84–85; Griffiths, *The Print Before Photography*, 210, 415–416. This, however, does not mean that producing woodcuts were always cheaper than intaglio printing, see a case study of prices by Bowen and Imhof, 'Reputation and Wage,' 161–195.

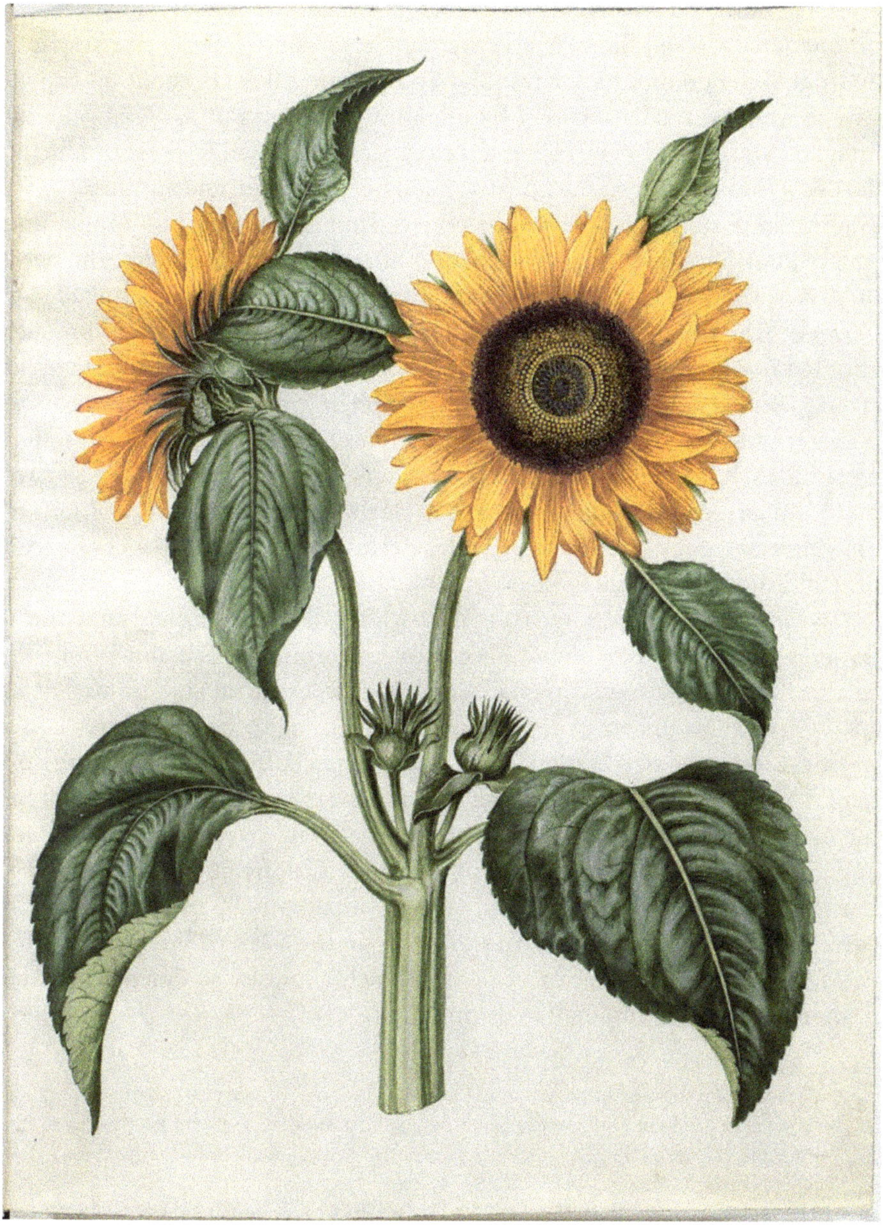

FIGURE 37 Sunflower with multiple heads from the Gottorfer Codex (1649–1659), SMK, National Gallery of Denmark, Copenhagen, KKSgb2950/30

hand-colouring. Even lower-quality colours would have further added to the production cost and hence made the copy even more luxurious. The paper and parchment supports for printing and painting flowers could be expensive as well, especially those of high quality.[21] The overall material value of intaglio-printing and watercolour as primary artistic techniques for producing florilegia thus reinforced the flowers' quality of being rare and curious.

More important, however, is that the techniques of intaglio-printing and watercolour have similar properties and affordances that enabled the early modern image makers to achieve certain visual effects when depicting lifelike flowers.[22] The rest of this chapter further zooms in on how the two techniques afforded florilegium image makers to depict *ad vivum* flowers at the stages of *drawing* and *colouring*. Despite the varying technical processes, the stages of drawing and colouring flowers are critical in both intaglio-printing and watercolour for achieving lifelikeness in flower depictions. The two stages also have more similarities than differences in both techniques when the definitions of drawing and colouring align more with certain concepts and practices instead of established artistic mediums.

*Drawing*, in this chapter, focuses on the aspect of visualising and depicting a flower on paper and parchment instead of concerning a discussion of historical drawing materials, such as chalk and ink.[23] Here, the intaglio-printed flowers are treated as drawings to be coloured, albeit highly lavish drawings. In a way, drawing is the step hidden from the viewers in both printed and watercolour florilegia. Prints reproduced the drawn flowers on the plates. Throughout the mark-making process, the original drawings were either transformed into engraved or etched lines, if a drawing was directly done on a plate, or largely lost or destroyed if done on a separate piece of paper to be transferred. In the cases of hand-coloured prints, the 'drawing' in the form of the printed lines was further covered by layers of paints, which could be so opaque that the printed lines became difficult to distinguish (Fig. 38).

---

21 For issues on the cost of papers in the early modern period, see particularly Bellingradt, 'The Paper Trade in Early Modern Europe: An Introduction,' 1–27. On parchment examples, see Oltrogge, '"Pro Lazurio Auricalco et Alii Correquisitis pro Illuminacione",' 189–198; Melograni, 'Manuscript Materials,' 199–221.

22 The term *affordance* is a critical concept in the study of material and materiality. It was recognisably written by James J. Gibson, 'The affordances of the environment are what it offers the animal, what it provides or furnishes, either for good or ill.' In other words, affordance is the properties of a material that determine how it can possibly be perceived and used. Gibson, *The Ecological Approach to Visual Per*ception, 127.

23 For general introductions on early modern drawing materials and techniques, see Meder, *The Mastery of Drawing*; Goldman, *Looking at Prints, Drawings and Watercolours*; Royal Collection Trust, 'Leonardo da Vinci's Drawing Materials,' video, 5:09, https://www.youtube.com/watch?v=-foym3CtleQ.

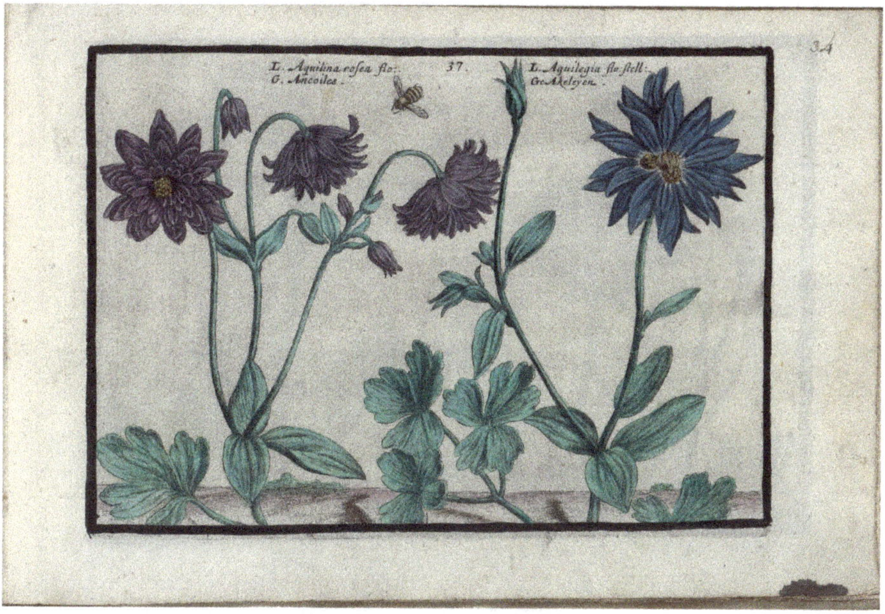


FIGURE 38 Hand-coloured columbine plate from a copy of the Dutch edition of the *Hortus floridus* (1614), Oak Spring Garden Foundation, Upperville, VA, 00004892

Similarly, drawings of flowers for watercolours were intended to be coloured with multiple paint layers. Unless a work is unfinished, viewers do not tend to see the drawing stage in a finished watercolour. Franciscus de Geest's (*c.*1635–1702/1712) *Hortus amoenissimus*, in this regard, offers some unique insights into the process of the drawing stage being covered, as several pages in the florilegium are unfinished.[24] For example, the page numbered 82 depicts six variations of geranium in pink, dark purple, white, blue, and magenta (Fig. 39). The white and magenta geraniums are unfinished. The blooms are painted, to the extent that there are shadings and highlights that give the petals volumes, but the stems and leaves are unpainted, with the exception of a stroke of green at the bottoms of both flowers. The unfinished parts show that De Geest drew the composition for each geranium with a brush and brown-green paint. Once covered with more paint layers in order to give them colours, the initial drawn lines vanished almost without a trace. Nevertheless, the completed coloured flowers in both printed and manuscript florilegia retain the structures of the

24 Franciscus de Geest, *Hortus amoenissimus*, Rome, National Central Library of Rome (Biblioteca Nazionale Centrale di Roma), MS Varia 291, http://digitale.bnc.roma.sbn.it/tecadigitale/manoscrittoantico/BNCR_Ms_VARIA_0291/BNCR_Ms_VARIA_0291/201. See Chapter 1 for an introduction to this florilegium.

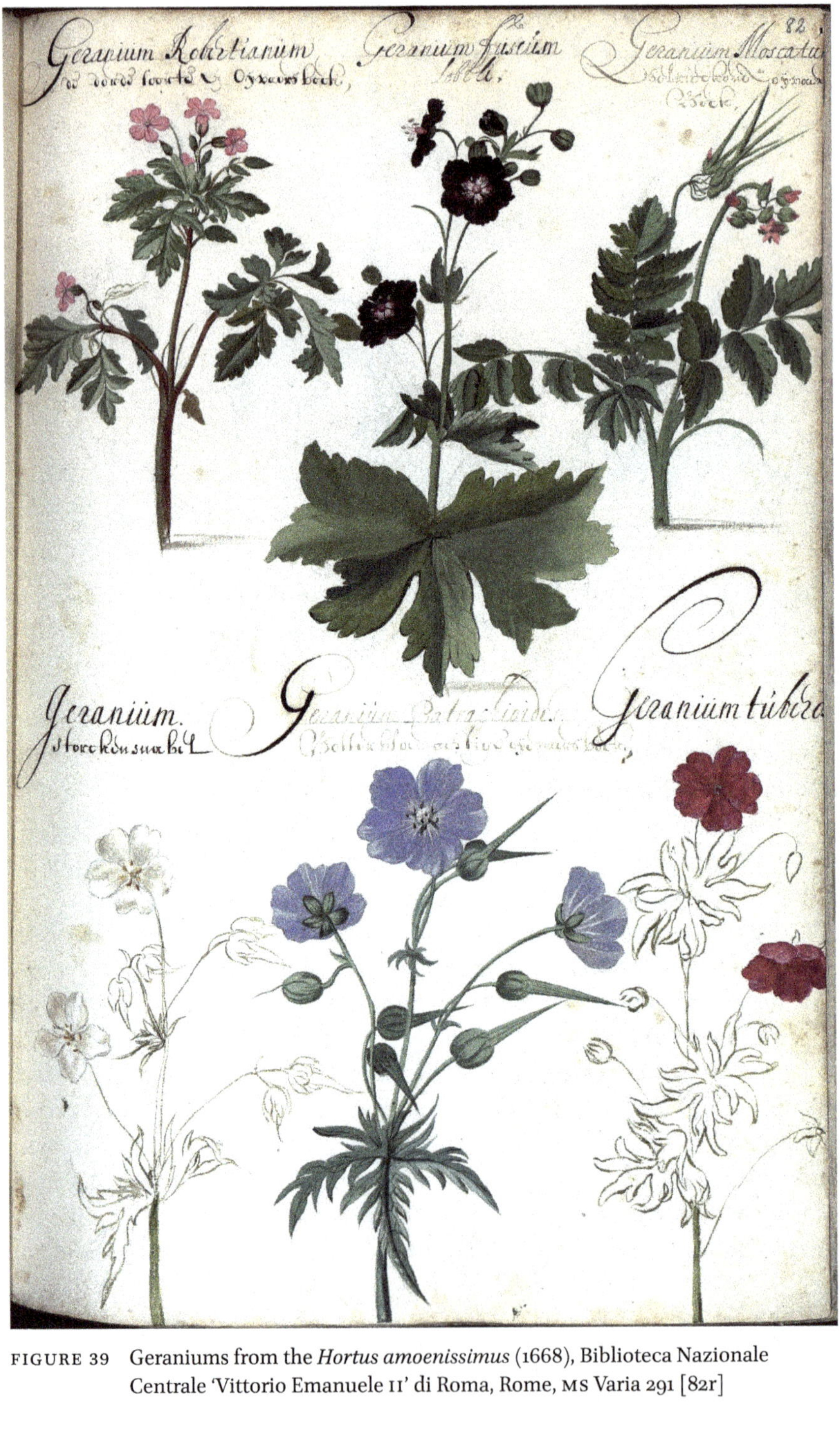


FIGURE 39 Geraniums from the *Hortus amoenissimus* (1668), Biblioteca Nazionale Centrale 'Vittorio Emanuele II' di Roma, Rome, MS Varia 291 [82r]

compositions on which the drawings built, and thus allow the covered stage of drawing to still be visually analysed and studied.

*Colouring*, on the other hand, refers to the efforts to add colours to the drawings. Intaglio printing is largely a black-and-white medium, even though colour printing was experimented with in the early modern period.[25] To add colours to printed flowers, as shown in Chapter 4, colour terms and descriptions were largely implemented to textually colour the images. Alternatively, professional colourists would be commissioned to hand-colour the flowers with paints, or the flower book owner could attempt it themselves. This chapter focuses on the printed examples that were materially coloured. Regardless of the mediums used for the drawn flowers and the skill levels of the practising colourists, the colouring material and technique to be discussed here for both hand-coloured prints and manuscript flower books is watercolour. In this volume, *watercolour* is an umbrella term for water-based mediums, which include what are categorised today as aquarelle or transparent watercolour and gouache or bodycolour.[26] There is a tendency for a clear distinction between transparent and opaque colours today, referring to the former solely as watercolour and the latter as bodycolour. The two are often perceived and practised as two rather different techniques, with watercolour being seen as the more prestigious one of the two in the public mindset.

Historically speaking, however, such a strict division is limiting and not fruitful. For the most part, seventeenth-century watercolour paints were simply mixtures of gum water (acting as the binder) and different kinds of pigments (acting as the colourants).[27] Gum water is acquired by dissolving in water a certain amount of gum arabic, the hardened sap of the Acacia tree species, and often with a little bit of honey or sugar added to help lessen the brittleness of the gum. The opacity of the paints largely comes from the properties of some pigments that are naturally opaque. Additionally, the ratio of gum water and

---

25 On colour printing, see notably Stijnman and Savage, *Printing Colour 1400–1700*.

26 A recent piece that explains in greater detail on the need of an inclusive definition of *watercolour* for historical works, extending to medieval illumination and portrait miniature, is Evans, *Renaissance Watercolours*, 8, 14–15. The authors of the essays for the book *Painted Prints* (Dackerman, ed.) refrained from using the word watercolour for the volume, but the described materials and techniques are the same as the watercolour medium investigated here. A useful study that delves into the development of watercolour on a longer time frame and shows the shift towards transparent watercolour in later centuries is Cohn, *Wash and Gouache*.

27 Glair (a liquid from beaten egg white) and (rabbit skin) glue were also regularly used as biners in the Middle Ages, see Thompson, *The Materials of Medieval Painting*, 54–57. By the seventeenth century, however, gum arabic became the major, if not the only, binder mentioned in seventeenth-century art treatises.

pigment can affect the transparency. The opaquer a paint is, the more pigment is used, thus making opaque colours more expensive and monetarily valuable than transparent colours for many hues.[28] Certain colours, on the other hand, exist only or mostly in liquid form as transparent washes. Saffron, for example, is a golden yellow wash that could be obtained by soaking the stigmas of the crocus saffron in water, with a little added rock alum.[29] Surviving copies of flower books more often contain opaque colours, but a mixture of transparent and opaque colours for different parts of a flower is also commonly found. A division between watercolour and bodycolour, within the genre of the florilegium, is thus not productive for analysing the capabilities and limitations of the watercolour technique for rendering *ad vivum* depictions.

## 3 Drawing Flowers Life-Size

A simple yet effective way for seventeenth-century florilegia to achieve lifelikeness was by making (close-to) life-size depictions of rare and curious flowers. Emanual Sweerts (1552–1612) wrote, in the preface of his *Florilegium Emanuelis Sweerti …* (1612), that the woodcuts depicting rare and curious flowers had been in reduced sizes:

> For this reason amateurs have not been completely satisfied, and so I have taken pains not only to show the nature of the plants, but also the size of flower and bulb, with its colour, as they ordinarily grow before our eyes.[30]

This life-size-ness, as emphasised by Sweerts, blurs the boundary between the image and the specimen of a flower. In some ways, this visual strategy of

28 Liming and washing were occasionally mentioned as two different painting and colouring techniques in the seventeenth century, more often in the English context and most notably in Jenner, *A Book of Drawing, Limning, Washing, or Colouring of Maps and Prints*, 20, 24, 36, 38; Norgate, *Miniatura: or, The Art of Limning*, 7. The distinction, however, emphasises mostly the qualities and status of the colouring materials. Limning uses more opaque colours. Washing uses more transparent colours and was often recommended for colouring prints only. Unlike today's preference, opaque colours were often more prestigious in the seventeenth century because of the material qualities of the paints.

29 Ter Brugge and Goeree, *Verlichterie-Kunde, of recht gebruyk der Water-Verwen*, 14.

30 '*Idcirco hac maiori forma non tantum genuinas figuras affequ, verum etiam ipsam florum, bulborumq; magnitudine & quantitat squibus vulgo crescere inueniuntur, delineare studui, additis declarationibus naturalium colorum.*' in Sweerts, *Florilegium Emanuelis Sweerti*, '*Ad candidum lectorem.*' Quote translated by Bleiler in *Early Floral Engravings*, xi. See Chapter 2 for an introduction to and more detailed discussions on this florilegium.

depicting flowers in life-size is similar to the *trompe l'oeil* that is prevalent in paintings.[31] *Trompe l'oeils* present painted objects so realistically that viewers may find it hard to distinguish reality from representation upon first encounters. Many *trompe l'oeils* display life-size or closer-to-life-size objects to add to the illusionism and artifice.[32]

When leafing through a florilegium, life-size flowers can give the readers an experience that is closer to encountering the living flowers in a garden or the preserved flowers in herbaria, which keep the flowers in their real-life sizes by pressing plucked specimens. This is especially striking when a flower is big and showy, such as the sunflower. The variety discussed below is the one that was usually marked as *proliferous* (multi-heads) or *minor* in seventeenth-century florilegia, which has smaller flower heads than the giant *major* kind. The sunflower *minor* (Fig. 15) in the Sweerts florilegium is around 30.1cm tall from the top of the head to the cutoff point of the stem; the head, including the yellow petals, is around 9.7cm in diameter. The sunflower *minus* (Fig. 21) in the *Hortus floridus* (1614) by Crispijn van de Passe II (1594/1595–1670) has the measurement of 13.2cm in height (from the tip of the top leaf to the bottom of the stem where the border cuts them off), and the head is around 5.6cm across the major axis of the oval, as it is depicted in a quarter view.[33] The sunflower *prolifer* (Fig. 36) in the *Hortus Eystettensis,* compiled by Besler, is larger. The flower is 43.3cm tall, and the largest head in the picture is 15cm in diameter. While bearing no caption, a sunflower variety with multiple heads (Fig. 37) appears in the Gottorfer Codex. From the top of the head to the cutoff point of the stem, the flower is measured at around 38.5cm in height, and the head depiction in a frontal view is 16.5cm in diameter. These measurements may seem to be on the (very) small end of the sizes to which the flower can grow. However, the *minor* is a smaller variety, and the varieties found nowadays have undergone a few centuries of breeding and cultivation to make them ever bigger and showier.[34] As a result, the above-mentioned sunflower depictions would have been recognisably life-size in the seventeenth century.

The life-size-ness of flower depictions in seventeenth-century florilegia often corresponds to the characteristic of many of them being big books in

---

31 See especially the use of *trompe l'oeil* as discussed by Kaufmann and Kaufmann, 'The Sanctification of Nature,' 43–64.

32 For example, see analysis of Pieter Claesz' breakfast still life that presents such visual effect in Grootenboer, *The Rhetoric of Perspective*, 35.

33 Van de Passe II, *Hortus floridus*. See Chapter 1 for an introduction to this florilegium.

34 This is often addressed by people working on reconstructions of historical gardens. An old but still useful booklet is Oldenburger-Ebbers and Heniger, *Ornamental Plants in 16th and 17th Century Gardens.*

their physical measurements. Florilegia are usually described as folio-size books, which is a large format in the early modern period.[35] Several of the printed collections have images that were printed with plates around the size of modern A4 (29.7 × 21cm) paper. The bound books are frequently larger, at least by a few centimetres, because the papers on which the images are printed are larger than the plates.[36] For example, there are two tones in the background of the sunflower page (Fig. 15) in Sweerts' florilegium. The grey-tinted area within and around the black rectangle frame is the plate, as the grey tone is caused by the ink on the plate that did not get completely wiped clean; the white border is the unprinted area of the paper. This sunflower plate measures around 34 × 21cm, and other plates in the flower book are usually around 33 × 21cm. The *Florilegium novum* is of a similar size to Sweerts' flower book, with most plates around 27 × 18cm, slightly smaller than an A4 paper. The largest in size, both in the number of plates included and in the physical measurements of the plates, of the printed florilegia in the seventeenth century is the *Hortus Eystettensis*.[37] Measuring around 39.7 × 48cm for the sunflower and for most of the plates in the collection, Besler's plates are larger than standard A3 (42 × 29.7cm) paper of today.

Manuscript florilegia often have similar measurements as well. De Geest's *Jardin de rares et curieux fleurs*, which is painted on parchment instead of on paper, measures around 29.3 × 21.2cm, almost exactly the size of A4 paper. His paper collection of *Hortus amoenissimus* is much bigger, with measurements of 44.7 × 28.3cm, around the same size as A3 paper. The Gottorfer Codex is at the large end of the manuscript florilegia. The sheets have measurements of around 50.5 × 38.5cm, almost double the size of A3 paper.[38] Not all flower books are large books, however. The plates in the *Hortus floridus* are relatively small, for example, at almost half the size of the plates in Sweerts' florilegium. Adopting a landscape format instead of portrait as all the above-mentioned collections, Van de Passe II's plates are around 14 × 21cm, which is around the

35 There is a range of measurements among the papers that were used for printing folio-size book, see examples in Gaskell, *A New Introduction to Bibliography*, 73–75.

36 The final book sizes usually differ from copy to copy due to trimming during the bookbinding process.

37 The *Hortus Eystettensis* consists of more than 350 plates.

38 The Gottorfer Codex most likely was (slightly) trimmed down during the process of bookbinding. Similar to printed florilegia, most sheets in manuscript flower books were trimmed. By doing so, the sheets in a collection became more-or-less the same size for binding them into a codex. The trimming is sometimes highly obvious because the flowers are cut off, in combination with the clean edges of the paper or parchment that is often an indication of the support having been sliced.

size of A5 paper. While it was not the rule, there was a clear tendency for the image makers within the genre of the florilegium to work on large formats.

Simply, a large picture plane allows more to be filled within and can hence contain life-size depictions of flowers for most species and varieties, and this is closely tied to the supports that were used for the two artistic techniques. Intaglio-printed and watercolour florilegia are works on paper and/or parchment.[39] While these two supports have completely different production methods and oftentimes different properties, they have a few shared qualities.[40] In particular, compared with the easel-painting support of wood panels, papers and parchments are thin and do not take as much storage space. Compared with the other easel-painting support of canvas, papers and parchments are firm and do not need to be stretched in order to be displayed.[41] This quality of being thin, yet firm, enables a pictorial collection to contain a high number of plates and leaves, space-saving enough to be bound into one, or a few volumes of codices that could be easily stored on shelves.

It is perhaps the 'limitation' of paper and parchment, however, that was more integral to the making of life-size images of rare and curious flowers. Early modern paper and parchment sizes could be big, but there was a limit to how big the picture plane could go. Historical papermaking had complex procedures.[42] In short, early modern papers were made by dipping a mould and deckle into a vat of pulp, most often fermented linen, recycled from old clothes and rags. The mould was a rectangular wooden frame attached to a wire mesh; a deckle was another wooden frame without wire that went on top of the mould. When dipped into the vat, some pulp went into the rectangle inside the deckle and onto the wire on the mould. The water dripped out of the mesh on the mould when taken out of the vat, leaving the pulp to form within the deckle a sheet of paper, before it was dried and further processed. The sizes of the papers were determined by the sizes of the mould and deckle, and they

39 Printed florilegia have only been found on paper supports thus far, but parchment printing was a known practice for extraordinary occasions, such as the printing of the Polyglot Bible by the Antwerp printer-publisher Christophe Plantin, see Voet, *The Golden Compasses*, vol. 1, 37.

40 See the general process on parchment making in Cheese, 'From Pelt to Painted Page,' 73–94.

41 Stretching is necessary during the process of producing a piece of parchment, but not when using it as a support.

42 On hand papermaking, see notably Barrett, *European Hand Papermaking*; Center for the Book, University of Iowa, 'Chancery Papermaking,' video, 11:42, https://youtu.be/e-PmfdV_cZU.

could only go up to a certain size.[43] The papers that were used to print the text and images of the *Hortus Eystettensis* were the largest size available during the time of its printing, making it one of the largest printed books in the first half of the seventeenth century.[44]

Parchments, on the other hand, were made from animal skins. The hide of a sheep, goat, or calf, the three most commonly used animals for parchment making, was soaked and dehaired before it was stretched on a frame to be further scraped to rid the last of the hair and to dry.[45] The sizes of the parchments were determined by the sizes to which the animals could grow. While sheep, goats, and calves were sizable animals, they were not gigantic. Papers and parchments could support huge or monumental sizes of images, such as wall maps, by connecting multiple sheets to form larger ones, but this was not a practice for florilegia.[46] Early modern papers and parchments being large but not massive provided a clear boundary for printing and painting. Image makers could pictorially fit many rare and curious flowers onto the picture plane according to their real-life sizes, not much smaller nor much bigger, in a way similar to fitting pressed flowers in herbaria, which are usually in folio sizes as well.

For printed flower books, the removal of text, in combination with using paper as support, contributed to giving plates a bigger picture plane to depict flowers in a life-size way. Florilegia are among the earliest examples of big flower and plant illustrations in print, which became more prevalent in the eighteenth century and onward, largely due to separating text and images on book pages.[47] After all, the same paper supports were used for printing herbals, which are also big books in folio sizes, and folio-size woodcuts were used to illustrate books in the sixteenth century.[48] Woodcut is a relief-printing technique, and the printing of woodcuts for images and the printing of (lead) types for text were operated in the same way. This means that text and images

---

43 Kinross, *A4 and Before*, 9–15.

44 More on the papers used for this florilegium, see Carroll, 'The "Hortus Eystettensis": How Paper Type Can Help Decipher Printing History,' 391–400.

45 See a similar procedure in action from the Pergamenum21 Project, University of Namur, 'Parchment making,' video, 9:44, https://youtu.be/_TmyEiVUTlg.

46 For examples of prints in large sizes, see Silver and Wyckoff, *Grand Scale*.

47 A renowned example of such big book with folio-sized plates is Merian, *Metamorphosis insectorum Surinamensium*; for bibliographical descriptions and a survey of surviving copies, see Van Delft and Mulder, *Maria Sybilla Merian. Metamorphosis insectorum Surinamensium.*

48 For example, many of the woodcuts in *De historia stirpium commentarii insignes* (1542) by the German physician-botanist Leonhard Fuchs, which was briefly discussed in Chapter 5, are folio-size that take up a full page in the book. Most herbal woodcuts, however, are (much) smaller.

could be printed together and at the same time, and there was the possibility to reuse and recycle the woodblocks to illustrate different publications.[49] In sixteenth-century herbals, images of flowers and plants thus tend to be smaller than the *forme,* which is the physical component of a page layout with types and blocks set in it for printing the text and images.[50] For example, the woodcut for the *Solis flos minor* (Fig. 14) first appeared in the *Plantarum, seu, Stirpium historia* (1576) by the Flemish physician-botanist Matthias Lobelius (1538–1616), and was the model for Sweerts' sunflower (Fig. 15).[51] The sunflower is around 12.7cm tall, with a head that is around 4.2cm in diameter, much smaller than the aforementioned depictions in florilegia. The woodblock that carries Lobelius' sunflower image is not much bigger, measuring around 13 × 8.2cm. Printing intaglio plates, on the other hand, was done on a different press. This resulted in more labour required to print a page, as text and images needed to be printed separately. However, this also allowed the possibility to make bigger images than the *forme*, especially when no text was intended to co-occupy the same page, as long as the plates could still fit within a sheet of paper. Printing text and images separately hence enabled image makers of intaglio-printed florilegia to work bigger.

In addition to, or alternatively to, working on larger picture planes, image makers often employed the visual technique of 'cutting' in order to depict the blooms in a life-size way, especially for flowers that were tall. The previous chapters have discussed that many florilegium images feature the upper parts of the flowers and eliminate roots and most of the leaves. By pictorially cutting away most parts of a flower, image makers not only sidestepped the challenge of fitting the whole flower onto a picture plane, but also directed the attention to the blooms where the rare and curious characteristics of petal count and colour variation resided. Depicting flowers life-size was not a practice solely to the florilegium. Many life-size images are included in the *Libri picturati A16–30*, a large image collection of flora and fauna, with some of its watercolours serving as models for the woodcuts for the herbals of Lobelius and his colleagues Rembert Dodoens (1517–1585) and Carolus Clusius (1526–1609).[52] The collection was bound in albums, with the sizes of the album papers larger

---

49 On the reuse and recycling of botanical woodblocks, see Moran, 'Preserving the Cutting Edge,' 393–419; Chen, 'A Woodblock's Career,' 20–63.

50 For an introduction on the technical processes of typesetting the forme, see Febvre and Martin, *The Coming of the Book*, 56–60.

51 See Chapter 3 for a discussion on the relationships between the sunflower woodcuts and engravings.

52 On this collection, see especially De Koning et al., eds., *Drawn after Nature*. A study by Helene Wille provides a detailed comparison of the woodblocks for Lobelius' book and the watercolours in the *Libri Picturati*, see 'De albums van Karel van Sint Omaars,' 15–18.

than the physical measurements of the watercolours. For example, while the album paper measured around 52 × 37cm, the watercolour of the common orange daylily (*Hemerocallis fulva* L.) in volume A22 (Fig. 40) is around 45.5 × 32cm, which is slightly larger than most of the folio-sized florilegia.[53] The large sheet of paper provided a spacious picture plane for the image maker to capture much of the flower without shrinking it down drastically, making it close to life-size for this species. Most of all the depicted plant parts, including the roots, the stem, the leaves, the buds, and the bloom, fit within the sheet of paper, with some tips of the roots and leaves going off the page. The stem, however, is folded twice so that it can fit the (close to) life-size length of 110cm (excluding roots) into the picture plane.[54]

The orange daylily in De Geest's *Hortus amoenissimus* (Fig. 41) is also life-size, but only for a portion of the full flowering plant.[55] De Geest's orange daylily measures around 40cm tall and the orange blossom 9cm, painted on a 28.3 × 44.7cm sheet of paper, and is close in size to the first folded part of the stem that holds the bloom and buds in the watercolour in the *Libri picturati* (around 41cm tall and the bloom 7cm). The bloom of the daylily in the *Libri picturati*, however, appears to be smaller. Despite the orange of the flower being the most eye-catching colour on the sheet, the buds and blooms need to compete with the green leaves and white roots for attention. By cutting away the other parts of the flowering plant, De Geest brought attention to the bloom and made its life-size-ness more noticeable. Such cutting especially enabled smaller-sized florilegia, most significantly for the *Hortus floridus*, to still contain many life-size images. Van de Passe II included a plate with a red/orange lily (*Lilium rubrum bulbiferum*; *Lilium bulbiferum* L.) (Fig. 42). While it is a different species from the daylily (*Hemerocallis fulva* L.), the flower is similar in appearance and in size to the daylily, but usually taller. This particular plate in

53 Number A22.026 in De Koning et al., eds., *Drawn after Nature*, 216–217.

54 The orange daylily can grow to be one meter tall today according to the Royal Horticultural Society, 'Hemerocallis fulva,' https://www.rhs.org.uk/plants/8622/hemerocallis-fulva/details. This folding is a similar technique found in many early modern herbaria for fitting a dried plant onto a page. However, this shared technique should not lead to the conclusion that the sheets in the *Libri picturti A16–30* were copied from dried plants, as the connections and uses between such image collections and herbaria remain a rich area for further investigations and assessments.

55 '*Lillium non bulbosum obsoleto Colore rubens. Lelyen van eenen dage, sonder clisters*,' in De Geest, *Hortus amoenissimus*, 118. This flower is identified as the *Hemerocallis fulva* L. and the brown daylily by Sam Segal in Menghini, ed., *Hortus Amoenissimus Omnigenis Floribus, Plantis, Stirpibus*, 79. The very bottom of the daylily has a different green paint from the rest of the stem, which might have been added in later centuries. Either shorter or longer, the flower is still about life size.

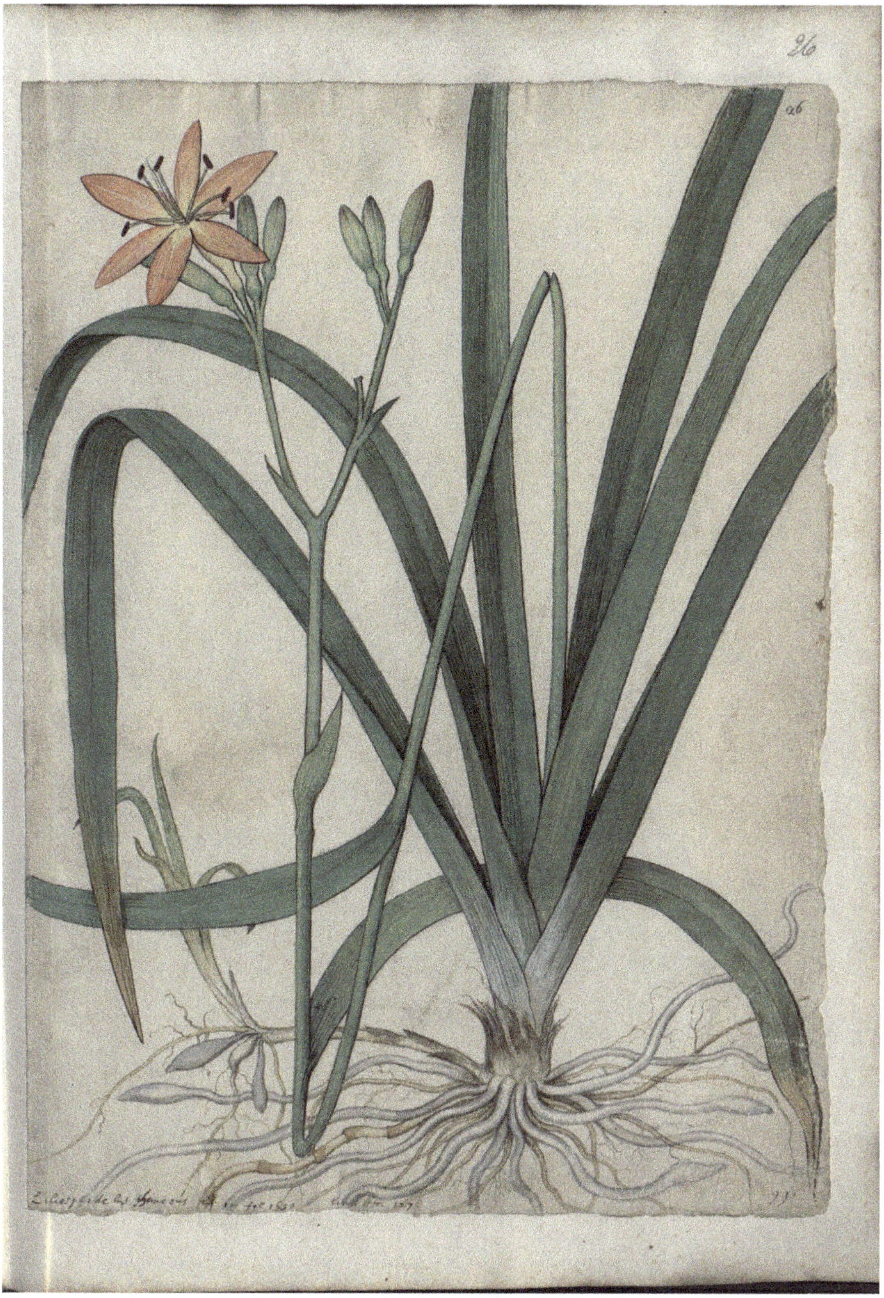

FIGURE 40 Orange daylily in the *Libri picturati A16–30* (second half of the sixteenth century), Jagiellonian Library, Krakow, *Libri picturati* A22

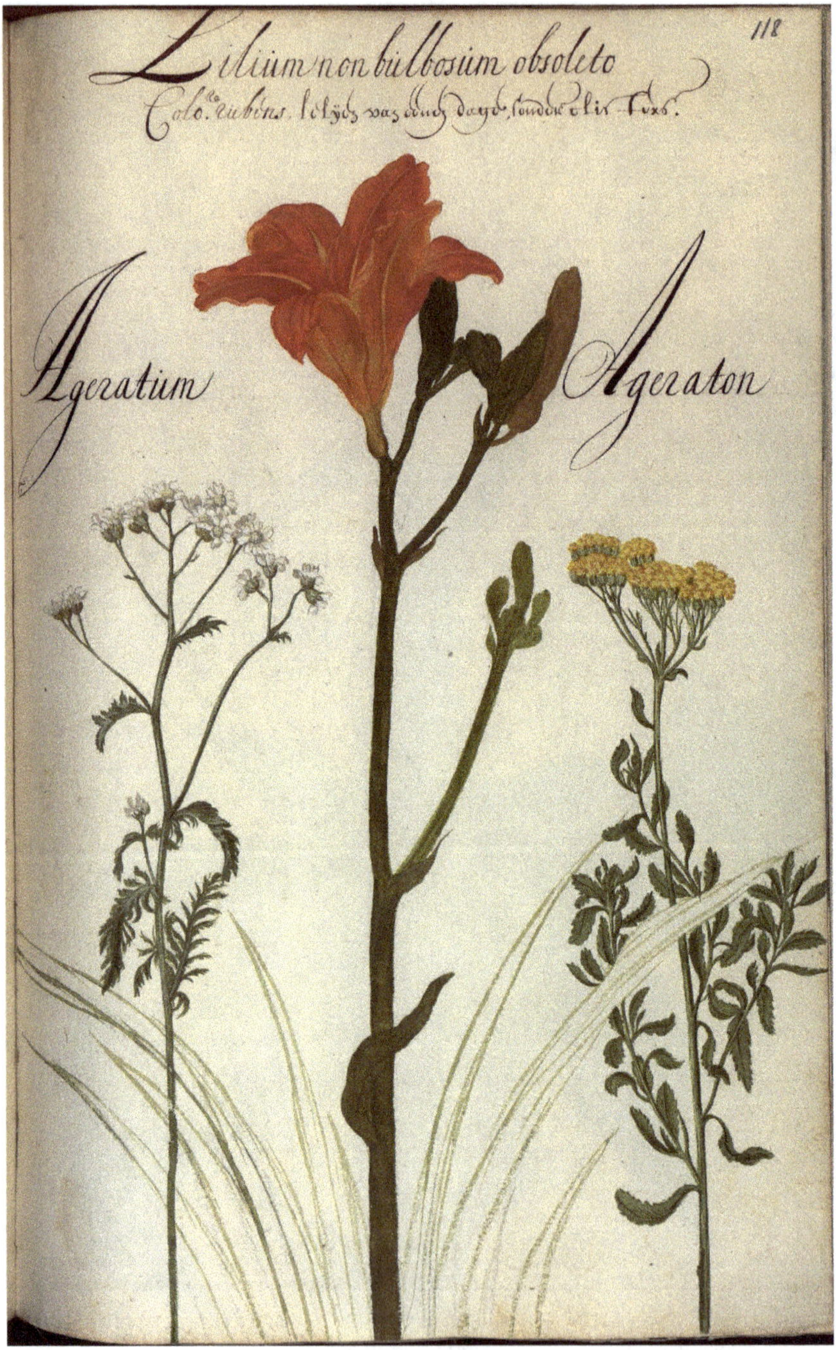


FIGURE 41 Orange daylily from the *Hortus amoenissimus* (1668), Biblioteca Nazionale Centrale 'Vittorio Emanuele II' di Roma, Rome, MS Varia 291 [118r]

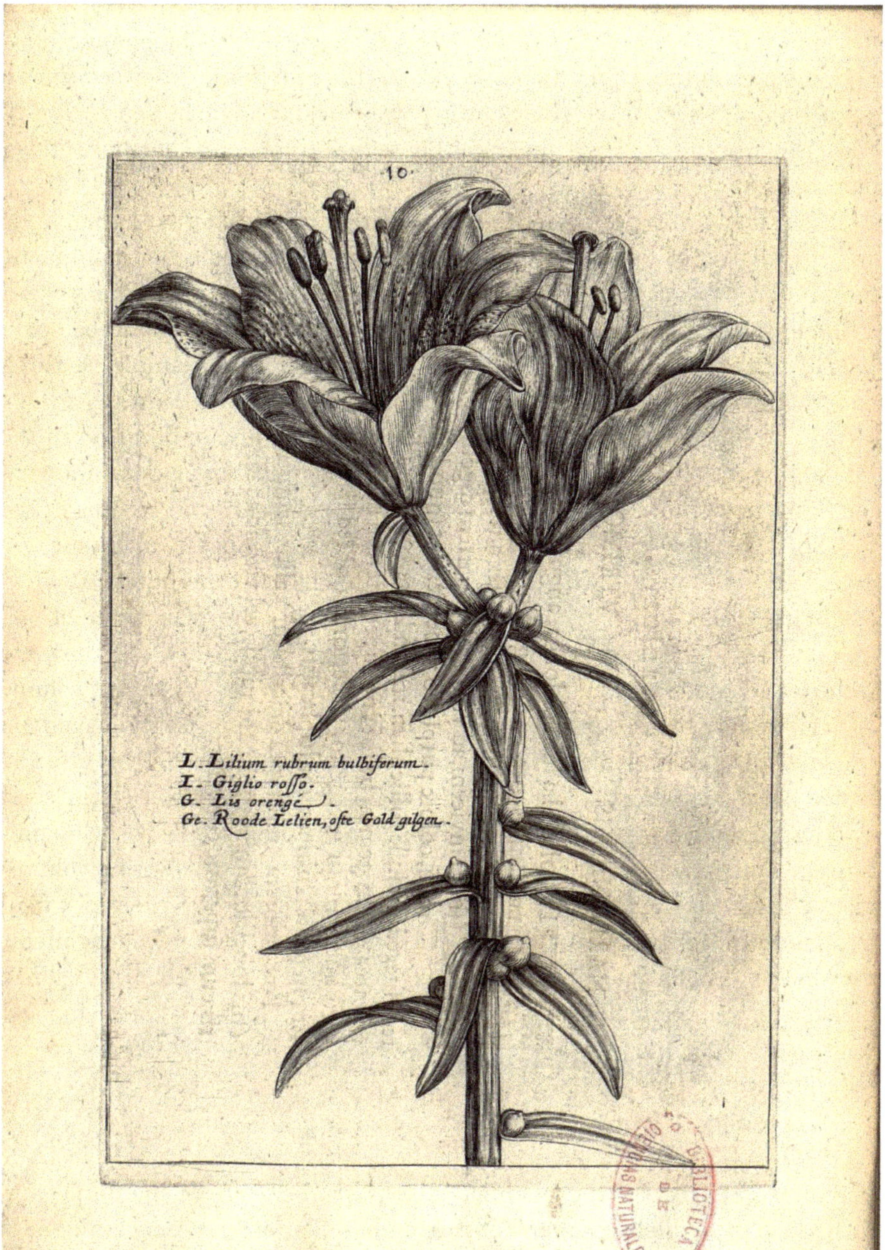


FIGURE 42 Red/orange lily from the *Hortus floridus* (1614), Biblioteca Real Jardín Botánico, Madrid, Spain, A PAS, https://bibdigital.rjb.csic.es/idviewer/13646/115

the *Hortus floridus* is 22cm in height and 14.3cm in width, much smaller than the papers in the *Libri picturati* and *Hortus amoenissimus*. The blooms, however, measure around 7cm, the same size as the one in the *Libri picturati* on a much larger sheet of paper. By cutting away most of the plant parts, Van de Passe II could still draw the blooms as big as they were in real life on a smaller picture plane.

This visual cutting of undesired plant parts and drawing only the bloom and upper part of a flower further allowed an image maker to manipulate the proportion of a flower without distorting its lifelikeness. In many ways, as long as the depicted flower is generally life-size, the flowers can be *better than life* with their rare and curious features magnified. For example, De Geest painted three variations of the pot marigold or calendula at least twice. One set can be found in the Oak Spring volume (Fig. 43) and another set in the *Hortus amoenissimus* (Fig. 44). The overall arrangement and compositions of the flowers are more-or-less the same on both pages, but the blooms in the Oak Spring volume are proportionally bigger than the ones in the Rome volume. The orange variety with double flowers in the centre measures 26.8cm in height in the Oak Spring volume, and the flower head is 8.7cm in diameter. The same variety in the Rome volume is overall taller at 35.7cm, but with the same width of 8.7cm for the head. The other two varieties in the Oak Spring volume also have larger blooms proportionally. The Oak Spring pot marigolds remain convincingly natural-looking, but the flower heads that contain their rare and curious characteristics of petal counts and colours are artificially emphasised by drawing them bigger. De Geest manipulated the proportions of the blooms to stems and leaves in most of the watercolours in the Oak Spring volume. Many life-size flowers in other seventeenth-century florilegia received similar treatments of subtle accentuation and exaggeration. By pictorially magnifying the blooms, the flowers' rare and curious features were not only captured in a lifelike manner, but made *better than life*.

## 4 Adding Colours to Flowers

The most important and challenging aspect for the genre of the florilegium to achieve is the lifelike colours of the rare and curious flowers. As the rareness and curiousness of many decorative flowers resided in their colour(s), delivering such colour information for the drawn flowers through applying watercolour visually preserved and highlighted this core characteristic of being colourful. This was especially a major issue for the black-and-white medium of intaglio-printed flowers to resolve. While it can be short, text is

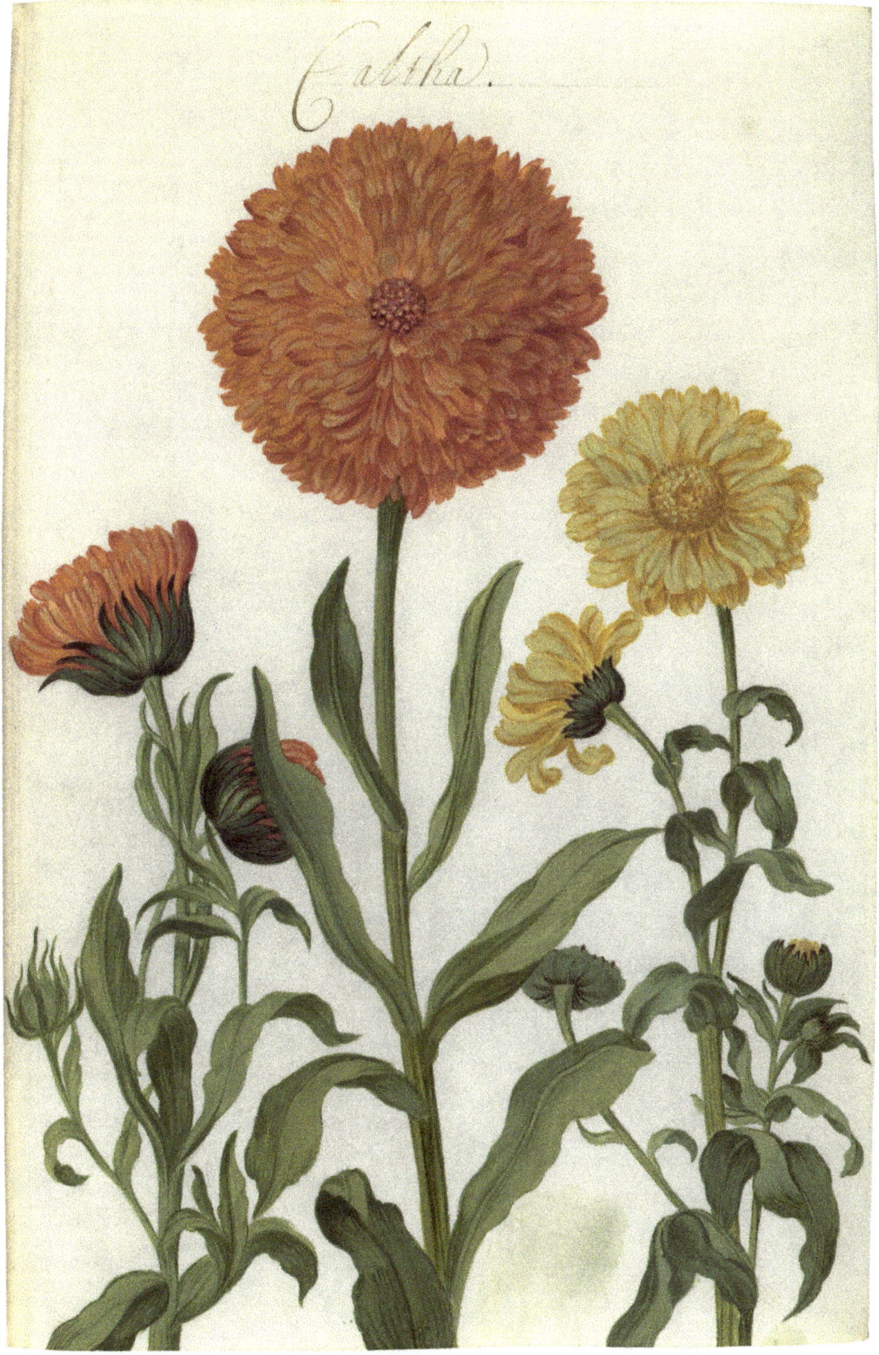


FIGURE 43 Pot marigolds from *Jardin de rares et curieux fleurs* (second half of the seventeenth century), Oak Spring Garden Foundation, Upperville, VA, 00004982

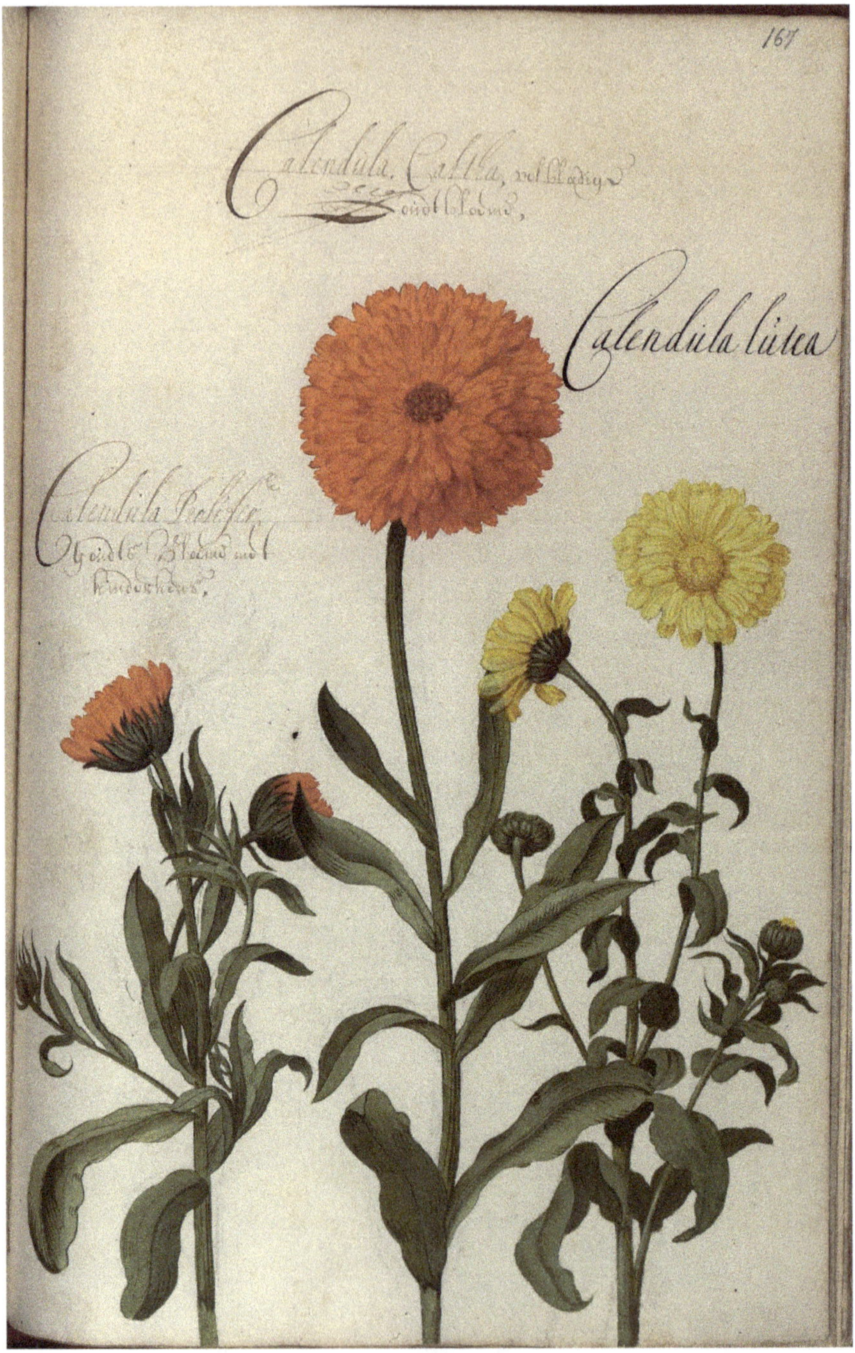


FIGURE 44 Pot marigolds from the *Hortus amoenissimus* (1668), Biblioteca Nazionale Centrale 'Vittorio Emanuele II' di Roma, Rome, MS Varia 291 [167r]

often a necessary device for providing colour information. Chapter 4 discussed how florilegium compilers utilised flower descriptions and colouring instructions to verbally colour flowers.[56] More essentially, the colour information is often engraved or etched directly onto the plates as captions, embedding the text within the picture plane. For instance, a plate in one of the reprints of De Bry's *Florilegium Novum* contains six varieties of *Viola* and three kinds of *Auricula* (Fig. 45).[57] The captions show *Viola alba, Viola flor. Lutea, Viola flammea Violacea, Viola Cerulea, Viola Martia purpur. plen. flo.*, and *Viola purpurea*, providing five different colour terms of white (*alba*), yellow (*lutea*), violet (*violacea*), blue (*cerulea*), and purple (*purpurea*). Manuscript flower books occasionally have terms or short descriptions of a flower's colour as well, such as De Geest writing in the caption for a variety of the anemone that the bloom is 'with a sulphur colour' in the *Hortus amoenissimus*, despite that the flower is already painted in yellow.[58]

The colour terms that are part of the plates provide simple yet useful information for colourists to apply paints to the flowers. It is important to note that the function of the colour terms is to point readers and/or colourists to the general direction of a colour group instead of codifying or standardising the colourisation of the flowers. The standardisation of colours has been a major point of discussion for images of plants and natural history, with increasing documentation of naturalists and image makers working on colour charts and systems that codified the colouring of flora and fauna in book illustrations in the eighteenth century and onward.[59] While colours can be an important factor in identifying and determining species, within the genre of the florilegium, it is the form and structure that provide the necessary details for identifying *which* flower is depicted. For instance, when early modern viewers saw a printed tulip in black-and-white, they would have been able to identify that the depicted flower was a tulip. A coloured tulip, either textually or materially coloured, further showed the viewers if the flower in the picture was a rare *Semper Augustus* or other tulip varieties. Colours, in this context, were not for identifying species, but for identifying one of the rare and curious characteristics in the flowers.

56 See sections 'Describing Flowers and Their Colours' and 'Describing How-To Colour Flowers' in Chapter 4.

57 This plate most likely first appeared in the reprint by Johann Theodor de Bry and Matthäus Merian for *Anthologia magna* (1626). Some copies of the 1612 *Florilegium novum*, however, include this plate.

58 *'Anemona met een solffer cleur der Bloem,'* in De Geest, *Hortus amoenissimus*, 21.

59 A noteworthy case study on this issue is Nickelsen, 'The Challenge of Colour,' 3–23.

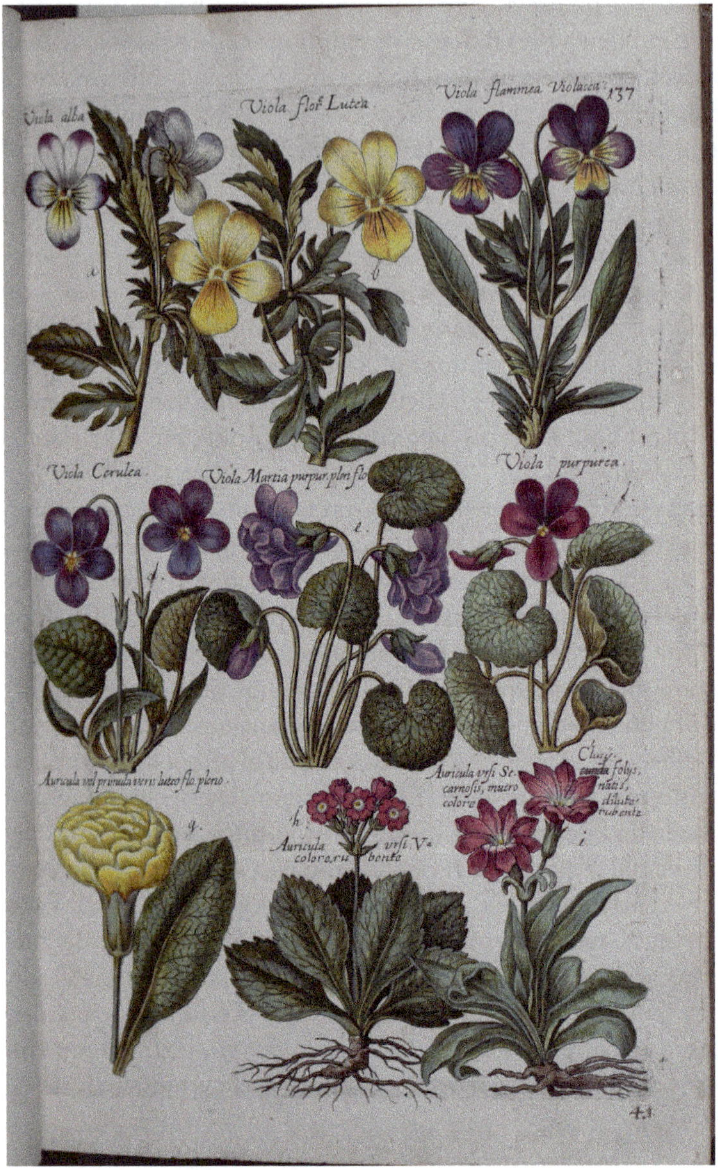


FIGURE 45 Hand-coloured *Viola* and *Auricula* plate from a copy of the *Florilegium renovatum et auctum* (1641), Oak Spring Garden Foundation, Upperville, VA, 00004681

The colour terms in flower books were not rigid in which colours they signify to the readers and colourists, as there is always a range of possibilities in tints and shades (or at times even hues) associated with the terms. For example, three hand-coloured *Viola cerulea* (Fig. 46) appear in a copy of

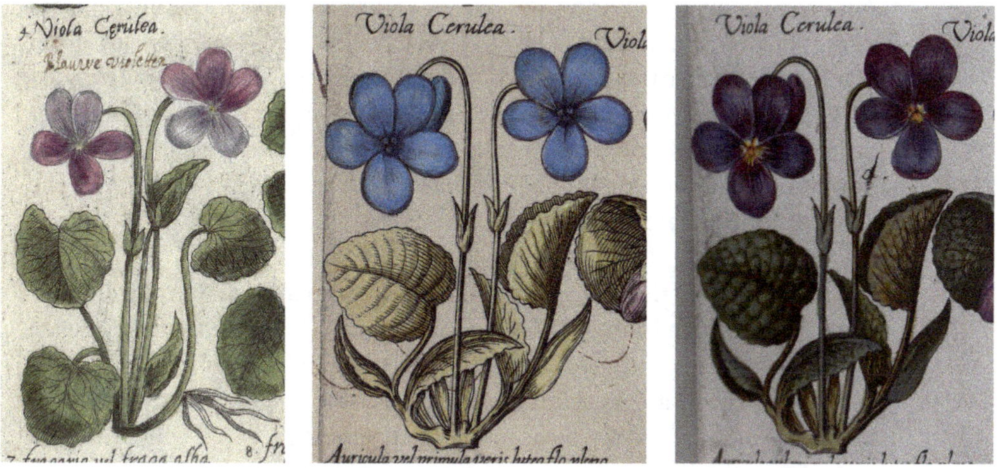


FIGURE 46 Different colourations of the *Viola Cerulea* from a copy of the *Florilegium Emanuelis Sweerti* (1612) (left), of the *Florilegium novum* (1612) (middle), and of the *Florilegium renovatum et auctum* (1641) (right), Oak Spring Garden Foundation, Upperville, VA, 00002978 / 00005057 / 00004681. Cropped and edited by the author

Sweerts' florilegium, a copy of De Bry's *Florilegium novum*, and a copy of the *Florilegium renovatum et auctum* (1641), a renewed and enlarged posthumous reprint of De Bry's flower book by his successor and son-in-law Matthäus Merian (1593–1650).[60] Although not identical on Sweerts' and De Bry/Merian's plates, the depictions of the small violet are close in composition. The three hand-coloured copies present three very different colourisations of the violet petals and leaves. However, even though there are varying degrees of naturalism in the colouring styles among the three versions, with the one in the *Florilegium Novum* flatter and less articulated in highlights and shadings, none of them is necessarily wrong or inaccurate. All three colourations fall within the realm of possible colours of early modern cultivars. Moreover, the same magnifying and accentuating of the flowers' colours could also be done by making the colours brighter, more saturated, or pushing towards a certain hue of blue, in order to make the violet's rare and curious features *better than life*. The colour terms in florilegia thus allow for a large margin of flexibility for representing the array of colours in rare and curious flowers.

60 Sweerts, *Florilegium Emanuelis Sweerti*, plate 2 (*Pars Secunda*); De Bry, *Florilegium renovatum*, plate unnumbered; De Bry and Merian, *Florilegium renovatum et auctum*, plate 137. The three copies discussed are all preserved in Upperville, VA, Oak Spring Garden Foundation, 00002978 (Sweerts), 00005057 (De Bry), 00004681 (De Bry and Merian). See Chapter 4 on the reeditions of these florilegia.

Achieving lifelike colourings of rare and curious flowers usually requires more application than a single layer of blue, yellow, red, green, or any other colour in order to give them volume and dimensionality. The colouring instructions in several editions of Van de Passe II's florilegium, once again, offer insights into the material practice of hand-colouring. To colour the columbine plate in the English edition of the *Hortus floridus*, Van de Passe II wrote, for the 'Aquilegia, with stellate purple flower,'

> The Coloure of the Starrey Columbines are also of a sad purple and may be painted with Indigo & lack, and minglinge a little white under it even to the middle of the leafe, and so towardes the end of it a more paler blew, the seed codds are greenish topt with masticot, the knopes or budds are white, the steale is of a pleasant green of verdigreece and white mingled together, the leaffe is blewish of mountaine greene and verdigreece mingled together, shadowed with sapgreene, and the edges topt with greene and white.[61]

The entries present only one person's specific way of colouring flowers. Some image makers and colourists had more elaborate approaches to handling the materials and techniques, while others had more simplistic ones.[62] Nonetheless, Van de Passe II's instruction describes colour materials and painting techniques that apply to the majority of florilegium images that fit into the larger watercolour practice during this period.

The columbine entry addresses three core issues for adding colours to flowers in a lifelike manner. First, there were many colour materials from which a colourist could pick to get close to the flowers' local colours, which are the colours of the floral objects when seen under natural light. Second, mixing colours produced even more colour varieties for a colourist. Third, painting

61 Van de Passe II, *A Garden of Flowers*, no. 37 (spring garden). The Dutch instruction is mostly the same, '*De coleur van de sterre Accoley bloeme is mede doncker purpur/ ende mach ghecoloreert werden met Indigo ende lack/ een weynich wit daer onder vermenght totten midden toe/ ende voortaen totten uytersten wat bleecker blau / de saethuelkens uyten groenen met masticot ghehoocht de noykens wit / den steel is licht groen van spaensch groen ende wit ghetempert. Del loos is uyten blauwen van berchgroen ende spaensch groen ghemenght / met sapgroen ghediept/ ende de tauten met groen ende wit ghehoocht*.' in *Den Blom-hof*.

62 The instructions by the French miniature painter Catherine Perrot for colouring the flower prints of the French printmaker Nicolas Robert, for example, uses more glazing technique with somewhat transparent colours for her entries. However, for the most part, the basic layering technique that is discussed in this chapter applies to her instructions as well. Perrot, *Les lecons royales*.

in layers enabled a colourist to create a volumetric effect. The hand-coloured columbine plate (Fig. 38) from a surviving copy at the Oak Spring Garden Foundation shows a rendition of this instruction. The blue columbine on the right is the stellate purple variety. Even though the petals on the flower do not look fully dimensional, they are painted with depth to present the volumetric form of the double flower, with the petals at the back being darker shades of blue than the ones at the front. It is possible that the petals would have been more purple than blue, but the red in the mixture may have faded. Despite the issue with the hue, this columbine demonstrates how painting in layers with multiple colouring materials can produce a flower that visually shows its fuller double form and blue/purple colour.

Van de Passe II's instruction gives glimpses of the many colour materials making up watercolour paints.[63] The diversity and variety of pigments and washes used for watercolour in the early modern period offered florilegia image makers and colourists an extensive stock of possibilities to depict life-like flowers in the colours that were either close to or better than how they appeared in real life. There were a number of colour materials listed in the columbine instruction, namely the pigments, including indigo, lack, massicot, verdigris, mountain green, and sap green.[64] The colours listed by Van de Passe II were common pigments for all kinds of painting techniques in the early modern period. Indigo is a blue pigment, lack (or lake) is a red pigment, massicot is a yellow pigment, and verdigris, mountain green, and sap green are three different kinds of green.[65] By tempering these pigments with gum water, as previously discussed, the image makers made watercolour paints of blue, red, yellow, and green.

There was an extensive array of pigments already available in the early modern period, which enabled the image makers to choose colours for the painting that closely matched the colours of the living flowers. The 1670 edition of the *Verlichterie-Kunde*, for example, includes a colour chart (Fig. 47)

---

63 There is ample literature on historical colour materials. Selected works that were helpful during the research include Harley, *Artists' Pigments c. 1600–1835*; Roy and Smith, ed., *Painting Techniques: History, Materials and Studio Practice*; Eastaugh et al., *Pigment Compendium*; Ambers, Hook, and Simpson, 'John White's Watercolours: Analysis of the Pigments,' 67–71; Panayotova, Jackson, and Ricciardi, eds., *Colour: The Art & Science of Illuminated Manuscripts*; Feeser, Goggin, and Tobin, eds., *The Materiality of Color*; Thompson, *The Materials of Medieval Painting*.

64 Chapter 4 discussed that some of the colour terms that were used for describing the flower colours are terms of colour materials.

65 Sap green is not a pigment, but a liquid wash made from the juice of ripe *Rhamnus* berries and commonly stored in bladders, Harley, *Artists' Pigments c. 1600–1835*, 79–81.

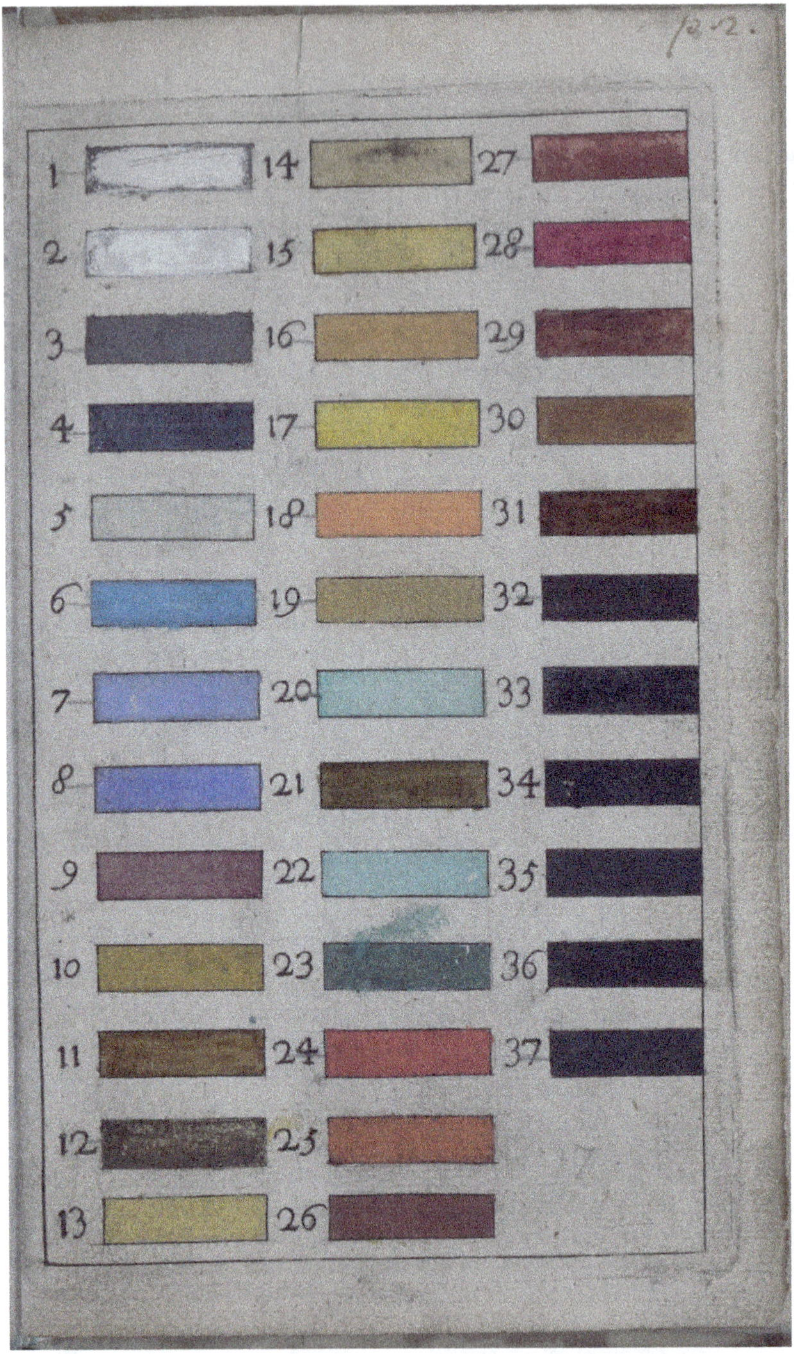


FIGURE 47 Hand-coloured chart in a copy of the *Verlichterie-kunde* (1670), KB, National Library of the Netherlands, The Hague, KW 1299 F 11 [3]

for the readers/users of the book to easily visualise the commonly used colours of the period that the book listed and described, including those given by Van de Passe II. The *Verlichterie-Kunde* is a treatise on the use and practice of watercolour, and the 1670 edition was revised and published by the Dutch bookseller and author Willem Goeree (1635–1711), based on the work of the Dutch colourist Gerald ter Brugghe (active 1616).[66] Goeree listed 37 colours and divided them into 7 groups. There are three types of white colour (numbers 1 to 3), six kinds of blue (numbers 4 to 9), ten varieties of yellow (numbers 10 to 19), four kinds of green (numbers 20 to 23), six types of red (numbers 24 to 29), three varieties of brown (numbers 30 to 32), and five types of black (numbers 33 to 37). The colours were made from different kinds of materials, including pigments derived from grinding and powdering cleaned earths and minerals, roasted metals, charred organic materials such as twigs and bones, and by extracting colourants to make washes from plant-based dyestuffs used in the textile industry.[67] This range of hues listed in the *Verlichterie-kunde*, which is by no means complete, could already represent a good number of colour varieties among rare and curious flowers.

Additionally, each of the colour materials can come in a range of tints and shades on their own. As most early modern colour materials were natural materials, their colouration could depend on the condition of where the raw materials were sourced and/or how they were processed. For example, ochres originating from different places can produce different tones of yellows. A higher grade of ultramarine would be a more intense and saturated blue than a lower grade, which tended to be a warmer blue as well. Each batch of the plant-based washes could be in a different shade because the colourants the dyestuffs contained could vary from batch to batch. Such inconsistency in colours could have been challenging to work with for image makers who wanted to maintain a certain quality or aesthetic throughout their corpus. However, such a wide range of hues and saturations offered florilegium colourists lots of options to closely match the paints with the flower colours of real life.

---

66 Ter Brugge and Goeree, *Verlichterie-Kunde*, between pages 2 and 3. The copy discussed here is preserved in The Hague, National Library of the Netherlands (Koninklijke Bibliotheek), KW 1299 F 11 [3]. The charts of most copies of this book are not coloured. For some insights into how early modern readers may have used the book, see Lehmann, 'An Alphabet of Colors,' 168–203.

67 The Teylers Museum in Haarlem houses a cabinet, compiled in the nineteenth century, with a large number of colour materials before they are processed into powdered pigments, many of which are the same type of materials that were used in the seventeenth century, such as yellow gamboge, see Pey and Homburg, *Een kabinet vol kleur*.

The columbine instruction further illuminates how colour mixing enabled image makers to create an ever-larger painting palette. This not only offered more possibilities to match the myriad local colours to the once-living rare and curious flowers, but the highlights and shadows for rendering the volumetric effect that made the images lifelike. Here, shadow refers to the shaded or darker portion of a picture for rendering dimensional objects instead of the cast shadow.[68] The instruction for the *Aquilegia* starts by stating that the colour of the flower is a 'sad purple,' or dark purple. This dark purple is not a material term; a colourist could not simply pick up a pigment labelled 'purple' to make a watercolour paint for this flower. Instead, the purple for colouring the columbine, as instructed by Van de Passe II, 'may be painted with Indigo & lack.' This means that the purple could be obtained by mixing together the two colours of blue indigo and red lake. Indigo is a blue pigment derived from the indigo plant and the lake is a kind of pigment derived from extracting red dye from madder roots or cochineal bugs, among other dyestuffs.[69] To mix indigo and lake into a dark purple for the columbine, a colourist would most likely temper the blue and red pigments with gum water, turning the mixture into a purple paint before applying it to the flower drawing.[70] Watercolours of this period do not emphasise the glazing technique for optical mixing, such as applying one layer of blue paint and one layer of red paint separately and letting the viewers' eyes perceive a purple through the layers. As many of the pigments have covering power and a certain level of opacity as part of their properties, a large number of colours would be mixed on a palette or in a container before being applied to papers or parchments.[71] The instructions usually specify when colour mixing through glazing would be the more suitable technique for achieving certain colours or visual effects. Through mixing paints, colourists could obtain colour groups of purple that better matched the local colour of the columbine.

68 For florilegium images, the shadowing simply aim at creating volume under a generalised light source. The visualisation of (cast) shadows can be a complex topic at times. For example, see Kaufmann, 'The Perspective of Shadows,' 258–287.

69 On the machinal process of making lake pigments, see especially Chapter 5 in Kirby, Van Bommel, and Verhecken, *Natural Colorants for Dyeing and Lake Pigments*, 69–106.

70 A notable manuscript that includes extensive swatches obtained by colour mixing is A. Boogert, *Klaerlightende Spieghel der Verfkonst*, Aix-en-Provence, Bibliothèque Méjanes, https://bibliotheque-numerique.citedulivre-aix.com/idurl/1/35315; see also the new platform and ongoing transcription, translation, and (technical) study of this manuscript on the platform Boogert Manuscript, https://www.amjournal.org/boogert-manuscript.

71 Mixing in a container before application by pouring liquid washes (which tend to be transparent) into a shell to create a new colour, instead of through optical mixing, is also advised by Jenner, *A Book of Drawing, Limning, Washing, or Colouring of Maps and Prints*, 28–30.

Colour mixing could further broaden the spectrum of a colour group. This kind of colour mixing is particularly useful for painting flowers that come in a variety of hues within the same colour group, such as purple flowers of all kinds, as well as for painting leaves, since they are mostly greens. The same blue indigo and red lake paints could produce a range of purple hues by adjusting the ratio of the paints. When more blue is added to the mixture, the paint becomes more blue-purple. When a mixture has a higher amount of red, it shifts to the red-purple spectrum. Two colours from the same group could also produce new colours within the group. To paint the blue-green leaves of the *Aquilegia*, for example, Van de Passe II instructed readers and/or colourists to mingle two green paints that were made from the pigments of mountain green and verdigris. Mountain green usually refers to a pigment derived from the mineral malachite, and verdigris is an artificial pigment, usually made by exposing copper (plates) to the vapour of vinegar or similar acetic acid. Mountain green comes with a range of hues, some of them warmer and more yellow, and others cooler and bluer. Verdigris, on the other hand, tends to be more on the bluer end of the green spectrum. Mixing a (warmer) green with a green with a blue tone shifts the mixture to a blue-green. This method of adjusting colour ratios when mixing colours within the same group enabled the image makers to represent the colours of the flowers with subtlety.

Mixing a colour with white, furthermore, produced tints that were useful for painting mid-tones and highlights. At the very end of the instruction, Van de Passe II wrote that the edges of the leaves may be 'topt with greene and white.' The diversity and variety of colour materials and the mixing of colours presented an array of hues, but they could sometimes be close in tone. Most pigments tended to have deeper and darker tones as well, with the pastel yellow colours of massicot and lead-tin-yellow (both of which are roasted lead-based pigments) being major exceptions. By mixing green with white, which usually refers to lead white in the early modern period, image makers could produce paler and lighter greens. Such lighter colours could be used for flowers that have pastel colours in nature, including light red flowers that are usually described as *pink* today.[72] Painting highlights with tints made by mixing a colour with white can also present a more cohesive effect, instead of having sharp lines by using white or the limited range of light-yellow pigments as highlights. Mixing colours with white created a pallet of nuanced colours for transitioning from shadow to local colour to highlight, which led to a lifelike depiction of volumetric and colourful flowers.

72 *Pinck* usually refers to yellow in the early modern period. See Harley, *Artists' Pigments c. 1600–1835*, 97–101.

Last but not least, Van de Passe II's instruction for colouring the stellate columbine exemplifies how a colourist could give flowers volume and dimensions with a few layers of local colour, shadow, and highlight, with a myriad of coloured paints. The function of the layers is to create volume on the flower. The instruction for the stellate columbine mentions around three layers, most explicitly on painting the leaves. As closely rendered in the Oak Spring copy of the columbine plate (Fig. 38), the first layer is a base of a blueish green. A layer of shadow in a deeper green is applied with sap green, according to the instructions. The third layer is the highlight, painted with a tinted green. The first layer gives a bloom or leaf its local colour. Without further rendering, the flower would have its main colours, but it would be flat. The shadow and highlight bring three-dimensionality to the flower. For printed flowers, sometimes a translucent colour wash for the local colour would be the only layer, and the volume is achieved through the engraved or etched lines, such as hatching and cross-hatching that shade the flowers. By shading and highlighting different parts of the leaves, the different tones of greens in close hues give the leaves some three-dimensional visual effect in the stellate columbine.

This simple layering technique is particularly useful in painting lifelike double flowers with many petals. The pot marigold on parchment (Fig. 43) by De Geest shows a clearer example of how shadows and highlights gave the form of the orange double flowers. Similar to the stellate columbine, the pot marigold is painted with a few layers of local colour, highlight, and shadow. The bloom is painted with an orange to block out its overall shape. Two darker tones of orange are used to define the petals, with the lighter one mostly outlining the petals and the darker one shading the shadow on the petals. A lighter orange is applied to some edges of the petals to highlight certain parts, which helps heighten the contrast. This use of layers further reinforces that, in order to paint a lifelike flower that is volumetric and dimensional, a large range of tonal colours achieved through colour mixing was needed in addition to the spectrum of local colours. By utilising highlight and shadow in tints and shades of the local colour, colourists could give definition to the forms of the rare and curious flowers.

## 5 Conclusion

The material and technical properties of intaglio-printing and watercolour in the early modern period afforded the visual effects for creating images of flowers *ad vivum*. Life-size and colourful depictions presented lifelike flowers that faded the boundary between image and object. The drawing supports of paper and parchment provided sizable picture planes that could contain most flowers without the need to miniaturise them. By visually cutting away roots and the majority of the leaves, image makers could capture the flowers' likeness in the sizes close to their living forms. Adding colours to the drawn flowers further brought the images to life, turning a black-and-white image with outlines into a colourful object with shadings. The many types of pigments and the mixtures of watercolour paints provided seventeenth-century colourists with an extensive palette with which to capture the colours of the flowers as accurately as possible. The painting technique of layering shadows and highlights on top of local colours rendered the flowers voluminous, bringing the flat drawings of the petals to dimension. Through the processes of drawing and colouring, image makers and colourists could accentuate and idealise the flowers, making the blooms bigger and the colours more vibrant. The subtle pushing for the rareness and curiousness often made the flowers not only lifelike, but better than their living forms. The *ad vivum* bloom depictions in florilegia recorded and preserved living collections of rare flowers and plants for many *liefhebbers*, and provided a chance for more people to have a rare flower collection, even if it was through images instead of living specimens. Whether or not a flower was drawn and painted from life, its lifelike appearance, often with enhanced qualities, was everlastingly captured on paper or parchment.

# Conclusion

## 1 A Customisable Genre Embracing Different Forms of Knowledge

This volume set out to understand why luxurious picture books of flowers were made in the long seventeenth century, besides serving as status symbols. Through investigating how florilegia were made, I have shown the importance of early modern makers in producing these books that are rich in materials. Foreign flowers and decorative plants that travelled long distances to areas to the north of the Alps, cultivated varieties with new and/or unusual colours, and double flowers with (many) more petals than their native forms made them desirable to collect as rarities and curiosities in gardens. From cultivating to depicting these flowers, makers of florilegia collectively endeavoured to immortalise a diversity and variety of ephemeral flowers. The florilegium as a pictorial genre recorded and preserved this aspect of early modern horticulture along with the knowledge of rare and curious flowers.

Seventeenth-century florilegia as objects of knowledge have a double meaning. On the one hand, they transferred and communicated different kinds of knowledge, from plant studies to gardening techniques, through images and text on their pages. On the other hand, they embody the many forms of knowledge that worked behind the scenes in order to produce the images and text on the pages. This book has unpacked some of these intertwined forms of visible and invisible knowledge between the pages. The six chapters of this volume have shown the many ways in which early modern makers—the gardeners, the florilegium compilers, and the image makers—contributed their knowledge and expertise to cultivate, document, and pictorially organise floral rarities and curiosities. This book has also illuminated how some seemingly simple tasks and activities performed by the three groups of makers were important in producing flower books as objects of knowledge.

The making of the florilegium genre took the practical and gardening knowledge of the *gardeners* to grow and cultivate rare and curious flowers from seed to bloom, as well as to keep them alive for those with access to the flowers to admire, study, and/or pictorially capture. By preparing different kinds of grounds and fertilising the soil, the gardeners fostered supportive environments for flowers to acclimatise to and thrive in the humanmade setting of early modern gardens that were often much different from the flowers' native habitats. By offering year-round care, including weather protection, the

© JESSIE WEI-HSUAN CHEN, 2025 | DOI:10.1163/9789004735149_009

gardeners' diligent labour and knowledge of when to supply what aid according to a flower's needs brought the seeds to seedlings, to small plants, to buds, and to blooms. Bulbous and tuberous flowers, in particular, required much more extensive care than rooted ones, with a much longer growing time (in nurseries) from seeds to bulbs and tubers before they could be (trans)planted into a garden ground.

The careful gardening knowledge made the living floral rarities and curiosities available for early modern *liefhebbers* to collect and for the genre of the florilegium to depict, record, and visually preserve. Florilegium images are embodiments of such gardening practice and knowledge, of the manual processes that were critical for sustaining flower book production in the long seventeenth century through depictions of blooming rare and curious flowers.

The making of the florilegium genre took the plant and artisanal knowledge of the *florilegium compilers* to identify, select, and order images of floral rarities and curiosities, as well as deliver and communicate various kinds of information related to the flowers' rareness and curiousness through text. Florilegium compilers participated in the curiosity-collecting culture of the early modern period—a backdrop for much development of natural historical knowledge—through assembling image collections of flowers with rare and curious characteristics. Compilers, especially those who put together intaglio-printed collections, excerpted flower depictions from the woodcuts and referenced the writings of sixteenth-century herbals, extracting visual and textual information about rare and curious flowers from authoritative plant studies of the time. Instead of following one of the several systems of ordering flowers and plants utilised by physician-botanists of sixteenth-century herbals, florilegium compilers often opted for a seasonal order of the flowers according to their blooming times. The seasonal order is a result of attentive observations of seasonality in the gardens, and offered readers of florilegia another way of accessing nature through books that were more in tune with the cultivated environment of curiosity gardens.

Compilers also used different kinds of writing to supplement information that was difficult to visualise, especially through the black-and-white medium of intaglio-printing. Through adding descriptions of a flower's place of origin, colour (s), and such, similar to the writings of sixteenth-century physician-botanists, florilegium compilers presented a more comprehensive picture of the floral rarity and curiosity in the scholarly language of Latin. By adding instructions on how to colour flowers, compilers turned to the writing style that was commonly used as one way to teach the hand-colouring practice to *liefhebbers* in vernacular languages. Through verbally creating vivid pictures of colourful flowers, compilers emphasised the knowledge of colours that was a critical feature of making these flowers rare and curious.

The making of the genre took the artistic and material knowledge of the *image makers* to present the rare and curious features of the flowers in a lifelike manner through various visual strategies, often emphasising the features to make the flowers more idealised than in real life. By depicting flowers in the moment(s) of bloom, image makers selected the most appropriate mode of representation to capture flowers when their rareness and curiousness were at their fullest. By visually cutting the flowers, keeping mostly the blooms and some leaves instead of a full plant including roots, image makers could fit many flowers onto the picture planes of paper and parchment that were close to their sizes in real life, providing a viewing experience similar to seeing living blooms in the gardens or pressed ones in herbaria. By knowing the many kinds of colour materials and how to mix them to create a range of hues and tones, image makers could represent and create myriad colours that match the flowers' colouration. With simple layers of base colour, shading, and highlight, image makers could render the volumetric forms and vibrant colours of floral rarities and curiosities through the medium of watercolour.

Image makers also accentuated the flowers' rareness and curiousness through different artistic manipulations. By drawing flowers in a manner that their petals and leaves are unfolded and arranging them in front of a mostly empty background, image makers employed techniques that were common for preparing living specimens to be pressed and laying out dried plants on the pages to be studied in early modern herbaria with dried plants. Such visual pressing enabled image makers to present the flowers' full forms with colourful petals more prominently, both of which were challenging to retain in the process of pressing for dried specimens. By slightly altering the proportions of the blooms and leaves, image makers made the blooms bigger to better showcase the many petals without losing the flower's overall lifelikeness. By colouring flowers with paints that were a shade more vibrant or in slightly different hues, image makers accentuated the already rare and curious colours of flower varieties to make them better than life. Through making bloom depictions with a number of visual strategies, with some comparable to those utilised in herbaria, florilegium image makers (especially the makers of the manuscript ones) preserved the critical features of double flowers and vibrant colour variations that made the flowers rare and curious.

The making of the florilegium genre was built on the diverse and varied forms of knowledge of the makers. While some individuals assumed every role of the maker—as a gardener, compiler, and image maker—others were commissioned to work mostly as the image maker for different patrons who took on the compilers' role to put together the living collections on which their florilegia were (usually) based. As a result, the genre can be incoherent because the different makers had different kinds of knowledge and interests. The topics discussed in the six chapters of this volume are thus tendencies and never rules, as regularly mentioned throughout the piece. For example, gardening knowledge only appears in the *Florilegium Emanualis Sweerti …* (1612) by the Dutch printmaker and nurseryman Emanuel Sweerts (1552–1612) and some copies of the *Hortus floridus* (1614) by the Dutch printmaker Crispijn van de Passe II (1594/1595–1670).[1] The *Hortus floridus* is the only florilegium that engaged with every form of knowledge that was investigated throughout this volume, but only the French edition contains all these forms. The Latin, Dutch, and English editions each contain selected types of visual and textual information.

The incoherency of the genre, however, precisely speaks to the strength of the florilegium. The making of florilegia in the seventeenth century was not concerned with establishing a defined genre with consistent rules or guidelines. Rather, through assembling images of flowers that had the qualities of being rare and curious, the florilegium was a device for people to customise collections that accommodated many forms of knowledge that were most interesting and/or pertinent to them, with the additional pleasure of being visually beautiful and materially sumptuous. The genre of the florilegium not only embraced a diversity and variety of rare and curious flowers, but also a diversity and variety of forms of knowledge in the long seventeenth century.

## 2 A Time to Make

Another important aspect that regularly surfaced throughout this book is the theme and different handlings of *time* in the historical making of florilegia. It took seasons, if not years, for gardeners to cultivate rare and curious flowers and see them bloom. It took years to decades for collectors to acquire specimens and assemble a living curiosity collection. While it could be a much shorter period, it took time for a compiler to read through existing herbals and botanical treatises and compile and extract text for florilegia. It also took time for image makers to produce each sheet and depiction of flowers; it took

1 Sweerts, *Florilegium Emanuelis Sweerti*; see Chapter 2 for an introduction to and more detailed discussions on this florilegium. Van de Passe II, *Hortus floridus*; see Chapter 1 for an introduction to this florilegium.

time to prepare materials, draw, engrave, bring, and/or paint the pictures. It took time to gather leaves of bloom depictions into a manuscript volume or to add plates to a later edition of a printed florilegium, both of which could take years to decades as to the assembling of living collections. It also took time to turn the image collections into codices, to bind the leaves or sheets into book blocks and adorn them with covers, even though this step is not discussed in this volume.

Between the pages, florilegium images capture the time of the flowers' lives when they present their rareness and curiousness most prominently. The range of bloom depictions is not only for capturing the morphology of rare and curious flowers, but their most shining moments. The common ordering of pages according to the flowers' blooming seasons is a time-based order, an uncommon strategy among all other methods of organisation. Even though the genre and its makers did not claim so, this recurring theme of time—of experiencing the often slow-pacing but always delightful wonders of nature in gardens and on papers and parchments—is perhaps what is unique to the genre of the florilegium among other types of images and books of natural history.

This aspect of time was especially brought forward to bright light during research through my working with hands-on and performative methods, as discussed in the Introduction. Making things requires time and patience. While some steps can be sped up to an extent, a sufficient (or more ideally an abundant) amount of time is necessary to acquire and better understand makers' knowledge. As I have shown, knowing the historical making of flower books led to better knowing the significance of the florilegium as a genre, and this was only possible because there was some stretch of time to research various aspects of making with enough time, albeit a longer period would have been more ideal.

As prefaced in the Introduction of the book, I employed three kinds of making in order to gain insights into the practical and embodied knowledge of different practices. *Historical reconstruction/reenactment* refers to following historical recipes to make pigments, washes, and paints that were commonly used for early modern watercolours. *Historical remaking* refers to reproducing selected pages from seventeenth-century flower books using early modern watercolour materials and techniques. *Historically inspired making* turns to current-day and creative practices that have historical roots, such as painting and pressing flowers and maintaining gardens, to engage with hands-on activities without worrying about adhering to specific techniques used in the early modern period.

The methods of historical reconstruction/reenactment and remaking offered closer material, tacit, and temporal experiences to the period's practices. They enabled me to better understand the early modern processes of making, preparing, and transforming artistic materials and techniques into printed and painted pages. The material, procedural, and embodied knowledge allowed me to better comprehend and interpret historical sources and evidence as well. Implementing these research methods has been largely advocated by historians from many fields through channels including articles, blog posts, books, and videos. Sharing the same enthusiasm, I aim to use this volume to attest to the usefulness and value these methods can bring to historical research.

I further appeal to researchers to consider *historically inspired making* as a way to think through the hands. As mentioned, many observations and questions throughout the volume came from my playing with making colours from plants that were not used for early modern paints, as well as from drawing inspiration from the past to create a twenty-first-century reinterpretation of the seventeenth-century florilegium. Chapter 2, on growing and cultivating flowers, and Chapter 5, on the shared artistic manipulations in florilegia and herbaria, also came from my trying to grasp the practices on a general level without anchoring them in a particular historical period. Yet, by simply making colours, painting and pressing flowers, and assisting gardeners to maintain and upkeep a garden, *making* already unveiled many forms of (practical) knowledge that could then be transposed to the seventeenth century, with the historical sensitivity of a historian.

Most historical reconstructions/reenactments and remaking start with skill building for researchers to learn about how to work with (period) materials and techniques; hands-on classes with (demonstrations of) artistic techniques are also taught in art history departments or museums to better equip students on conducting object- and technique-based research.[2] These makings at a foundational level, however, are often considered as a step stool to other more advanced methods. I argue that they can be valuable in themselves for generating research questions and making connections. While the precise materials, techniques, tools, and other details in a practice might have changed throughout the centuries, there remains essential (embodied) knowledge for a maker to have. The skill levels of makers differentiate their mastery of the practice, but accessing makers' knowledge on a broad level does not have to be intimidating and can be achieved by simply getting one's hands into the action.

This 'amateur' approach might be even more suitable for certain historical topics than a hyper-focused reconstruction/reenactment or remaking.

2 Such as the course *Van papier tot plastic* (From paper to plastic) for first-year bachelor students in the art history department at Utrecht University and the Summer Institute for Technical Studies in Art (SITSA) at the Harvard Art Museums.

As I have demonstrated, the forms of knowledge in seventeenth-century florilegia were wide-ranging, even though many of them may not be regarded as deep into one field. The maker of the *Florilegium Emanualis Sweerti*, for example, is recognised to have worked in two very different fields, even in the seventeenth century, as a printmaker and a nurseryman (and sometimes as a rare flower/plant trader as well). Assessing purely on an aesthetic and artisanal skill level, Sweerts as a printmaker might not be the most masterful one.[3] However, Sweerts' unique perspective and knowledge as a maker came from participating in two trades. Sweerts and many early modern figures, including the *liefhebbers* that were the main readership and ownership of florilegia, had extensive (practical) knowledge on a more general level instead of deep, specialised knowledge. The genre of the florilegium is not the only group of historical objects and topics that sits in this junction of multiple forms of knowledge. Working with historically inspired making on a novice level can be a non-stressful way to engage with hands-on work that can lead to making more connections and combinations that may be uncommon, but still exist, to help a researcher better grapple with the many crossroads found in the past.

Turning to makers' knowledge that can come from combinations of different fields, practices, and expertise, the history of gardening and horticulture provides a fertile ground for further research, aside from the technological developments (such as hothouses and greenhouses) that have received much attention. Gardens were an important part of natural history in the early modern period, and they are deeply connected to environmental history. Chapter 2 discussed some of the gardening techniques as written in *De Nieuwe Nederlandse Bloem-Hof* (1700) by the Leiden arborist Henrick van Oosten. Flowers are not harmless, and neither are the methods for cultivating them. Many of the described early modern techniques, including those on how to grow trees, are no longer used today, or at least discouraged: they are no longer considered the most environmentally friendly approaches. Most of the foreign flowers in florilegia were either contained in gardens, or became naturalised to areas to the north of the Alps, but accounts of introduced flowers and plants becoming invasive species due to careless growing that caused major problems for local ecologies appear time and again throughout history. Additionally, even though the flowers in florilegia are not often the most notorious examples, the importation of plant matters to Europe in the long seventeenth century was

3 Many studies commented on the 'poor quality' of this florilegium, such as Blunt and Stearn, *The Art of Botanical Illustration*, 101–102; Segal and Alen, *Dutch and Flemish Flower Pieces*, 1019.

not separated from what would become the systematic cultivation of plants by violently exploiting people, lands, and other natural resources in the colonies.[4]

The history of gardening and horticulture is not only about what took place within the garden walls, but a collation of makers' knowledge that stemmed from the grounds of many places around the world.[5] There is, naturally, significance in looking at historical techniques within the garden walls to reconstruct historical sites and for living (history) museums. Such works have successfully (re)created spaces that give the general public glimpses of the historical past in an immersive and experiential manner, which has helped in drawing attention to and raising interest in heritage sites.[6] Furthermore, gardening and horticultural history can contribute to a better understanding of how human efforts and activities have constructed and destructed natural environments from the ground up. Writing about such efforts on a grand scale for ecological restoration, Laura J. Martin traced the twentieth-century shifts in the United States from nature preservation, conservation, to restoration of ecological systems.[7] The different terms all strive for caring for the natural environments and organisms that face(d) destruction due to human activities, but the dilemmas and struggles lie within balancing how much human intervention should go into co-designing wildernesses. While not everyone can be an active ecologist, making decisions for 'cultivating' wildernesses, many more people have gardens. Zooming in on these humanmade ecologies, horticultural practices and history can participate, and have often done so, in restoring nature on a small but accumulative level with a large number of participants.

In this vein, it is especially important to also access makers' knowledge by working with current-day makers, in this case gardeners, who (unintentionally) practice historically inspired making on a regular basis. An open dialogue between today's gardeners and historians can be mutually beneficial, with the makers shedding light on the practical knowledge to confront environmental challenges at present and the researchers on the historical knowledge of how people have (not) done so throughout the ages. An example is the task

4 An example that pinpoints many of such issues is Schiebinger and Swan, eds., *Colonial Botany*. See also, Schiebinger, *Plants and Empire*.

5 While there remains much for exploration, an edited volume by Michael Conan and W. John Kress offers many interesting studies as starting points, *Botanical Progress, Horticultural Innovation and Cultural Changes*.

6 A hands-on example is Greene and Temple Lombardi, *Vegetable Gardening the Colonial Williamsburg Way*. Reconstructing a historical garden, however, usually involves many more aspects of decision making, see Harney, ed., *Gardens and Landscapes in Historic Building Conservation*.

7 Martin, *Wild by Design*.

of dividing a perennial plant. In his gardening treatise, Van Oosten sometimes advised his readers to divide a flower, so that they could get more of them in number. This suggestion was for obtaining more of the floral rarities and curiosities in the early modern period. Today, dividing plants is also common advice for (amateur) gardeners, but emphasising how doing so can help in creating and/or maintaining a luscious garden in a more sustainable way because it reduces the need for constantly purchasing new plants from garden centres that require more resources to grow. Aside from learning about specific period techniques for reconstructions, knowing the historical past of a practice can further help makers see that how to do things has never been static, but constantly developing and adapting to the changing environment, and to learn from what was good to (re)adapt for the future.

What seventeenth-century florilegia have immortalised, after all, is not a one-off interest in the distanced past. Growing and breeding cultivars, or horticulture in general, thrived and became a major industry in the Netherlands, now exporting many modern cultivars to other corners of the world.[8] The Keukenhof garden attracts hundreds of thousands of visitors from around the world every year, amazing people with its plant beds filled with millions of cultivars of tulips, narcissi, grape hyacinths, and many more flowers.[9] These (travelling) cultivars have different colours and features that are rare and curious to people today, usually even more vibrant in colours, ever-larger in sizes, and with evermore petals than the historical varieties.[10] The current-day varieties of floral rarities and curiosities have both positive and negative impacts on the lands in which they grow, as their historical counterparts did in the early modern period. The everlasting flowers between the pages of seventeenth-century florilegia witnessed the horticulture of the period, one page from the ever-changing forms of human effort to cultivate the natural environment that continues today.

8 Gebhardt, *Holland Flowering*.

9 There were over 1.4 million visitors in 2024 according to the official website, Keukenhof Holland, 'Keukenhof Concludes Successful Anniversary Season,' https://keukenhof.nl/en/news/keukenhof-sluit-een-succesvol-jubileumseizoen/.

10 There are many specialised horticultural or garden societies today that have registries of cultivars, usually from the past century, and each society normally focus on one to a few kinds of flowers. An example is the Historic Iris Preservation Society, 'Historic Iris Gallery,' https://historiciris.org/gallery/.

APPENDIX 1

# Making Colours Series

This is the archived version of the blog post series 'Making Colors' that reflects on my working with performative methods for hands-on research while writing this volume. It is initially published on my personal website, https://jessieweihsuanchen.com/tag/making-colors-series/. The series is reformatted into this appendix, with minor updates to the text, such as changing to British spelling and the citation style. The videos originally viewable on the website are replaced with screenshots. Even though the multimedia aspect is lost in this version, this appendix ensures that the blog posts will remain accessible after the website eventually stops working

The materials, tools, and equipment for the historical remaking research of my project are supported by the grant for the Stacy Lloyd III Fellowship for Bibliographic Study from the Oak Spring Garden Foundation.

## 1 Experimental Series on Historical Remaking

Posted on 3 February 2020

At the beginning of my project, I reflected on the fast pace of going through a large number of botanical watercolours during my research trips. With every page, I only had a few minutes to study the image(s) and took as many notes as I could in order to grasp the collections within the allotted time. As I embark on historical remaking to investigate the materiality of seventeenth-century watercolour this academic year,

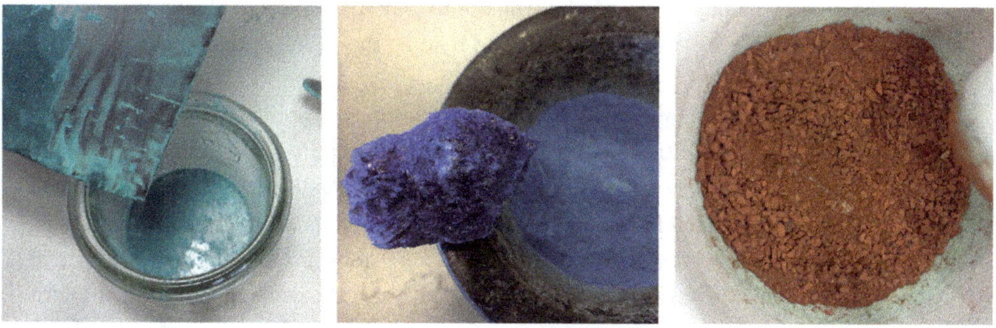

FIGURE 48 Collecting verdigris (left); crushing azurite (middle); crushing red ochre (right)

I find myself working on a much slower pace, which can be uncomfortable at times. Since last October, I have been spending a lot of hours in our Kunstlab (ArtLab) at the university to experiment with making pigments from natural materials (Fig. 48), such as minerals and plants. Now, I have just mentioned that I consulted many early modern flower books and watercolours. The goal of studying these images in person and up-close is to ask questions and hopefully get answers to the materiality of historical watercolours. If I already have such sources, what, then, is a historian (in contrast to a conservator) doing in a lab to make pigments?

The basic concept of watercolour, or water-based media in the broader sense, is fairly similar to what it was in the early modern period. However, pigments are much more finely-ground with modern machinery, and manufactured watercolour tubes and cakes today have different formulas and pigment sources. For example, the toxic lead white, corrosive verdigris, and expensive authentic ultramarine are no longer used as ingredients, even though the same colours with synthetic pigments can still be purchased. Additionally, painters would have made their own paints (and sometimes pigments as well) instead of buying ready-made ones. This requires a high level of skill and allows customisation of one's painting supplies. It is thus necessary to produce some pigments from scratch to better understand the material properties and qualities of historical colours.

As I experiment with making colours in the lab, I see a good opportunity to also experiment with my writing to communicate the making process. To some, the making process is the ultimate joy and more valuable than the product; to others, it can be boring and unnecessary to appreciate a piece of work. Writing about the making process can be both as well. Therefore, in the next few months, I will be working on a series of short pieces that discuss several issues I have leaned or observed in the lab. Each will be accompanied with the process of (one of) my experiments. In these pieces, I do not attempt to answer any question, for that is what the dissertation will. Rather, they will resemble fieldnotes, but in a more organised manner. The main goal of the series is not to argue, educate, or even inform, but to play with different elements—such as formats and visuals—to make writing and reading the making process a more engaging experience.

Since my project concerns with the application of watercolour in seventeenth-century botanical illustrations on a macro level, rather than the practice of specific painter(s), I address my research method as *historical remaking*. What one should call a performative method for empirical research is a heavily-debated and ongoing issue. Historical reconstruction might be a more familiar term for most.[1] I chose historical remaking instead to emphasise on the procedure and the generality of early modern

1 There is a wealth of literature on historical reconstruction as a method. Two pieces of writing that inspired me to apply this methodology to my research are good ones to start.

watercolour practices, as well as the inevitable ahistorical challenges. The term is helpful to keep my questions (semi-)focused and my decision-makings clear, but is not definitive.

Similarly, my approach to historical recipes is, comparatively, more material-oriented than text-based. I do not always 'reconstruct' pigments by following one recipe book, but consult different sources for different colours or purchase ready-made natural pigments. Oftentimes, I would test a material without reading a lot of the existing literature. The natural materials are my primary sources during my remaking experiments, and working with them usually raises many questions. I then delve further into literature to learn if people in the early modern period had similar questions and if they have found answers or solutions. Or why if they did not encounter my problems or challenges when dealing with the materials. For these aforementioned reasons, this series does not represent the full spectrum of applying historical reconstruction/remaking as a research method. It shows the processes that are specific to my project, but with a potential to join larger debates in the future.

## 2 Time and Patience in Historical Remaking

Posted on 17 February 2020

How long is one minute? There is the objective answer of sixty seconds, but there is also the subjective personal perception. One minute can feel like a blink of an eye, but it can also last an eternity. When working in the lab, I often experience both perceptions within a workday. The days are always too short, and I regularly wish that I could live in the lab to work into the evenings. However, during each experiment, time often seems to have stopped moving. One experiment can appear to take forever, and I constantly have to prolong the allotted time for the tests. The temporal aspect of making is amplified, and patience is much required to conduct this slow research.

Taking time to manipulate a material is imperative to observe how matter changes its behaviour gradually. The quality of the product usually relates to the time spent on the process as well. Almost every experiment I have done on grinding pigments from various natural sources circles back to the issue of time, when the tools I use to grind them remain consistent. Minerals, such as lapis lazuli, take much longer to grind to fine particles due to their hardness, whereas earth pigments like ochres are less time-consuming because they are generally softer. These observations do not come as a surprise, but going through the motion of grinding makes experiencing time much

---

See Staubermann, ed., *Reconstructions*; Fors, Principe, and Sibum, 'From the Library to the Laboratory and Back Again,' 85–97.

more intense than reading about it in historical recipes. A recent experiment might be able to visibly demonstrate what time does to change a material.

It is known that over-grinding of certain minerals can result in the pigments losing their brilliance and turning the colours dull. However, how long it takes for a pigment to pass the threshold and become over-ground can vary drastically depending on the type of the mineral and the quantity of the pigment. I wanted to see how easy it is to over-grind malachite using the simple technology of a glass muller and a granite slab. To do so, I put on a batch of malachite powder and ground it with water for ninety minutes. The powder was prepared from a previous experiment, in which I simply broke the stone into small pieces, crushed the pieces into coarse powder, ground the powder with the muller and slab until it no longer produced a cracking and scratching sound, and removed impurities through levigation. I took a little pigment out of the batch every ten minutes to see how the sizes of the particles changed during this ninety-minute process.

Time-lapse is a fascinating film technique which condenses a long sequence into a short period of time. The video (Fig. 49) of me grinding malachite gives a deceptively brisk working pace that appears to be full of energy and excitement. In reality, the process was agonisingly long and taxing. Ten minutes into grinding, I was already losing my patience. The ten-minute marks never arrived fast enough, and there were still so many ten minutes to go. When grinding, there was no highly noticeable visual and tactile difference in the texture of the powder-water mixture. It was smooth and easy to be pushed around. However, the colour of the mixture gradually became ever-so-slightly paler in the first half hour of grinding, and it was most observable when placing a small sample of the mixture onto the plate at the end of each ten-minute mark (Fig. 50). The colour of the mixture stopped changing around the forty-minute mark, but I continued to complete the full ninety minutes of grinding.

I left the pigment samples on the plate to dry and called it a day in the lab. Two days later, I returned to the samples and bound them with gum water to turn them into watercolour paint. Each sample became a small swatch painted over a strip of black India ink to test its opacity (Fig. 51). Immediately, the powder that did not undergo further grinding (0 minute) shows that the particles are too coarse to make a good paint. The sample with ten minutes of grinding time shows an improvement in the paint quality, but still not good. The swatches painted with pigments ground for thirty minutes or longer are more similar, in their colours and the smoothness of the paint surfaces.

Despite the change in colour to a paler green, I do not think that I have over-ground the malachite. The darker greens have appealing colours, but the paints did not behave well. The higher grinding time resulted in finer particles, but more grinding did not further alter the colour of the pigments in a highly visible manner. Rather, pigments with higher grinding time can be more evenly mixed with gum water to make paints with a

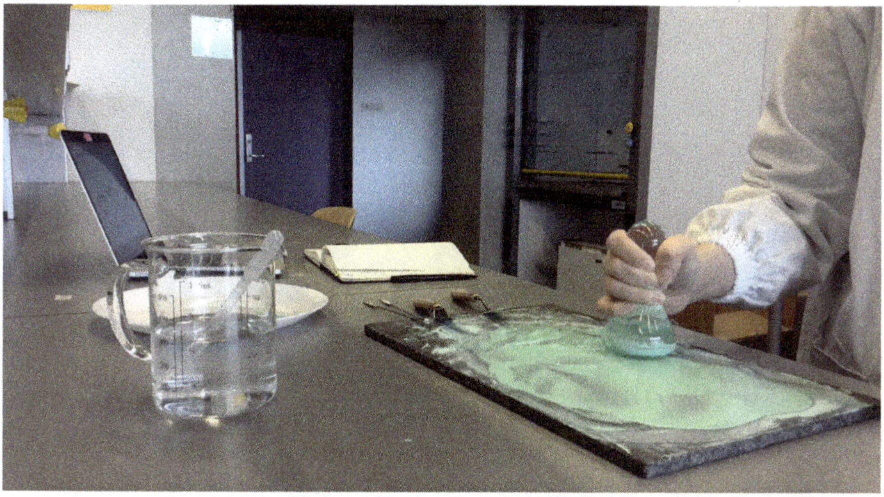

FIGURE 49 Time-lapse of the malachite grinding experiment. (Screenshot of the video that is not available for this archived version of the blog post series)

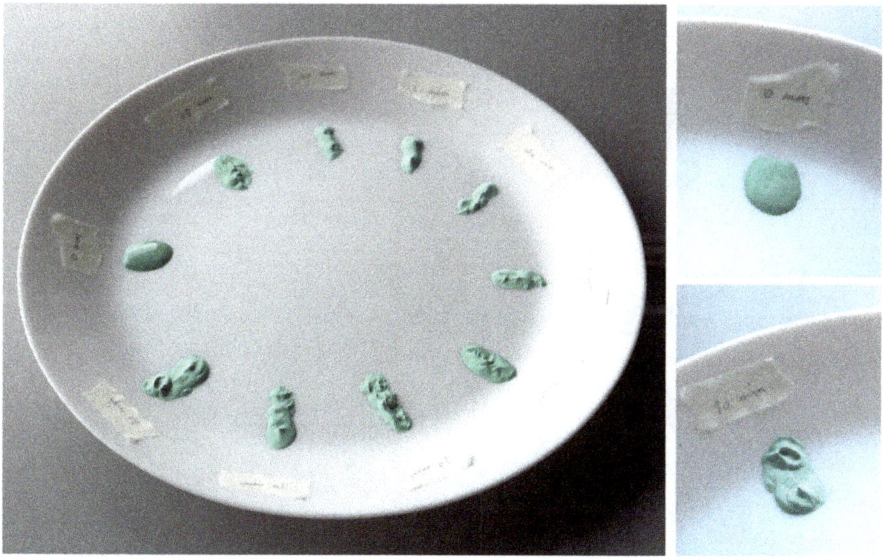

FIGURE 50 Samples of pigment ground for different lengths of time

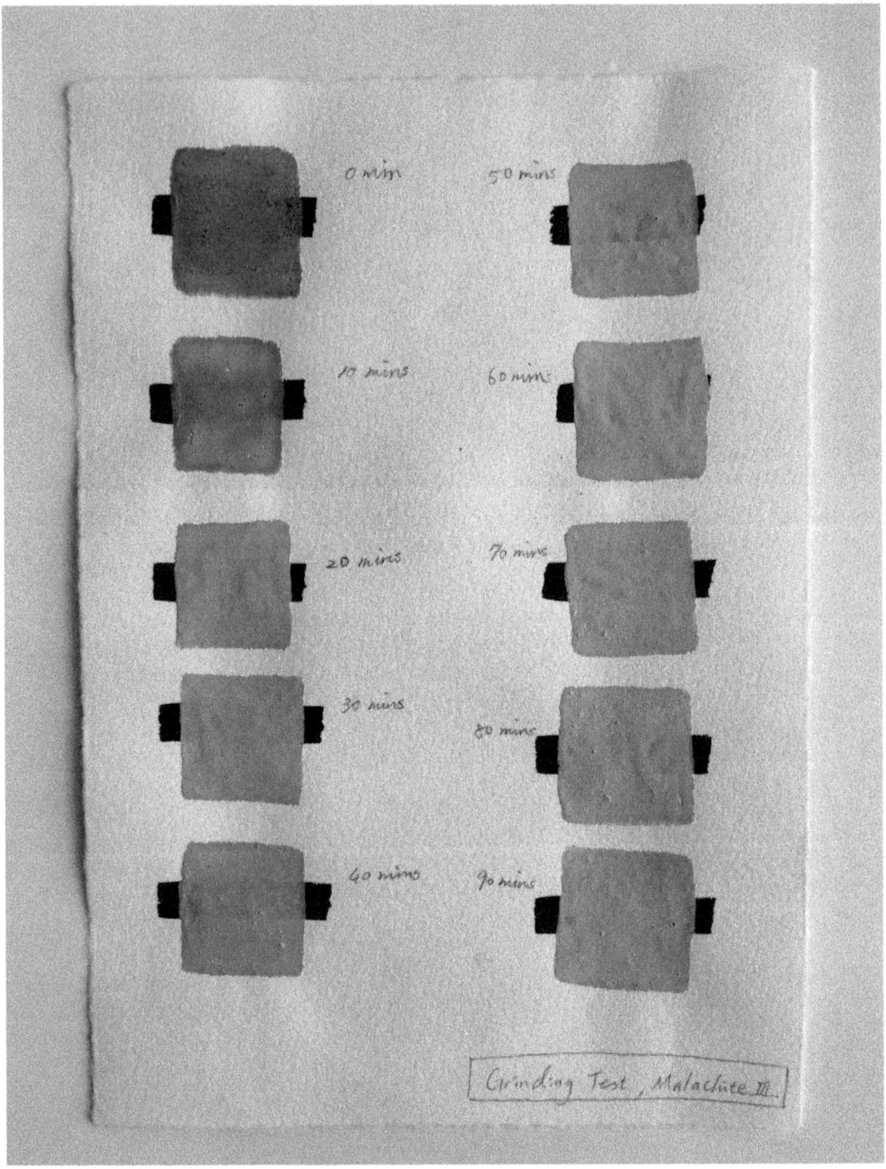


FIGURE 51 Swatches painted with watercolour paints made from pigment samples ground for different lengths of time

better flow. Time, in this experiment, is a critical element to produce usable pigments, and it needs to be accompanied by the patience of repeating the same motion for a minimum of thirty minutes.

## 3 Repetition and Boredom in Historical Remaking

Posted on 26 March 2020

As part of my research and a goal of my project, I have been building a watercolour reference set that consists of colour studies using the paints I made (Fig. 52). This will allow me to have some tangible references to compare the material qualities between the homemade paints and the ones found in historical flower books. It will also be useful for teaching workshops and classes in the future. To compile the set, I went through a selection of early modern treatises and listed the colours mentioned in these books. I then cross-referenced with the most authoritative secondary literature and got around fifty colours from seven colour groups—black, brown, blue, green, yellow, red, and white—that were most commonly used in early modern watercolour. For each colour, I did several tests with freshly made colours, and repeated the same tests with reactivated colours after they had dried.

Historically, artist supplies, including pigments and brushes, could have been purchased through suppliers like apothecaries, but artists would have to make their own paints. To do so, one usually takes a pigment and mix it with a binder. For example, to make yellow ochre for oil painting, the practitioner would take the pigment and mix it with linseed oil.[2] To make yellow ochre watercolour, one would take the same pigment, but bind it with gum water, which is a combination of water and gum arabic (and often also with a little bit of sugar or honey). The mixture is mulled with a muller on a slab for some time to ensure the pigment is evenly bound with the gum water. The historical application of this yellow ochre watercolour, however, is more similar to our modern understanding of gouache because it was often used as a thick and opaque paint, rather than giving the light aquarelle effect that most people would picture when hearing the word watercolour.[3]

The word *wash* would be closer to the modern-day aquarelle. For instance, the English treatise *A Book of Drawing, Limning, Washing*, printed by M. Simmons for Thomas Jenner in London in 1652 made this distinction.[4] The treatise refers to painting

2 The University of Delaware produced a brief and useful video to introduce this process, visit https://www.youtube.com/watch?v=Yu7olO1oqXw.

3 For more information on historical water-based medium, see Cohn, *Wash and Gouache*.

4 Jenner, *A Book of Drawing, Limning, Washing or Colouring of Maps and Prints*.

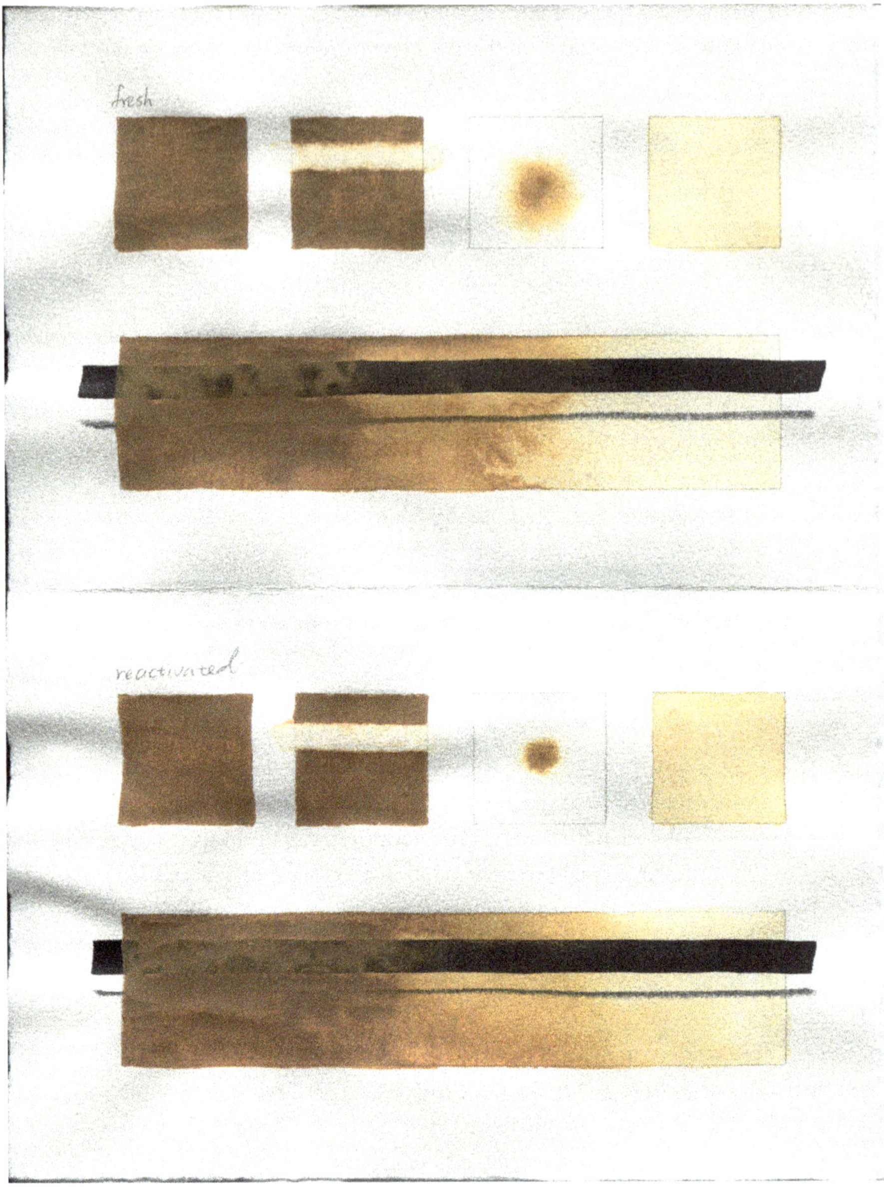


FIGURE 52 Colour study of yellow ochre to learn about the material properties of the paint through several tests

with translucent and watery colours as washing and opaque and thick colours as limning. Modern-day transparent watercolour achieves washing through watered-down paints, but the washes still contain finely ground pigments. Some historical washes, however, are more similar to dye, and they need to be used within a comparatively short amount of time because they cannot be dried and reactivated. In many cases, practitioners would have used both opaque paints and translucent washes in a watercolour. As a result, some of the colour studies on my list are washes that require different methods of making.

For the paints with pigments, I produced several pigments from scratch by grinding stones and making lakes (lake pigments will be the topic of another post in this series), and procured the rest from suppliers that sell quasi-historically-made pigments.[5] Normally, to get a good quality paint, a practitioner needs to find the right ratio for gum water and the selected pigment. Some colours require more gum water, whereas others need less. I decided to standardise the pigment and gum water ratio for all of my colour studies, as well as the mulling time, which means that some colours would be bad in quality.

Intentionally setting myself to fail achieves two things. First, it saves time because it will otherwise take multiple tries to find a good ratio for every colour, which would have drastically added to the research hours with more than fifty colours at hand. Second, failed and bad-quality colours are more useful for me as a researcher (although the practitioner inside me often screams in dismay during this process), as they show a wider spectrum of material properties and effects that the historical masterworks in museums and libraries do not. The paint-making process is rather straightforward. For each colour, I mixed 1 portion of pigment and about 1.5 portions of gum water, and mulled for fifteen minutes (Fig. 53). I did the tests with the fresh paints right off the slap and cleaned up the tools before moving onto the next colour. This set of motions took, on average, forty-five minutes, and was repeated close to fifty times.

Repeating this standardised process many times transformed my understanding of the pigments from abstract text that I read about in books to personal descriptions generated by my own experience. After a while, it became intuitive skills that my body could perform without too much thinking required. I started to pick up patterns and identify the potential issues of working with earth, mineral, metal-based, and lake pigments. Despite the subtle nuances, pigments made of similar materials—such as earth pigments that include ochres and umbers—behave more or less the same. Recognising my own accumulation of material knowledge through repetition was exciting, but the process was less so. For some, mulling paints is relaxing and therapeutic; for me, it was unfortunately boring.

5 All of my ready-made pigments come from either Kremer Pigmente or Verfmolen De Kat.

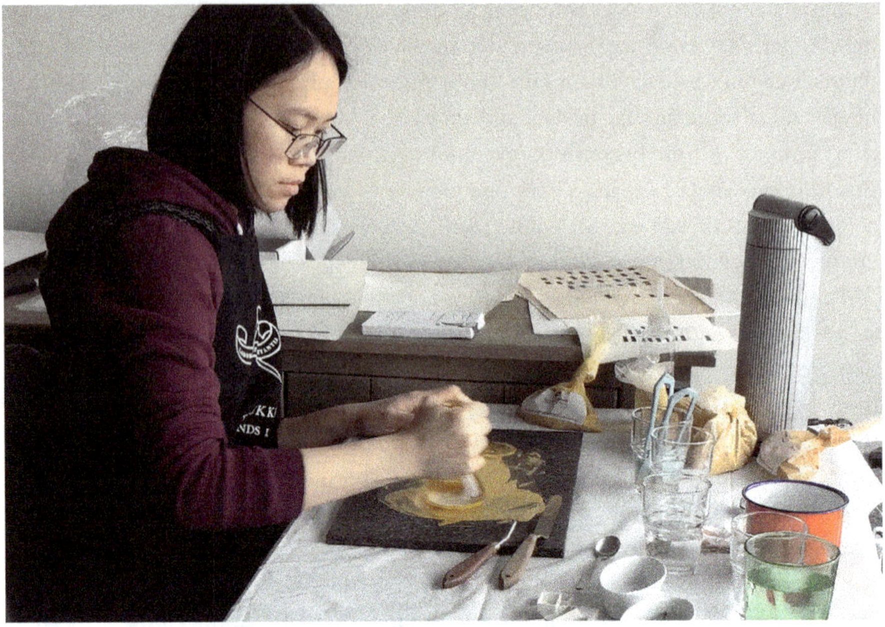

FIGURE 53 Mulling paint in home lab, showing only two minutes of the process. (Screenshot of the video that is not available for this archived version of the blog post series)

This boredom in the making process, however, is perhaps a more important feature to consider than excitement. Engaging in historical remaking as a research method is fun, but artistic practices are also hard work and come with a less glamorous side of things. On some days, it was difficult to even drag myself through the mulling motions. On other days, however, the boredom became productive. My muscle memory guided me through mulling paints, so my mind was free to be lost in my thoughts to raise more questions to be further researched. For example, I asked myself how apprentices dealt with boredom related to similar repetitive tasks in early modern workshops.

Moreover, I wondered how many artistic or technical innovations were results of people being bored with their old or current approaches of doing. Recalling my experience from art school, oftentimes when people master a technique, they do not simply repeat the same thing because it becomes taxing and boring. Instead, many people would explore a new material or technique. Finding new and different ways of making not only livens the process, but also offers a possibility to take one's artistic creation to the next level. Boredom is, in this sense, a prime motivator for innovation in artistic practices.

Returning to mulling the yellow ochre between my glass muller and stone slab for the reference set, this collection will never be complete if I want to be exhaustive, as

I cannot test every pigment that there is in the world. Nonetheless, after hours and hours of mulling paints, I am close to the finishing line of compiling the materials references that are essential for this project. With two binder-full of colour study sheets, I am curious to revisit some early modern flower books, hopefully in the near future, and find out what more I will notice with my newly accumulated and growing material knowledge of early modern watercolour.

## 4 Labour and Pain in Historical Remaking

Posted on 17 April 2020

> 'Oh my, that's heavy.' This thought comes up every time I pick up my granite slab for grinding pigments and mulling paints.
>
> 'Why won't this malachite break?' I asked myself impatiently when breaking the stone into small pieces.
>
> 'My fingers hurt.' I complained after mixing colours for hours.

The three examples of my honest responses demonstrate some of my physical reactions during the process when conducting experiments. An important aspect of working with historical remaking as a research method is to understand the level of labour and the performance of one's body in different situations. For example, when printing an image with a hand press, I had to put my entire body to work in order to pull the heavy handle that lowers the platen to get an impression (Fig. 54). Operating a printing press is hard labour, and it manifests in the size of the mechanical device. Similarly, one easily notices the labour that goes into blowing glass and casting metal because much attention is paid on the movement of the body in order to work with the tools and materials, as well as to keep the maker safe.

Comparatively, working on colour making and painting is much less physically demanding. What kind of labour is involved when making and painting with watercolour? This was a question I had never given any thought when studying in an art school. Back then, the working process was—while not mindless—an intuitive act. The attention was usually directed to obtaining a satisfying result, such as a desired visual effect. Discussions and critiques centred around the concepts and techniques of our works, but not the labour behind the pieces. Returning to watercolour as a researcher, however, I started to analyse decisions and actions that used to be performed naturally without too much thinking. I realised that labour, which often is accompanied by pain, was one obvious element that I overlooked as a practitioner.

Because I work in a lab, instead of a studio or a workshop, the space needs to be cleaned and all the materials and tools need to be put away by the end of a work day.

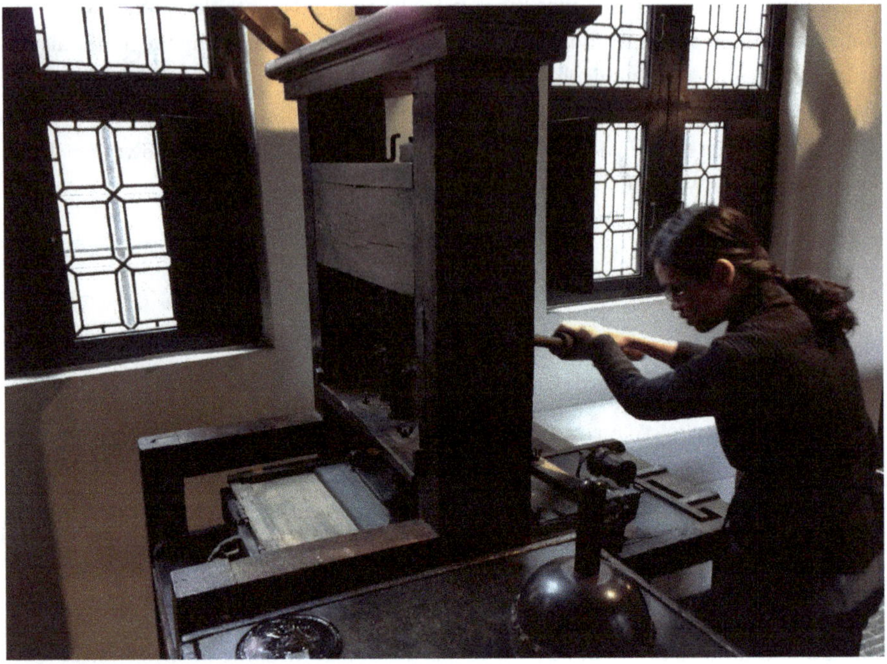

FIGURE 54 Operating a hand press to print a woodcut

Setting up my paint mulling station and transporting the granite slab in and out of storage became a routine (Fig. 55). This step may be ahistorical, as a proper workshop might have a designated space for mulling paint. However, even when such a space is available, there is always the preparation of materials and the cleaning of tools before and after paint mulling. The weighty slab magnifies the labour in the preparing stage before the actual work of paint mulling can start.

Sometimes, excessive labour is caused by ineffective procedures. To produce a *green pigment with malachite*, the first step was to break the stone into small pieces before they could be crushed into powder (Fig. 56). On the first day of this experiment, I did not have a hammer, and all I could find in the lab that could serve as an alternative was a metal rod. To procure a proper tool meant to lose several hours in the lab, so I made the decision to bash the stone with the rod. This became a typical scenario of working harder but not smarter, as I most likely spent as much time, but more physical labour, to break the stone into little chunks, and calluses developed on my hand due to the feeble tool that was not designed for this purpose. The calluses are a signpost that historical recipes sometimes advise specific tools or devices for certain tasks for a reason.

Calluses are unpleasant, but the pain induced by overworking would be the hidden danger of seemingly easy labour. On the surface, painting with watercolour does not seem to require much effort. It is a clean and neat practice, and the needed tools are lightweight. Nevertheless, holding a brush puts tension on fingers, and working for

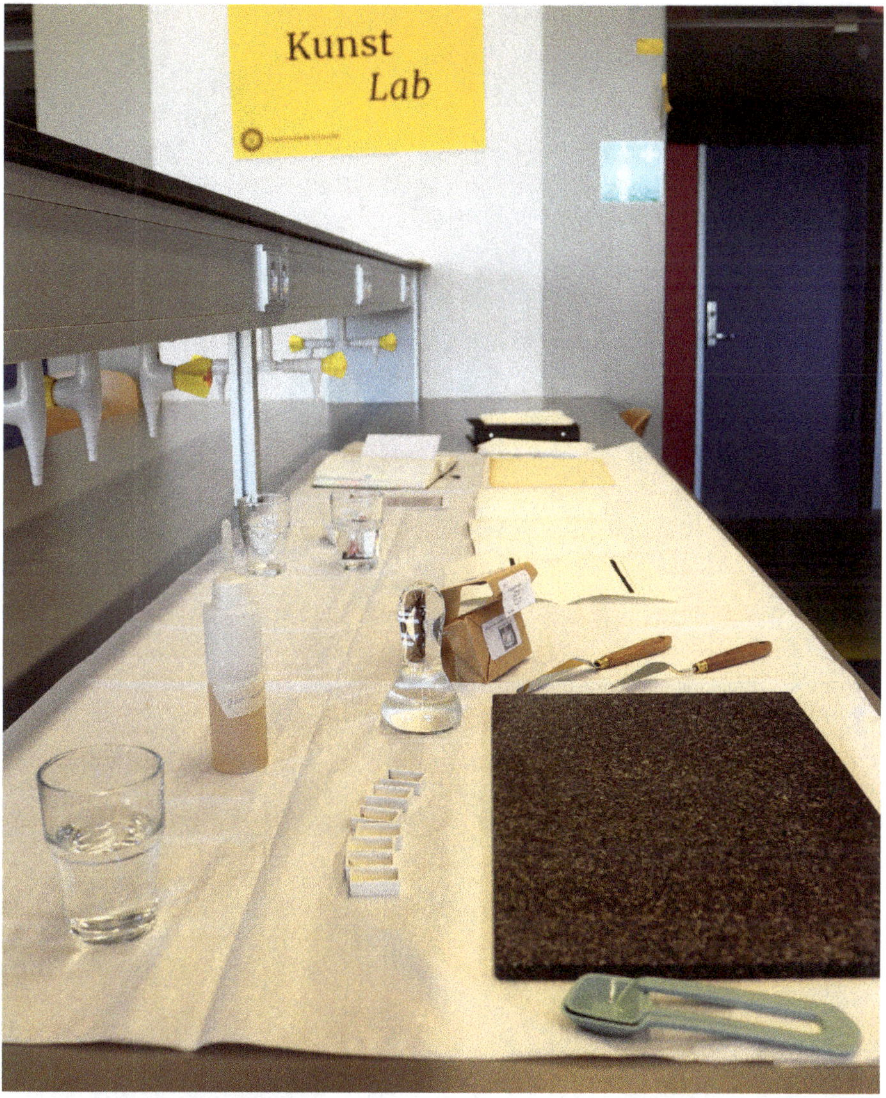


FIGURE 55 Paint mulling station in the Kunstlab, with the grey granite slab in the foreground of the picture

long hours could lead to pulled muscles. One of my latest experiments was systematic colour mixing (Fig. 57). To mix the 1,326 colours from the selected fifty-two colours for my project with each other took me ten days and a minimum of eighty hours to paint a total of 3,978 dots. This intensive use of my fingers in a relatively short time caused some mild pain in my fingers.

These physical discomforts from remaking experiments are familiar, as they have always been part of the making process, only now they have been brought to the

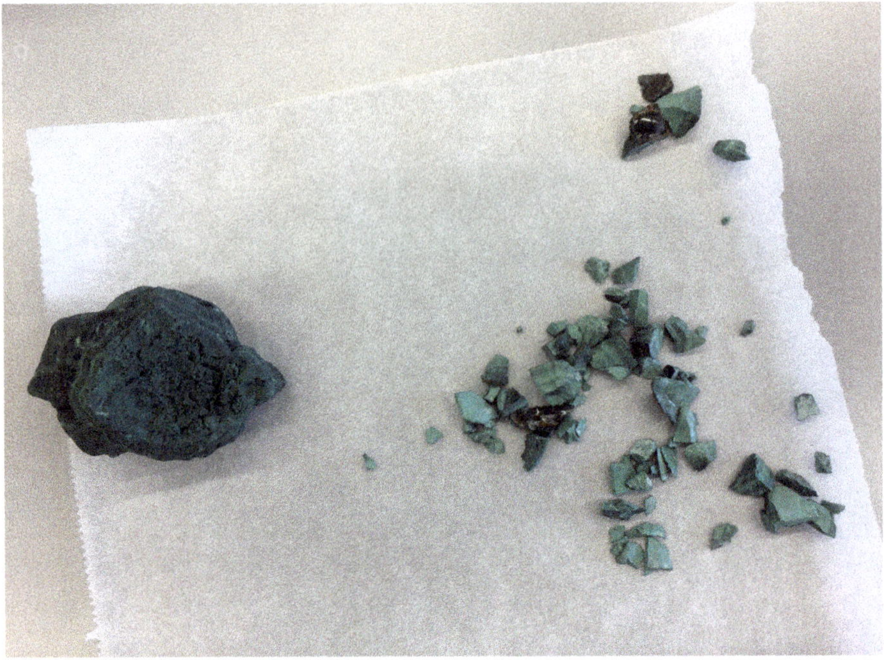

FIGURE 56 Breaking malachite into small piece

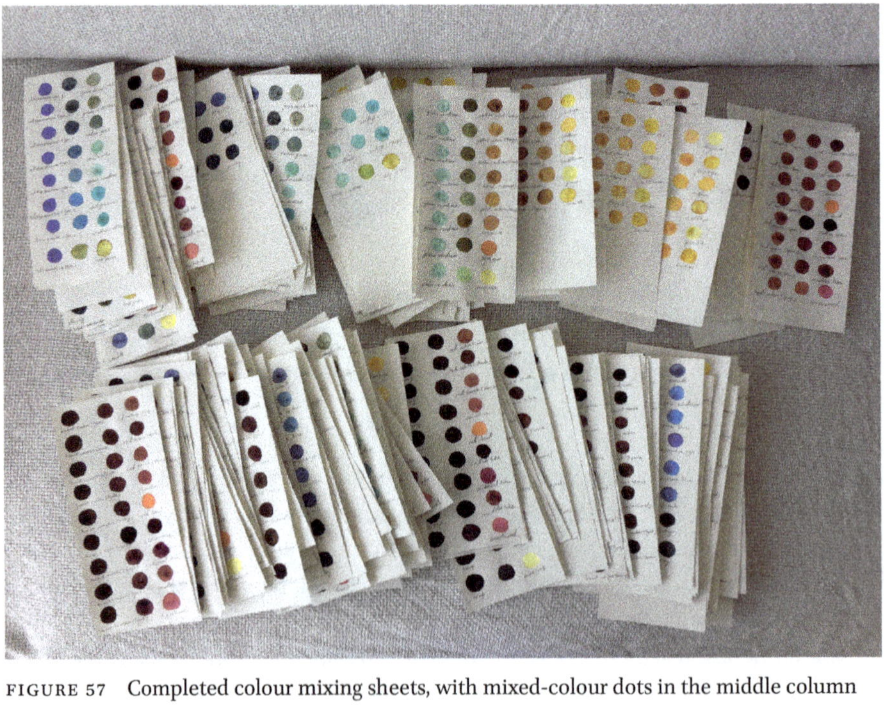

FIGURE 57 Completed colour mixing sheets, with mixed-colour dots in the middle column

foreground. I recall that there were two weeks that I could not hold a pencil after a big project; stories of students suffering the same pain or even seeking medical help due to overuse of hands and fingers are also common. Safety measures to prevent burns and cuts that could be potentially life-threatening would be one of the first things to mention when learning a new technique. I am always alert and careful when working with knives to avoid turning my artistic endeavours into lessons of anatomy. Yet, the pain and injury from accumulated small labour did not seen to find its way into discussions in classrooms, nor in historical recipes. With my new awareness of labour and pain, perhaps possible descriptions that deal with similar experiences would become more identifiable when revisiting historical sources.

## 5 Senses and Skill in Historical Remaking

Posted on 30 June 2020

*This blog post was first published on website of the Microscopic Records under a slightly different title and format on 29 June 2020.

Brazilwood, buckthorn berries, logwood, madder, and weld. Between the months of February and April, my major research activity was boiling these plants. Why? I experimented with a number of recipes for making ink and lake pigments in order to explore the materiality of colours in the early modern period on a small scale. One of them, taken from a mid-seventeenth-century recipe collection, reads as follows:

> How to make Red Inke. Take [Brazill] and seeth it in Vinegar and small beer, in an earthen pot, and put a little Allum therein, untill you taste it to be strong on your tongue, and when it is boyled, straine the [Brazill] through a clout, cleane from the water, and let the water stand and cool for your use.[6]

Generally speaking, the first steps in making ink and lake pigment are similar. Immerse a plant in some amount of liquid, usually water and/or vinegar. Heat the mixture for some time to extract colourant from the plant. Add alum in different quantities as needed. Ink requires only a small bit of alum to add some thickness to its consistency. Lake pigment, on the other hand, requires a larger amount because the dyed alum is

6 Recipe from Jenner, *A Book of Drawing, Limning, Washing or Colouring of Maps and Prints*, 30. The given recipe is the description for 'How to make a Purple Colour' with logwood. The same instruction applies to making red ink, as the recipe for the red ink does not provide all the steps. Instead, it states 'Boyle Brazill as you doe the Log-wood [...]'.

the main substance that constitutes the pigment.[7] The red ink recipe describes many of such generic steps and features.

### 5.1 *Senses in Text*

This recipe is particularly interesting because it specifically mentions the sense of taste. Among all the senses, taste was one that I ignored when creating pieces during my art school years. Working primarily with modern art supplies of all kinds, we were instructed NOT to put anything in our mouths because most materials, while not necessarily highly poisonous, are hazardous to the human body when consumed. The red ink recipe challenged my own understanding of the making process in a confusing way. What does it actually mean to 'taste it to be strong on your tongue?' How strong is 'strong?' What kind of taste would it be? Sour? Bitter? The listed ingredients (Fig. 58) are either edible or relatively safe to ingest in minute quantities. Nonetheless, remaining sceptical and not daring enough to risk my life in the worst-case scenario, I forewent the step of tasting the mixture for both attempts of this recipe.

### 5.2 *Senses during Making*

Missing the sense of taste did not consequently compromise the experiments. Taste was not the only sense involved in this process. Even though other senses are not mentioned in the text, it became clear that they all played their parts to produce the red ink, and by extension ink and lake pigment in general. Sight, unsurprisingly, was a crucial sense that participated in most of the process. The eyes saw the changes of the colour and the consistency of the solution (Fig. 59). They checked the marks on the thermometer and looked out for overflow while boiling. Hearing was attentive to the bubbling sound of the solution. The ears informed me when to start to keep an eye on the decreasing amount of liquid remaining in the beaker (Fig. 60). Touch was alert to the heat to prevent me from scalding my fingers on the hot beaker. Smell did not really have a critical role in this case. However, the aroma of the brazilwood, vinegar, and beer mixture was so overwhelmingly present that it became the most memorable aspect of this making experience.

### 5.3 *Building Skill through the Senses?*

By conducting several experiments involving such and similar ingredients and procedures, I became familiar with the properties of the materials and the typical steps in the making process. After doing it several times, I noticed that how I worked with recipes changed unintentionally. During the second attempt of the red ink recipe, text dissolved into action, and taking notes required more effort because it became

7 Kirby, Van Bommel, and Verhecken, *Natural Colorants for Dyeing and Lake Pigments*, 69.

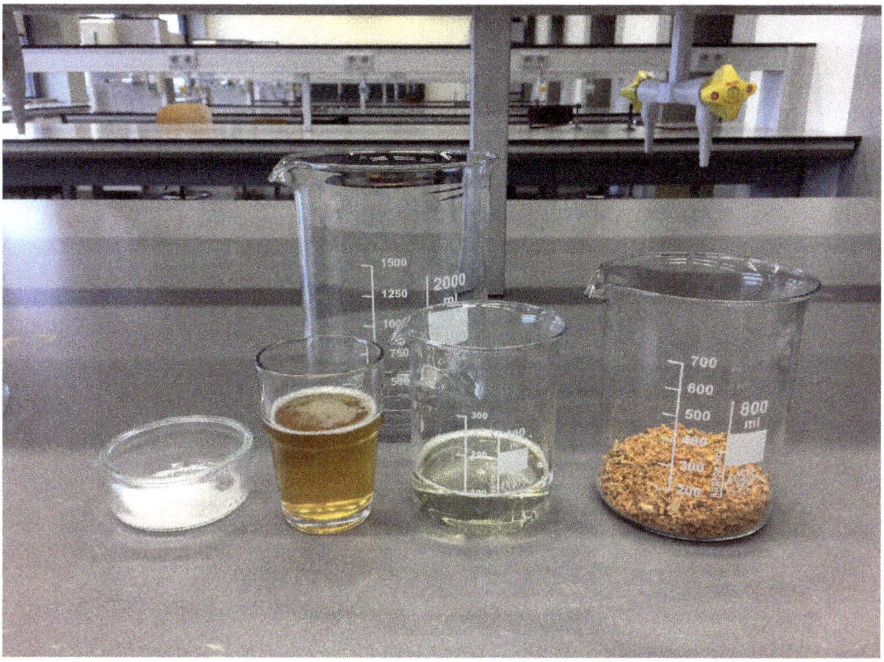


FIGURE 58 Ingredients for the first attempt of the experiment (left to right): alum, beer, vinegar, Brazilwood shaving

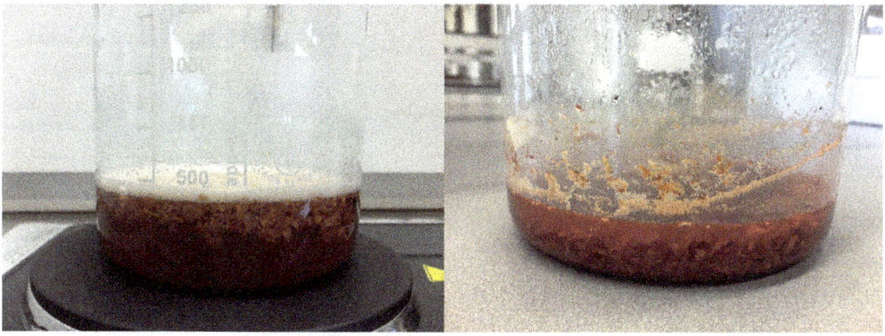

FIGURE 59 The colour of the mixture of the first attempt before (left) and after (right) heating

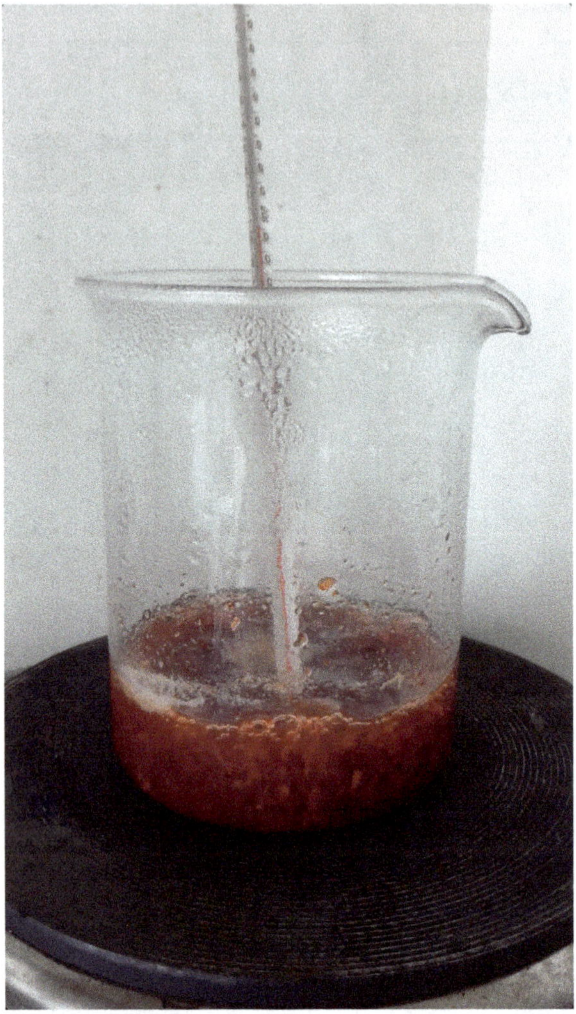

FIGURE 60
This video illustrates the role of the sound of the heating hot plate and the bubbling solution during the boiling process. (Screenshot of the video that is not available for this archived version of the blog post series)

counterintuitive. Instead of checking the text at every step, I simply reacted to what was happening in front of me on the burner. The sensorial experience from the previous ink- and lake-pigment-making experiments provided me with a frame of reference to more precisely expect how the materials would behave and transform; I had become fluent in the process of making this ink. Not textual instructions, but sensorial engagements informed me about what (not) to do.

Is this fluency what could be called *skill*? What exactly is skill?[8] Having a higher level of familiarity with the materials and the making process does not mean that

8 Many scholars have discussed senses and skill in different fields. This blog post does not so much engage with the literature since this series focuses on my own observations and

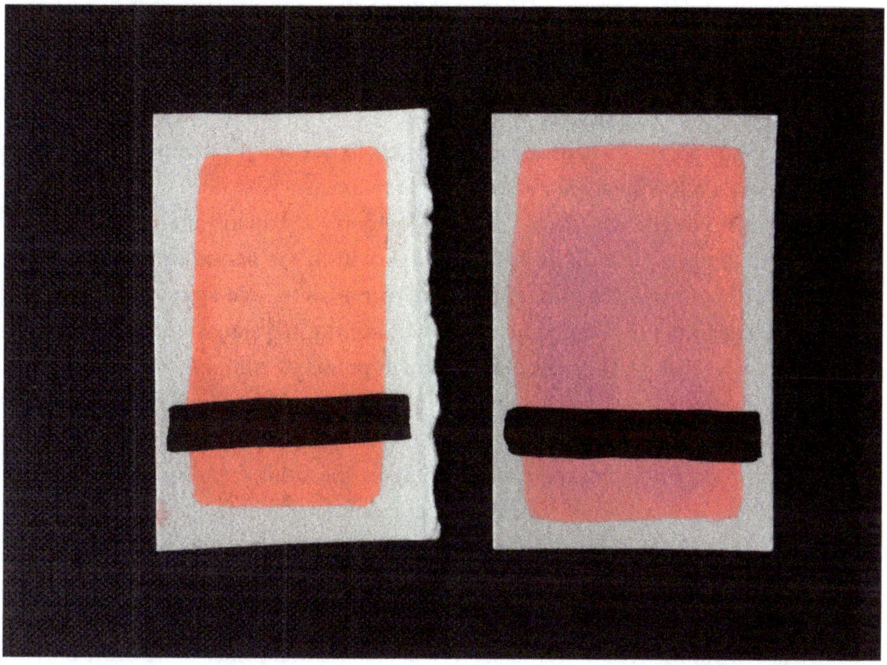

FIGURE 61 Swatches from the first (left) and second (right) attempts of making the red ink. Both had a syrupy surface that did not dry, which was then treated by lifting the sticky substance off with a wet brush to allow the swatches to be stored

I can then produce the ink without failing. My first attempt of the recipe gave me a sticky red ink that did not dry. By changing the ratio of the ingredients, my second attempt was less sticky, but the colour was too light to be useful (Fig. 61). The product did not show an improvement of my ink-making skill. However, I was able to identify the problems and reasons behind this failure much more quickly. With more attempts, eventually, I will be able to find the right measurement for a result that reflects an increase of the skill level. Perhaps, what skill means in making ink and lake pigment might not be the visible quality of the final product, but the awareness of the senses in performing the actions and identifying the issues. For me, remaking the ink and lake pigment recipes shed light on the simultaneous sensorial engagements in transforming these small-scale plant materials into art materials. By training my senses to be skilful in manipulating these microscopic records, the materiality of colours in the early modern period became more than simply terms and visual samples, but a holistic experience of the body and the mind working together in the making process.

---

questions from the experiments and the making processes. However, one book that has been serving as one of my conceptual frameworks in exploring senses and skill that needs to be mentioned is Ingold, *Making*.

## 6 Space and Resources in Historical Remaking

Posted on 2 August 2020

Among all the aspects of historical remaking, the space for conducting experiments may be the most ahistorical one. The recipes to reconstruct are directly from the past. Even though the interpretation of recipes is modern, historians and researchers are trained to carefully contextualise historical text within the period in which a recipe was written or recorded. The materials used for remaking—in spite of the fact that they are often grown, produced, or processed differently nowadays—usually have similar properties as historical ones to offer enough points of comparison. For example, despite the subtleties in shades, madder root is still a source of red dye; it does not become a yellow dyestuff instead.

The workspace for remaking experiments, on the other hand, usually does not resemble a historical studio or workshop in its appearance. This is particularly obvious when working in the Kunstlab of the university (Fig. 62). The lab is entirely modern, equipped with fluorescent light fixtures and laminate countertops. Nothing in this lab bears semblance to an early modern workspace for artistic practices. However, this ahistorical environment is useful to serve as a constant reminder to be critical in my remaking experiences. As a historian in the twenty-first century, there are certain facets of the past that I can never fully relive. Keeping a distance from the past through

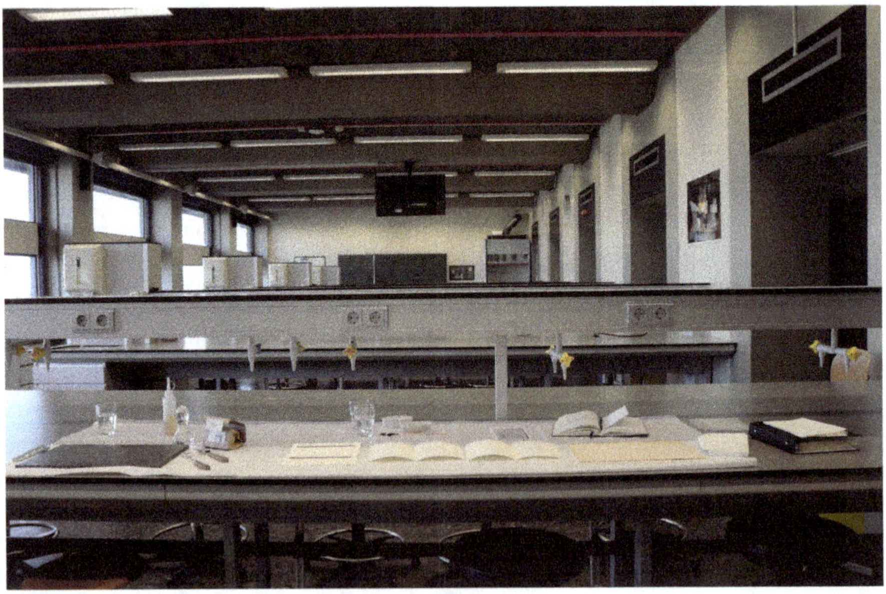

FIGURE 62 The Kunstlab at the Utrecht University

the setup of the Kunstlab is thus an effective way to prevent over-interpretation of my remaking experiments.[9]

A few months into routinely experimenting in the lab came the semi-lockdown in the Netherlands due to the Covid-19. Because of 'working from home,' or 'living at work,' which became the reality for many people, an awareness of how a (work)space affects our productivity and ways of working, sometimes drastically, is alive and well in our current collective memories. The condition of my workspace changed as expected. No longer did I have a big surface to spread my items and a large and quiet space all to myself. However, no longer did I have to bike for forty minutes to get to the lab and spend an hour minimum for preparation and cleaning. Where I lost the luxury of extensive workspace, I gained working time.

In a way, converting my shared living room into a home lab became a transhistorical experience of working in a maker's studio or workshop. Historical studios/workshops were (and still are) often part of the building in which a maker and their family lived, and they were usually repurposed space with many figures working in it simultaneously. The drawer and the folding table in my living room was transformed into a workstation, and reconfigured a few times depending on the needs of the experiments (Fig. 63). When my roommates and I were in the living room at the same time, we choreographed our movements to avoid getting into one another's way. Noises from my roommates' movements (and at times unpleasantly loud music from the neighbours) were distracting, but their occasional curiosity in my projects resulted in some unexpected new ideas to try something different in the follow-up experiments.

The most striking realisation, however, is perhaps the simple fact that I have a space which allows me to continue my hands-on research at home. A functional workspace,

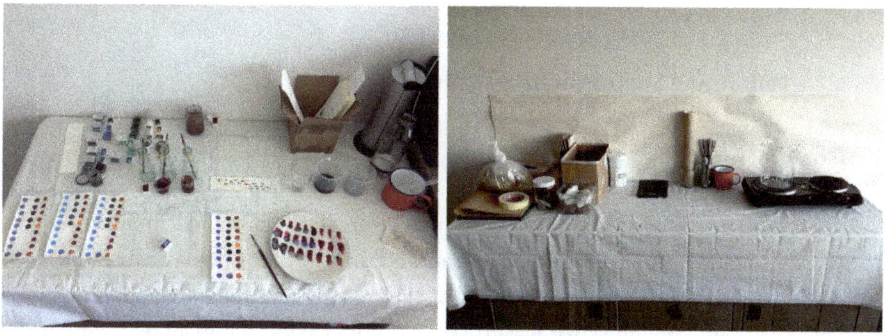

FIGURE 63 Left: temporary painting station in the home lab; right: temporary lake pigment making and textile dyeing station in the home lab

9 This line of thinking is heavily inspired by conversations with Thijs Hagendijk, who has briefly discussed it in his article 'Learning a Craft from Books,' 198–235.

albeit small, is a valuable resource. Even only a little more than a year ago, my living situation would not have permitted me to have a simple home lab, and this aspect of my research would have stopped until further notice. In fact, as one sometimes forgets, a project usually depends heavily on the spatial and material resources, which are often conditioned by the financial resources that are available to a researcher.

Archives, libraries, and museums supply consultation space and research materials. Even though diligently digitised collections have made it possible for one to work in more types of situations, they remove most of the material aspects of the objects that are instrumental to answer some research questions. In addition to workspace, equipment and supplies in labs, usually procured through project funds, enable researchers to test and experiment with different kinds of materials. Without access to try several combinations, the results could be limited. For my own research, the possibility of working with eight types of blue pigment, instead of only one or two, makes a substantial difference. In order to do so, especially with the pricier materials such as mineral pigments, I have to rely on the institutions that financially support my project.

This reminds me that much of artistic production is materialised through patronage in the broadest sense, of people commissioning paintings, sculptures, and buildings, and of people subscribing to books, to name a few. We are now seeing our modern form of patronage, with platforms like Patreon that help artists and creators find alternative financial resources to keep making things. Perhaps, we have never really been that far from the past, as Covid-19 and many other recent events have shown that we continue to repeat history (unfortunately way too often the terrible side of it), only in a different shape and form. The reliance on—and the gratitude to the privilege of having—space and resources to make things is not only a challenge I alone have faced, but an issue that might speak to many makers across time and place.

## 7 Failure and Success in Historical Remaking

Posted on 29 November 2020

The first lesson I learned when I began the historical remaking research of my project was embracing the mistakes, errors, and failures in the process and result. They have been an expected and integral part of the making and remaking process in the past year. Despite knowing so, it was frustrating nonetheless when an experiment did not lead to a successful pigment, wash, or painting. However, I find myself not only welcoming failed experiments, but often valuing them more than successful ones as I progress over time.

## 7.1 *Defining Success in Artistic Production*

What do failure and success mean within the context of artistic production, then? Of course, the definitions vary greatly depending on a maker's goal and intention. Broadly speaking, however, one can objectively (and sometimes subjectively) assess and determine if an artwork is successful from its use of formal elements and level of technical execution, as well as its effectiveness in communicating visual information and/or emotion. On the material end, a 'successful' pigment or other kinds of art supply can generally refer to functioning and high-quality products, usually with a greater level of permanence.

When conducting historical reconstruction or remaking experiments, I often apply similar criteria to assess the results. For example, I successfully made a deep and saturated red lake pigment with dye extracted from cochineal insects (Fig. 64). In the same week, my experiment with madder red (Fig. 65) failed because the process hardly

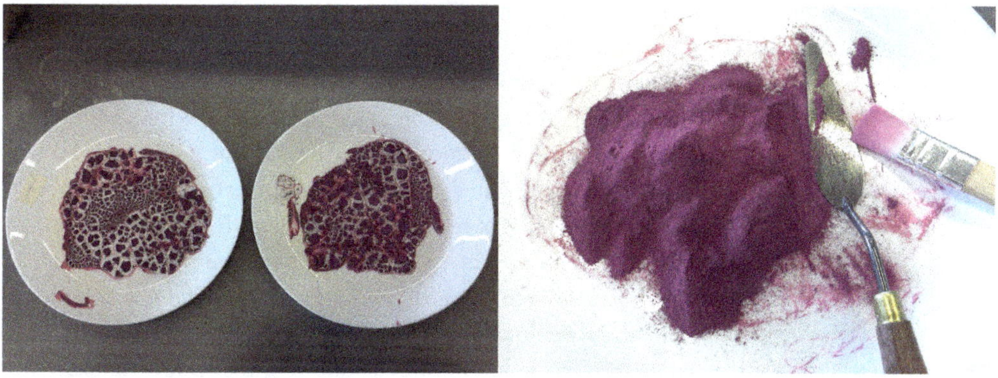

FIGURE 64 Result of the cochineal lake pigment

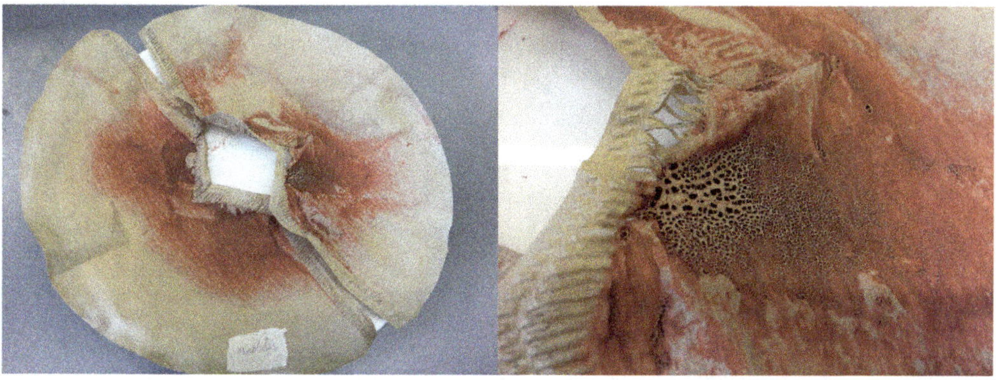

FIGURE 65 Result of the madder lake pigment

yielded any pigment; the quantity was too small for the colour to be useful for painting, and the recipe is supposed to produce a proper amount of pigment.

### 7.2 *Making a* Pastill

My greatest failure among all the experiments is perhaps the reconstruction of the *pastill* to extract ultramarine pigment from lapis lazuli stones. There are several historical recipes for this process, including one from Cennino Cenini's *Il Libro dell' Arte* (*c.*1400).[10] The art supplier Master Pigments tested this recipe with a successful result. This demonstration became useful for me to expect what a functioning *pastill* should do and have a concrete point of comparison.

For my own experiment, I followed the recipe from the *Miniatura; or, The art of limning* (1627/1628, 1648).[11] This book is one of the major sources for understanding the materials and techniques of early modern watercolour and the time period is contemporaneous to the botanical watercolours of my investigation. Norgate's recipe calls for a few more ingredients and has rather different descriptions for the steps, but the overall process is comparable to the one of Cenini's recipe.

The experiment took me two days to complete (Fig. 66). The first day involved solely breaking the stone. I first crushed a piece of lapis lazuli into chunks with a hammer, and continued breaking them into small pieces with a mortar and pestle. There are ways, both described in Norgate and by modern practitioners, to weaken the stone by heating it, but I could not do so because of the safety measures with which I was working.[12] With brute force, I was only able to get 22g of crushed stones out of the big piece, weighted 244g. I took the coarse powder to the granite slate and ground it with water under the glass muller for about an hour, and left it to dry overnight.

On the next day, I collected the ground powder. Following the recipe, I melted 6 parts of beeswax before adding—in this order—4 parts of colophony, 2 parts of mastic, 3 parts of linseed oil, and 5 parts of turpentine. Two ingredients, *Rosen* and *Razo de pino grasso* were eliminated from this experiment because they were not available to me. Finally, I added the lapis lazuli powder when the mixture was fully melted. The result of the *pastill* was far from ideal. The dough was excessively sticky; it never fully dried, and thus not possible to knead and extract the pigment. Two days of work and the cost of the materials gone, and I still had no ultramarine blue for painting.

10 Cennini, *The Craftsman's Handbook*. Transcribed text can be found here: http://www.note-access.com/Texts/Cennini/2a.htm.

11 Norgate, *Miniatura*, 68–69.

12 Norgate, *Miniatura*, 67.

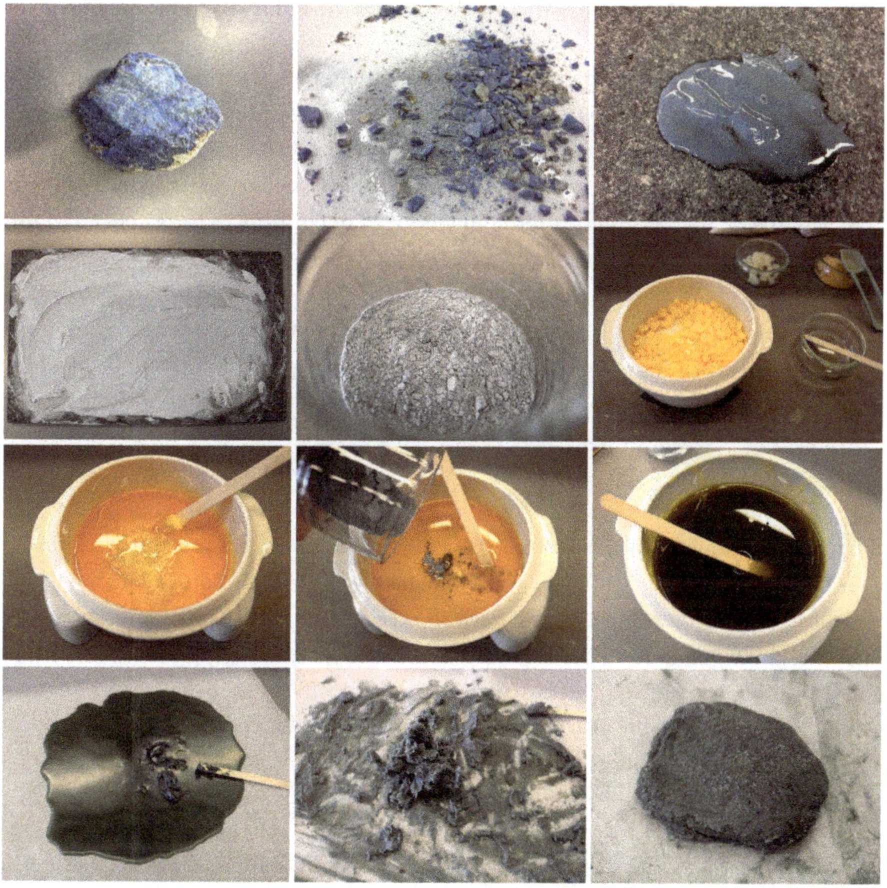

FIGURE 66 Photo sequence of the experiment process. From left to right and top to bottom: 1. lapis lazuli stone; 2. crushed stone; 3. grinding crushed stone into powder; 4. ground and dried lapis lazuli powder; 5. lapis lazuli powder; 6. melting beeswax; 7. adding colophony and mastic; 8. adding lapis lazuli powder; 9. all ingredients added; 10. cooling the mixture; 11. mixture too sticky and wet to knead; 12. failed *pastill*

## 7.3 *Shifting Perspectives*

Looking back, it became clear that the experiment was destined to fail. The stone I purchased has a high content of impurity; the ground powder was nowhere fine enough; there were missing ingredients in the *pastill*; the list goes on. Learning from this failure, I can expect that a future attempt might have a higher possibility of success after identifying and correcting the problems. It should be pointed out, however, that this evaluation of failure and success comes from the perspective of a supposed practitioner, with the eventual goal of the experiment a usable (if not quality) ultramarine blue pigment.

When reflecting on the experiment through the lens of a researcher, the value of this failed attempt became more prominent. It connected many dots that were previously implicit from reading historical text and secondary literature. It raised many questions, several of which align with existing discussions by researchers of early modern recipes. It sparked new possibilities, either hypotheses or curiosity to be tested. From this perspective, the *pastill* experiment was among the most 'successful' in engaging with different kinds of thinking. Historical reconstruction and remaking is particularly useful in the sense that multiple perspectives can be explored and experienced simultaneously. Consequently, the 'data' gathered throughout this process needs to be carefully interpreted to avoid conflation of these perspectives, but that should be done for any type of source.

### 7.4 *Looking to Fail*

Once I have trained myself to better distinguish these perspectives, the level of success in the end result became less relevant. Successful results of quality pigments are welcomed for further painting tests, while failed experiments provide other useful information. Going one step further, I started setting myself up to fail by not always following the 'proper' steps as instructed. The curiosity to see what would happen gradually outweighed the yearning for successful end products, but that is the subject for the next post to come.

## 8 Fun and Curiosity in Historical Remaking

Posted on 15 December 2020

Throughout my PhD project thus far, and particularly since I started the historical remaking aspect of my research, I have constantly asked myself, 'Am I having too much fun?' Often, the hands-on research does not appear to be work. I frequently experiment into late evenings because it is too exciting to stop exploring the material potentials of historically-oriented pigments and dyestuffs, despite the boredom, pain, and frustration that occasionally accompanies the processes. I have wondered if all this is children's play instead of serious research, as most of my experiments do not intend to answer specific questions. However, why can research not be fun and serious simultaneously? Why, generally speaking, do we tend to separate fun explorations from hard work?

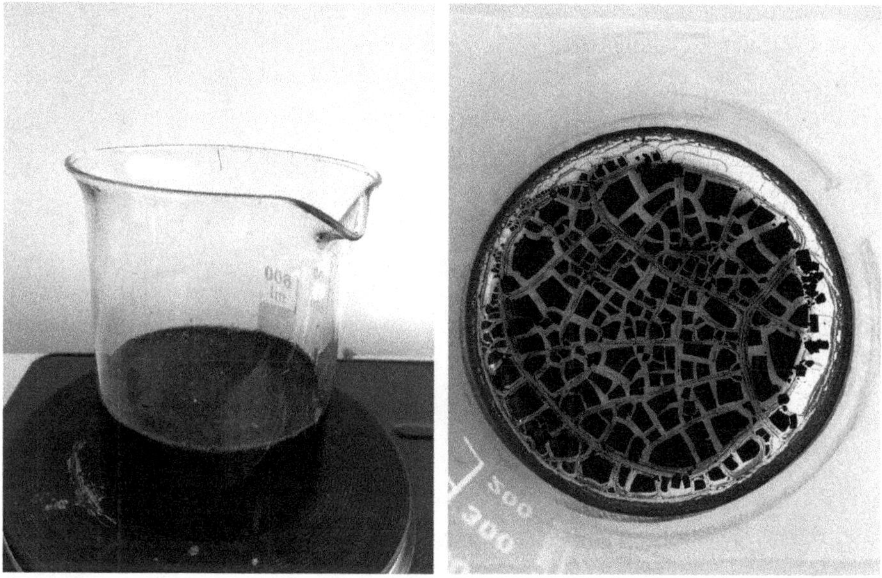

FIGURE 67 Logwood red dye (left) and deep purple lake pigment (right)

Logwood is one of the materials that has captivated me. The plant was commonly used for textile dyeing in the early modern period, but much less so for painting.[13] Similarly, the first time I encountered this plant material was during a workshop for reconstructing late medieval and early modern black dyeing technologies, for logwood red was the base colour for one of the black dye recipes. Compared to other dyestuffs, such as Brazilwood, logwood appears much less frequently in recipe books or manuals for historical watercolour. The main thing I was most curious about, however, is that blue dye can be obtained by altering its PH level. Can this dye make a blue lake pigment? To see what happens, I purchased some logwood shavings and substituted the Brazilwood in a recipe for making a red lake pigment with the same amount of logwood. The result was a deep purple pigment (Fig. 67).

Given that *blue* can be a vague term throughout history that sometimes covers colours in the violet spectrum, it was not entirely 'out of the blue' to obtain a purple pigment. What was interesting, however, is that the excess dye was still so saturated that it was difficult to spot the pigment (Fig. 68). Usually, after adding the substrate (mostly alum, but sometimes chalk or other powdered materials) to the dye and letting the

13 Harley, *Artists' Pigments c. 1600–1835*, 59–60; Eastaugh et al., *Pigment Compendium*, 248.

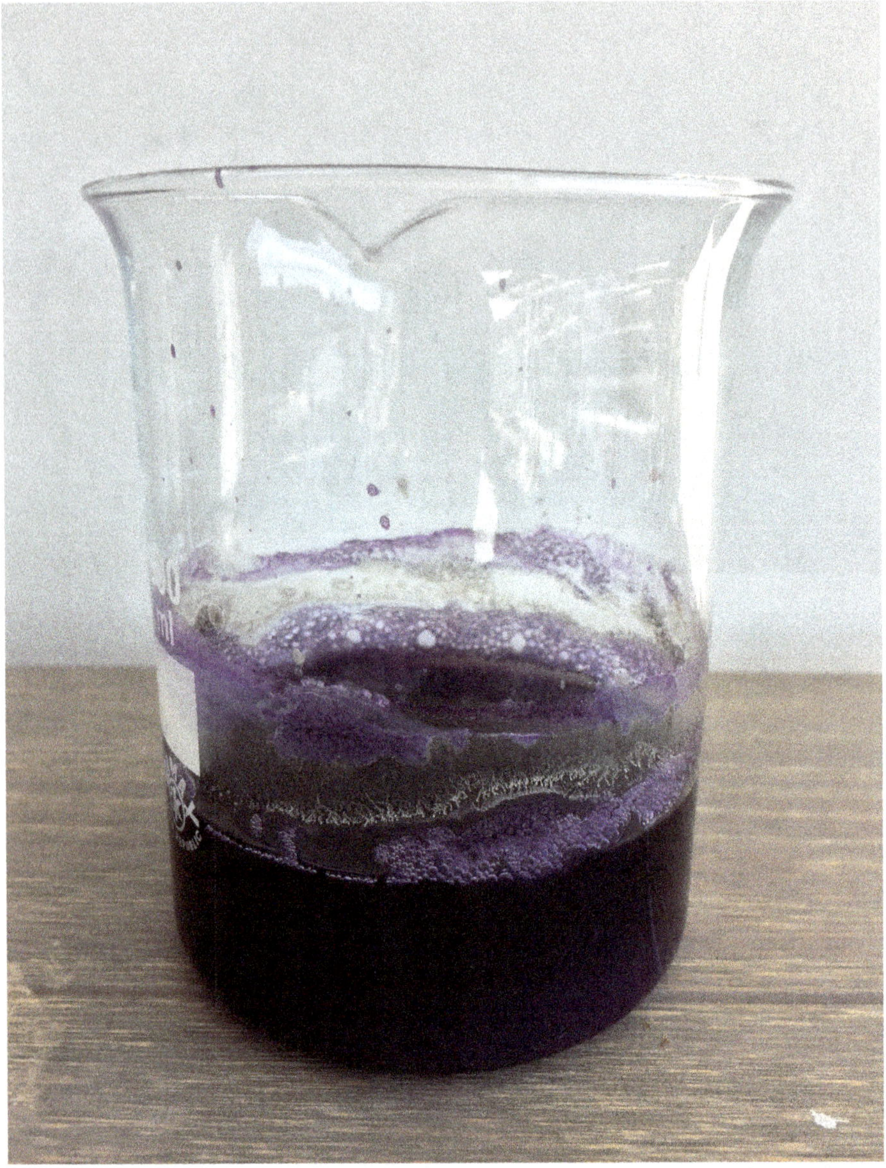

FIGURE 68 Logwood lake pigment with highly saturated excess dye that the particles are not easily seen

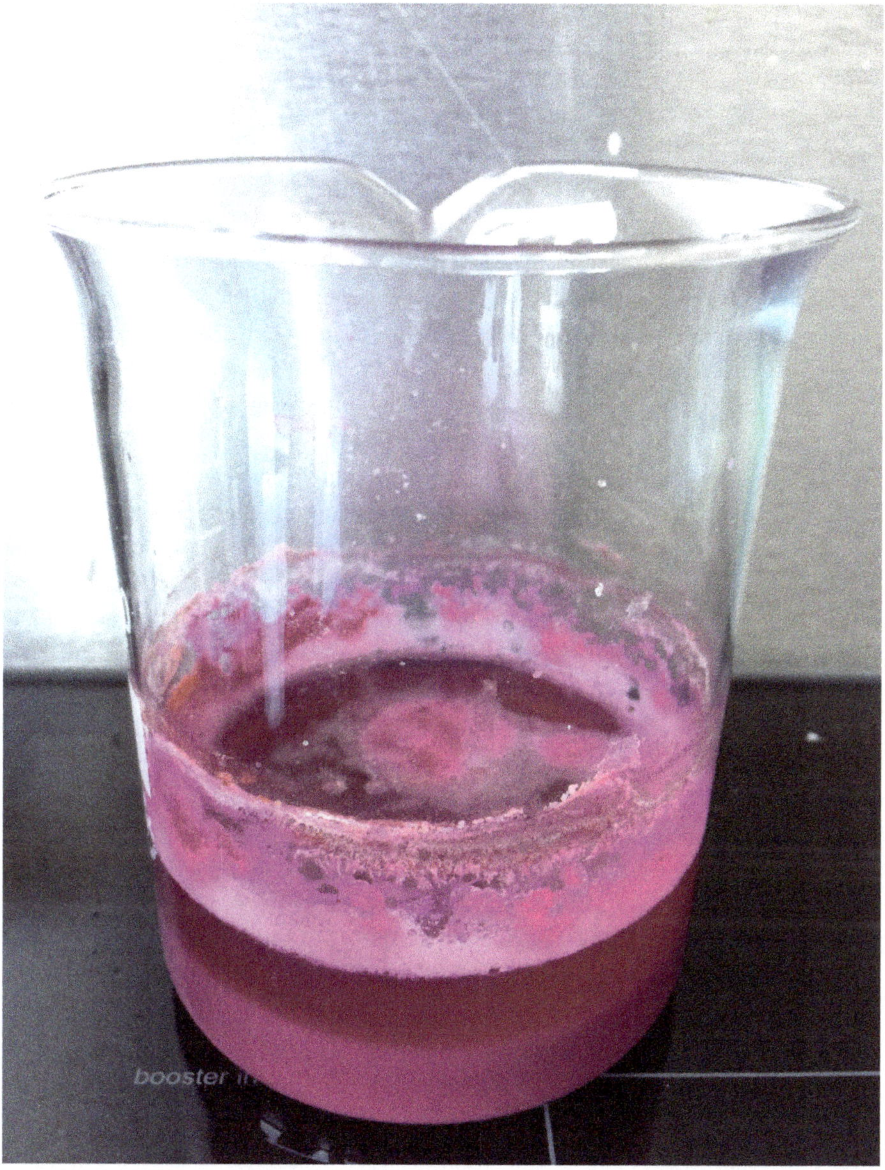

FIGURE 69 The separation of Brazilwood lake pigment and excess dye became visible overnight

mixture settle overnight, the dyed substance would sink to the bottom of the container. The substrate normally absorbs most of the colourant, leaving the dye a clearer liquid than it was (Fig. 69), which can then simply be tossed out. The excess logwood dye was so deep that throwing it away seemed to be a waste. What would happen if I made pigments by using the excess to dye multiple batches of substrate?

I prepared another logwood dye bath following the same recipe two weeks after the first experiment, and made the first batch of pigment (Batch A). The next day, I filtered and washed Batch A, and left the pigment to dry. The excess from Batch A became the dye for Batch B, the excess from Batch B became the dye for Batch C, and this process was repeated many times. After the eleventh time (Batch L), the excess was still darker than anticipated; I added chalk instead of alum into the dyebath for the last batch of pigment (Batch M) and produced thirteen batches of pigment from one dye. The idea was that each batch would be a tint lighter than the previous one. The results (Fig. 70) were not as quite straightforward, but the later batches are indeed lighter than the earlier ones. How about the quality of these pigments? Despite the nice range of purples in appearance, most batches are low in quality. The particles are coarse, and the lighter colours ones turn dull when bound with gum water (Fig. 71).

This string of logwood experiments has no concrete historical base in text or practice. They were entirely driven by curiosity and the mindset of 'what if.' As excess is not a proper dye that can produce quality, or even usable, pigment, the second experiment was impractical. However, the amazement and curiosity elicited by the (plant) material is perhaps worth mentioning. The early modern curiosity-collecting culture—as seen in the *Kunst- und Wunderkammern* filled with natural and humanmade material objects throughout Europe—was a big part of knowledge production of the period. Early modern alchemy shared a different kind of enthusiasm towards materials in their transformation and application. The empirical and heuristic processes of knowledge production became relatable on a personal level through the logwood experiments.

By now, knowledge production has developed so much that we can easily explain the phenomena that were considered 'wonders' in the past. While it is unequivocally important to know the historiography and updated understanding on the subject for good research, I sometimes ponder if several centuries of reasoning has made it difficult for many of us to remember that small but fundamental curiosities, often coupled with wonderment, have been a critical part of knowledge production. The fun explorations during my historical remaking research have been particularly instrumental in making me rethink the emotional side in the pursuit of knowledge throughout history.

## 8.1 *Afterword*

At the end of this series, and especially throughout this last blog post, you may have noticed that I have asked many big and small questions. Some of them were followed by an answer or explanation; some of them became more complex questions; and some

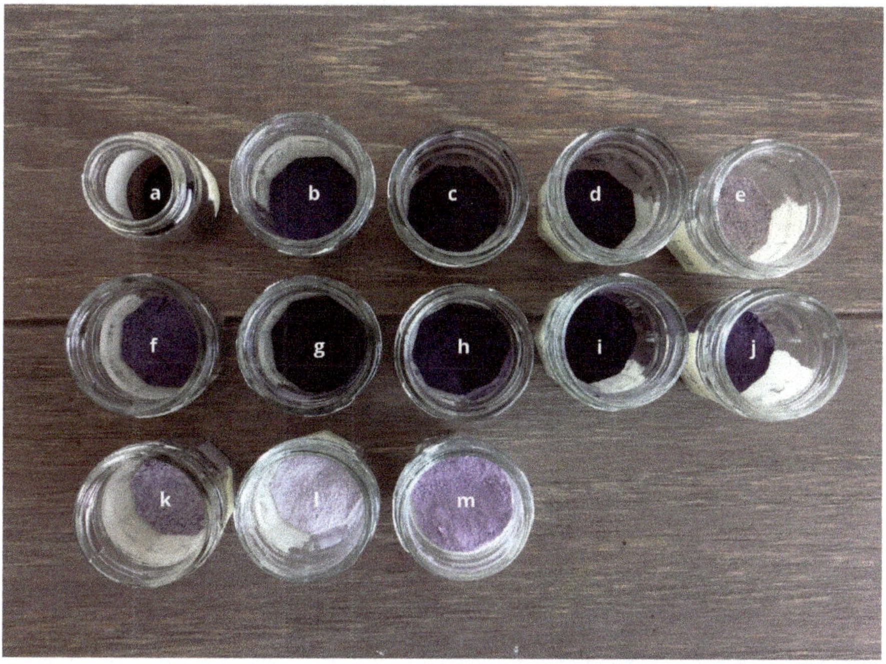


FIGURE 70 Thirteen batches of logwood lake pigment from one dye

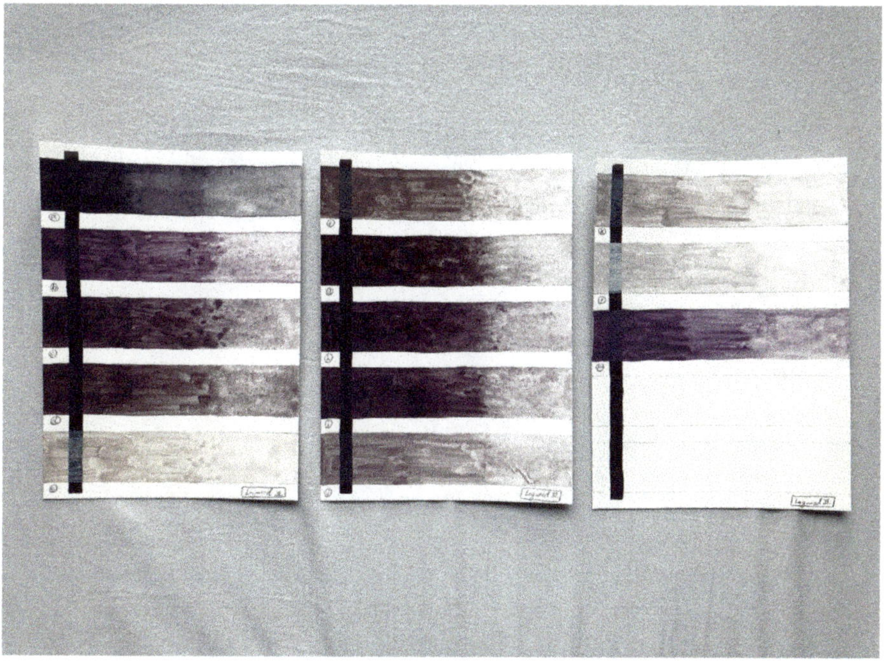

FIGURE 71 Paint swatches of the thirteen batches of logwood lake pigment

of them led to absolutely nowhere. This is reflective of my process of applying historical remaking as a research method. If I were to summarise my past year with hands-on research in one word, historical remaking means *possibilities* for me. It means the possibilities to observe, to engage, to appreciate, to be amazed, to understand, to confront, and to question historical (practical) knowledge production through bodily gesture. The experiments are similar to collecting sources in archives, libraries, and museums, and follow-up digestion and analysis is required. With more experimentation and contemplation down the road for my dissertation, the work continues.

APPENDIX 2

# Planting Pigments ColLAB

The Planting Pigments ColLAB was a series of one-day workshops that gathered participants to make historical and experimental plant-based lake pigments. The workshops were organised by Anna Svensson and me and took place on 17 February, 18 February, 30 March, and 31 March 2021. The ColLAB was part of the preparation to produce a multimedia work titled *Planting Pigments, Painting Plants*, partially based on my PhD dissertation research. The piece was created for an interactive exhibition for a string of sessions, 'A Green Turn? How plants are shaping the Environmental Humanities landscape' at the online conference STREAMS 2021 (International Conference for the Environmental Humanities) on 3–7 August 2021. The ColLAB participants were provided with tools and materials for making lake pigments by receiving a kit of supplies. Participants who wanted to be included in the multimedia piece shipped their pigments to me after the workshop.

*Planting Pigments, Painting Plants* sits at the intersection of plants, colours, and makings in the past and present. It draws on historical techniques to examine the small-scale production of pigments, paints, and dyes, as well as to consider the maker's responses to and the results of the creative process, among other topics of concern. This multimedia work comprises of three parts:

- A digital booklet.
- Physical artwork created using the watercolour medium and exhibited within the digital booklet.
- A collection of pigment and paint samples created by international makers from many backgrounds, also documented within the digital booklet.

This appendix is the archived version of the digital booklet, which was initially published on the *Planting Pigments* website, https://plantingpigments.sites.uu.nl/. The booklet is reformatted into this appendix, with minor updates to the text, such as changing to British spelling and the citation style. The original graphic design of the booklet that aimed to provide a certain aesthetic experience is lost in this version, but this appendix ensures that the content of the booklet will remain accessible after the website eventually stops working.

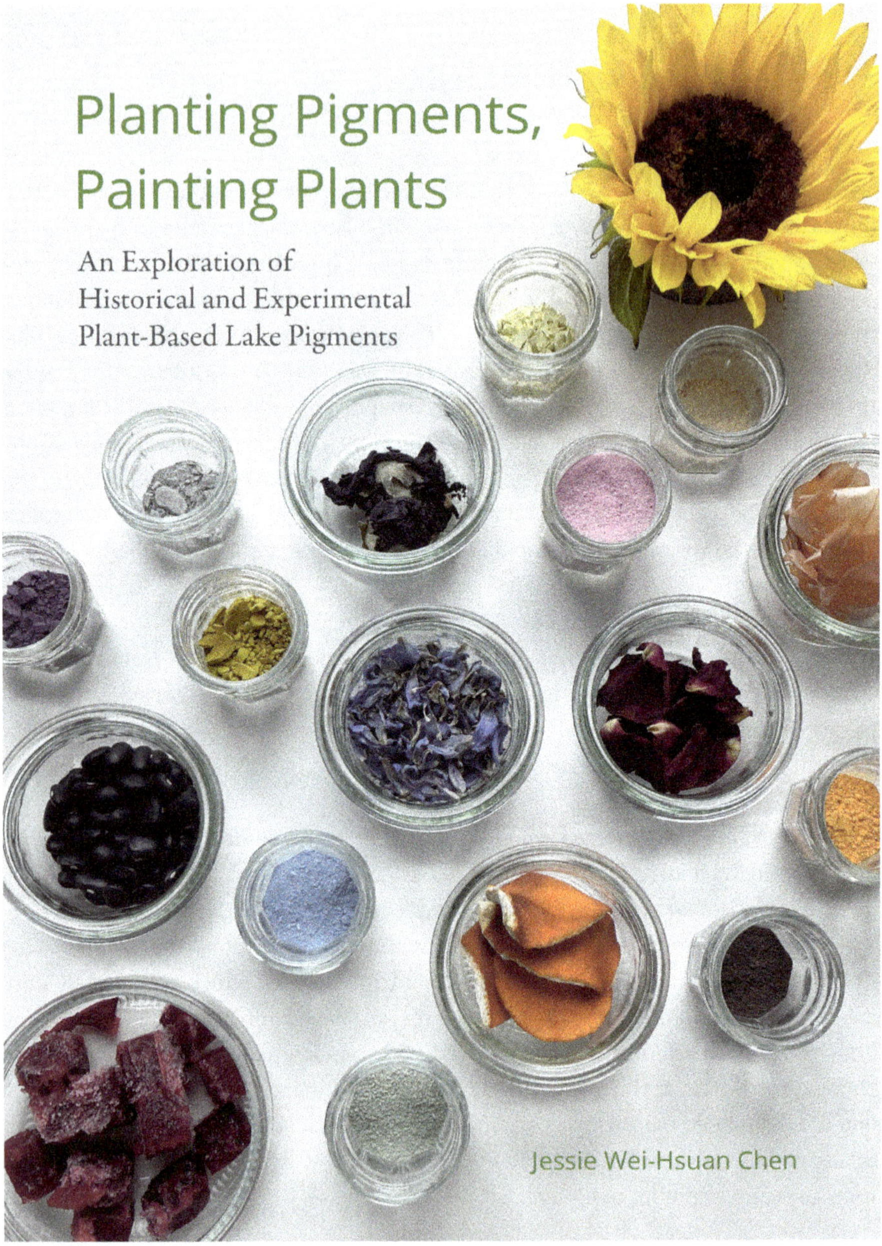


FIGURE 72 The original front cover of the booklet

## 1 Prologue

This booklet is a small book with a big identity crisis. It is not a master class on the theory and practice of painting materials and techniques. It is not an exhibition catalogue that thematically introduces and describes artworks. It is not a piece of gallery work displaying the art and craft of painting plants. It is not a how-to manual to show the step-by-step process of extracting plant-based dyes and making pigments from them. And it is not a sample book because the pigments in this booklet are not intended to be purchased nor further disseminated. Debatably, some might not even consider it a 'real' book, for it is digital-born and not intended to be printed and bound. Nevertheless, this booklet took inspiration from all the above-mentioned categories. What it can be best described as is a series of end results, but not conclusions, of experimenting with turning plants into pigments, into paints, and into paintings. Moreover, the colourful swatches of pigments and paints in this booklet are abstract visual representations of the conversations, discussions, and many more questions that arose from the making process.

This project stems from my PhD research, part of which delves into the practical making of botanical watercolours from the early modern period, which roughly refers to the period from the late fifteenth to the late eighteenth centuries. It tries to understand how image-makers in early modern Europe worked with the 'colour technology' of their time to visualise the plant world. Applying the performative research method that can be best described as historical reworking/remaking, I spent a lot of time at the university Kunstlab (Artlab) to work with historically-oriented art materials and techniques.[1] This particularly enabled me to safely handle toxic colours, such as lead white, that were essential in the past. The Covid-19 pandemic disrupted this lab access, and some aspects of the research have been halted, but being forced to experiment from a simple home lab brought into play different perspectives on a maker's relationship with various working environments.

Concurrently, two things came to the foreground while working from home. First, the interest in hand-making things grew throughout many places in the world during the pandemic lockdowns. Making lake pigments from dyes gained traction, particularly

1 For more on this research in progress, see the Making Colours Series (Appendix 1). On performative methods, see especially Hendriksen, 'Rethinking Performative Methods in the History of Science' and other articles in the same journal issue; Dupré et al., ed., *Reconstruction, Replication and Re-enactment in the Humanities and Social Sciences*.

among communities of natural dyers, a practice that has deep historical roots.[2] Albeit constantly changing their forms and shapes, many arts and crafts share experiences and qualities that transcend time and place. Second, the longstanding habit of sending bouquets or potted plants to express appreciation of a person's effort, to celebrate their accomplishment, to console them in sorrow, or simply to brighten their day became even more common. The question of what colours can be extracted as a visual reminder of the memories these plants commemorate grew in the back of my mind.

Plant-based lake-pigment making became a suitable means to bring historical and present-day experimental colours together. This booklet shows three different (sets of) watercolours done with paints made from plant-based lake pigments. The first single-sheet watercolour concerns itself with early modern materials and techniques, with the goal of achieving a specific artistic style through the maker's control of the paints. The second set of twelve small paintings took the opposite approach. Instead of manipulating the paints to create what I envisioned, the final style was a response to how the paints behaved in the moment of painting. The final watercolour is a combination of having and losing control. Through making lake pigments and painting flowers with them, this booklet offers a possibility of what can happen when plants, colours, and makings intersect.

## 2 (Re)Painting Colourful Flowers

The road to finding one's own artistic expression often starts with imitating the works one likes. The path for me to better understand the application of historical colour technology took me to copy a seventeenth-century painted parchment (Fig. 73). The reproduction and the original painted folio (Fig. 74) show a snowdrop, two winter aconites, two hellebores, some hepaticas, and some crocuses.[3] This folio is one of the 100 paintings in the manuscript that is titled *Jardin de rares et curieux fleurs* and attributed

2 Three online resources that were particularly useful in introducing me to many present-day practitioners are World Pigment Day by Ruth Siddall and Jo Volley (https://www.instagram.com/worldpigmentday), Wild Pigment Project (https://wildpigmentproject.org/), and the blog of Botanical Colors (https://botanicalcolors.com/category/blog/).

3 The album is preserved at the Oak Spring Garden Foundation in Upperville, Virginia. See catalogue entry in Tongiorgi Tomasi, *An Oak Spring Flora*, 84–87. There is another album by De Geest with very similar content and many flowers in same compositions that is preserved at the National Central Library of Rome (Biblioteca nazionale centrale di Roma) and is digitised for public access (http://digitale.bnc.roma.sbn.it/tecadigitale/manoscrittoantico/BNCR_Ms_VARIA_0291/BNCR_Ms_VARIA_0291/1). The 2012 printed facsimile of the Rome

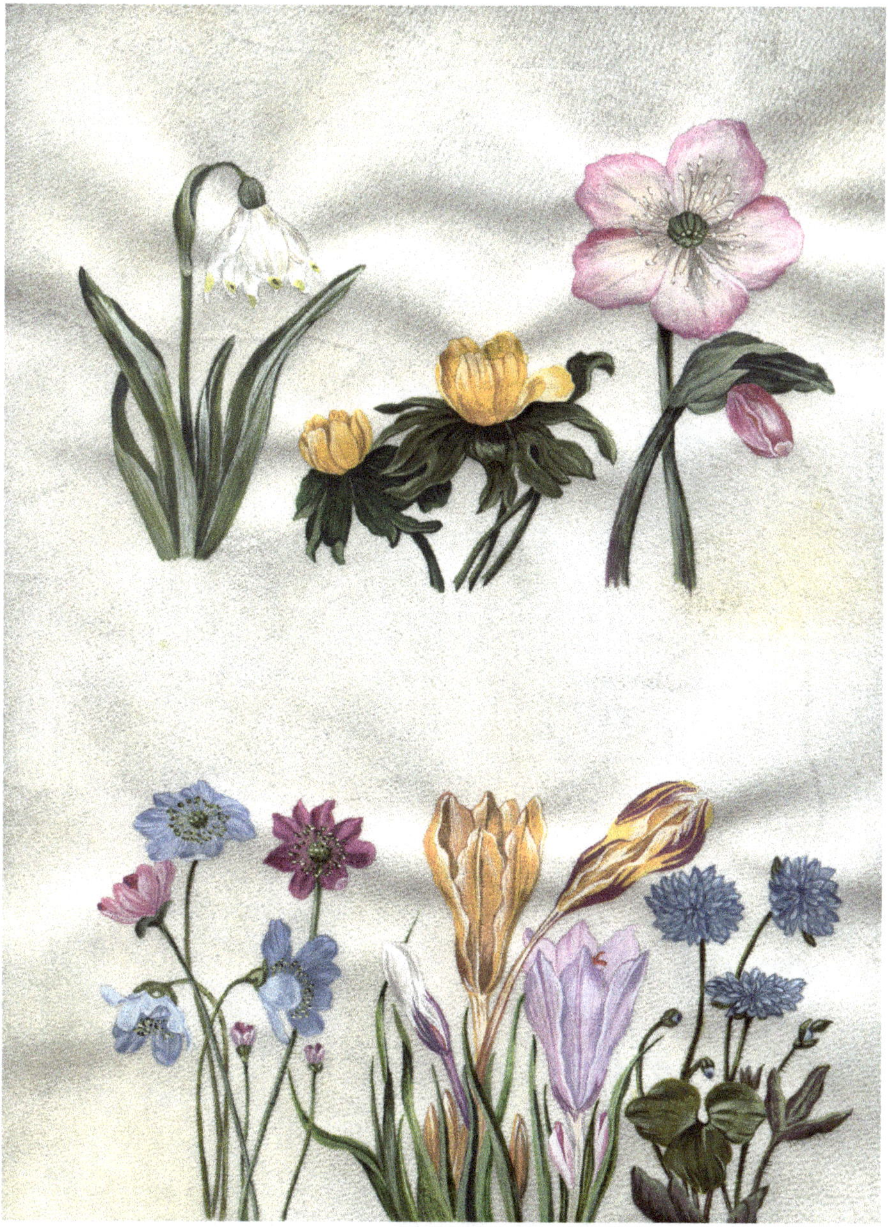

FIGURE 73 Reproduction of watercolour by Franciscus de Geest

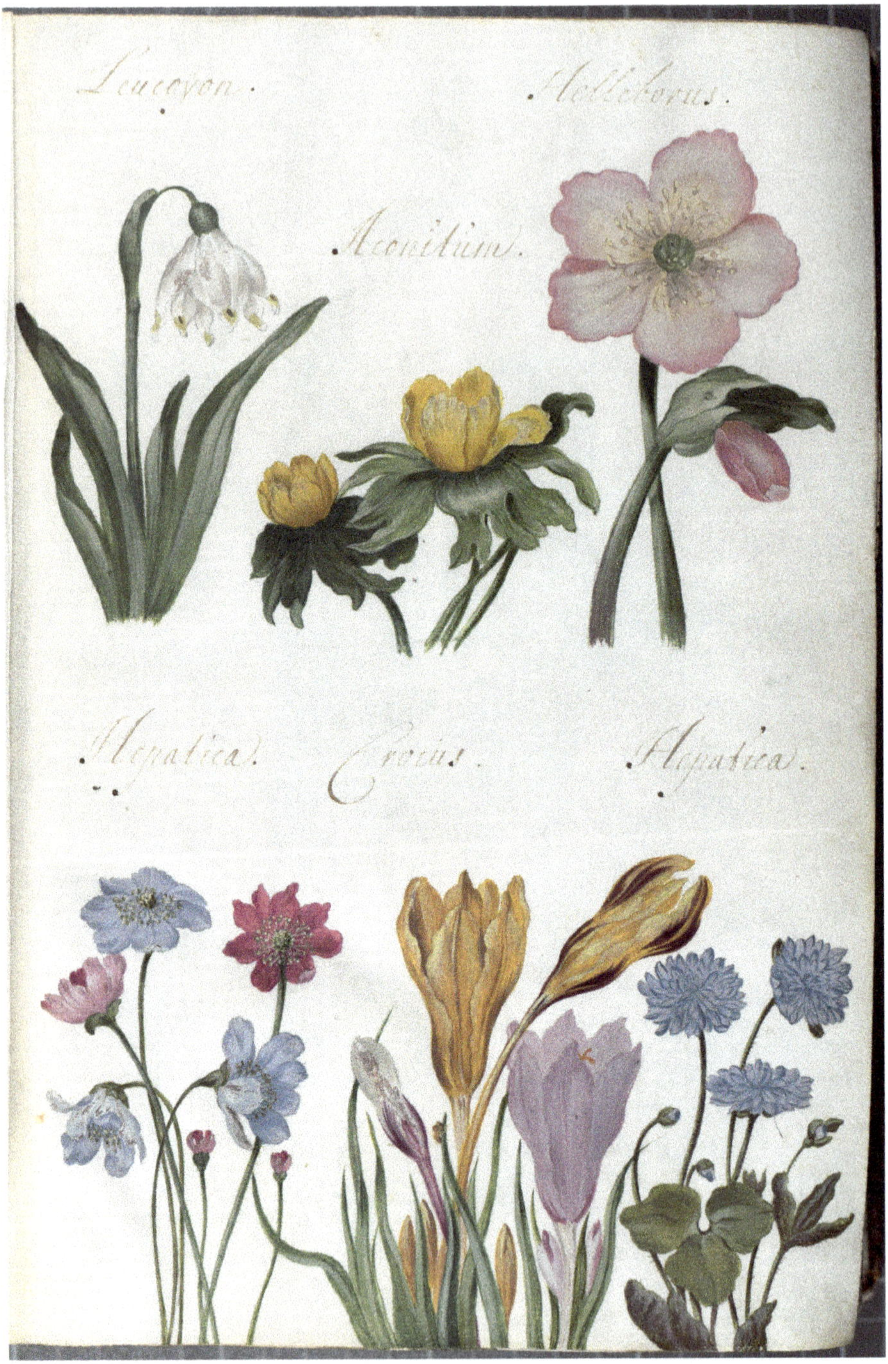


FIGURE 74 Original watercolour by Franciscus de Geest
IMAGE CREDIT: OAK SPRING GARDEN FOUNDATION, UPPERVILLE, VA

to the Leeuwarden-based Dutch painter, Franciscus de Geest (after 1626 and before 1638/39–*c.*1699).[4] Today, these recognisable plants are common garden flowers in the Netherlands and many European countries. Procuring them requires only a simple trip to a local florist or supermarket. As its title suggests, however, the depicted flowering plants in the manuscript were considered rare and curious in early modern Europe.

De Geest's book of flowers can be broadly categorised as a *florilegium* (plural *florilegia*). While it is not a strict label and the content of each book can vary, the florilegium is a genre of book featuring a collection or a selection of flower images. Literally meaning 'a gathering of flowers,' the term florilegium first described a collection of excerpts from other bodies of text in the Middle Ages, but gained an association with a 'flower book' in the early modern period. Florilegia were essentially picture books. Copperplate engraving (sometimes hand-coloured) and hand-painted watercolour/bodycolour were two major ways to create the images in a florilegium. Text often only appeared in the introductory section(s) to state the owner's possession of the plants and again in the captions for the illustrations. Florilegia of the seventeenth century focused on the plants' decorative quality, as opposed to their medicinal or utilitarian functions, which were the focus in early modern herbals.[5] Early modern florilegia also often had close connections to specific gardens that once existed and showcased the several colour variations of the same flower and the most valued cultivars of the garden owners.[6]

Regardless of whether it was in the form of a printed book or painted manuscript, florilegia were markedly expensive. The cost in producing and reproducing a volume full of folio-size-or-larger engravings was (and still is) much higher than printing a book without pictures.[7] Additionally, the emphasis on the wide range of colours of the depicted flowering plants meant that (professional) colourists had to be brought in, as hand-colouring was the common method to add colours to black-and-white prints in the early modern period.[8] The cost of the many people and artist's materials involved in its making process gave a printed florilegium, particularly a hand-coloured one, a

volume includes updated studies of De Gesst and his works, see Menghini, ed., *Hortus Amoenissimus*.

4 Huisman, 'Franciscus de Geest and His "Delightful Gardens",' 24–29.

5 For more information on the genre of florilegium, see Nissen, *Die botanische Buchillustration* 66–80; Chapter 2: Florilegia and Pattern Books in Saunders, *Picturing Plants*, 41–64; Blunt and Stearn, *The Art of Botanical Illustration*, 13–14.

6 See, for example, Sikkens-De Zwaan, 'Magdalena Poulle (1632–99)'; Lauterbach, 'Commerce and Erudition.'

7 A famously expensive florilegium is the *Hortus Eystettensis* (1613). See Barker, *Hortus Eystettensis*.

8 See Dackerman, ed., *Painted Prints*; Goedings, *Afsetters en meester-afsetters*.

high price tag. Similarly, manuscript florilegia were often commissioned over a long period of time, from several years to more than a decade. Although each folio may not have been exorbitantly expensive, the accumulated cost of more than 100 folios, which is typical for surviving florilegia that have not been dismantled, means that only people with certain financial means and privileges were able to afford such an item.[9] Florilegia were not the most democratising objects, but the early modern fascination with flowers which they embody and visually represent led to many gardening and technological developments (such as the greenhouse) that eventually made the garden flowers so easily available and commonly enjoyed today.[10]

To pictorially capture the colourful flowers in florilegia, the medium of watercolour is important for the making of both hand-coloured printed and manuscript volumes. However, the term 'watercolour' may be somewhat misleading for present-day readers. Today, watercolour is often treated as a synonym for transparent watercolour, painted using a technique of thin and translucent washes and with a luminescent effect. The opaque paints De Geest used to depict the flowers are usually described by the term bodycolour, or alternatively gouache, instead.[11] However, this rigid separation of watercolour and bodycolour is not necessarily productive when discussing the majority of works from the past, since both translucent and opaque paints were often used together in a single painting.[12]

In a more general sense, watercolour is a water-based painting medium, consisting of pigments mixed with a binder; a pigment is a powder of insoluble coloured particles, and a binder holds the particles together. Gum arabic (Fig. 75) is the most used binder for watercolour throughout the ages, and the only kind I used for this project. By adjusting the ratio of pigment and the gum arabic dissolved in water, the paint can become more translucent or opaque.[13] There is one major complication, however. Some of the translucent paints in early modern works that have a closer effect to

9 Two exhibition catalogues feature several essays that discuss the people who commissioned a few seventeenth-century florilegia that are most well-known to us today, see De la Fuente Pedersen and Kolind Poulsen, ed., *Flowers and World Views*; Bushart et al., ed., *Maria Sibylla Merian und die Tradition des Blumenbildes von der Renaissance bis zur Romantik*.

10 A useful starting point is Lee and Helphand, ed., *Technology and the Garden*.

11 See Cohn, *Wash and Gouache*, for an introduction on the materials and techniques of watercolour and bodycolour/gouache that is close to the present-day understanding of the terms from a conservation perspective.

12 Creators of the recent exhibition Renaissance Watercolours at the Victoria and Albert Museum (https://www.vam.ac.uk/exhibitions/renaissance-watercolours) share the same view. See particularly, Evans, *Renaissance Watercolours*, 8; 14–15.

13 Tallian, 'John White's Materials and Techniques,' 72.

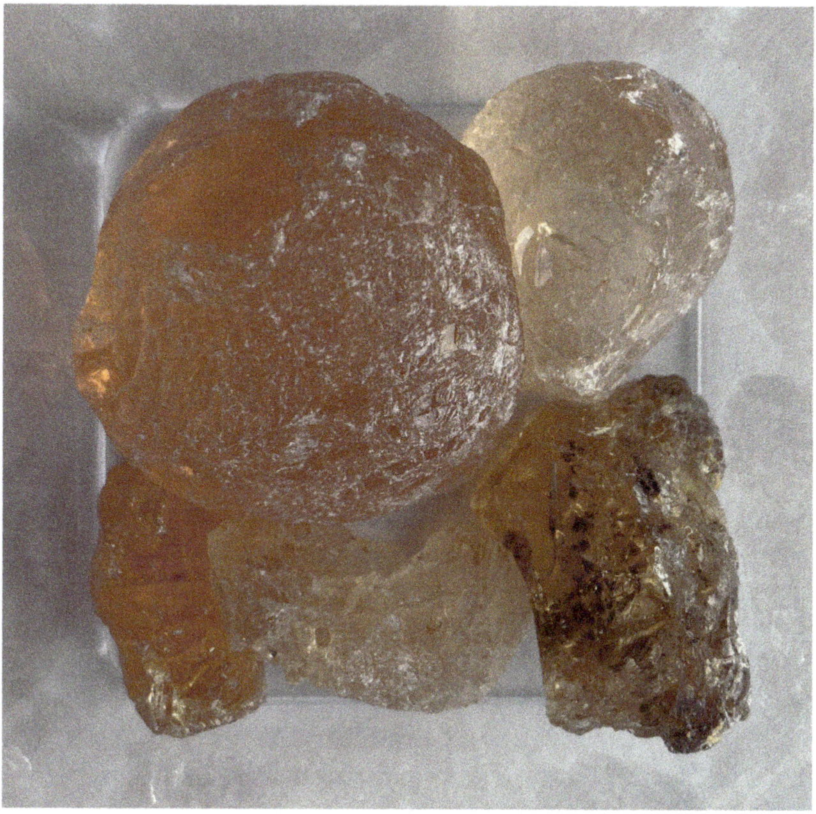

FIGURE 75 Gum arabic pieces, with the largest round piece in the picture measuring around 3.5cm in diameter in real life

transparent watercolour are more in line with the general definition of inks for writing and drawing (but not for printing). Instead of a pigment, an ink is usually made from soluble dye with a binder, which in most cases was also gum arabic. Sometimes there was even no mention of adding the gum, and only the dye was used for hand-colouring or painting as a glazing colour.[14] Within this booklet, 'painting with watercolour' includes using all three kinds of colourants: pigment-based paints, inks, and dyes (indicated as 'clothlet' instead of 'pigment' on the colour charts).

14 For instance, in an important seventeenth-century treatise on the art of illumination and watercolour, *Verlichterie-kunde* (1670), the recipe of 'Bresilie Verwe' (a colour made of redwood/Brazilwood) describes the steps of making a red ink and the recipe for saffron states that the yellow extracted from the plant can be used without gum. See Ter Brugge and Goeree, *Verlichterie-kunde,* 14, 24–25.

Determining what colours to use for painting the reproduction required educated guesswork. Early modern artists' manuals on illumination and watercolour provide useful information on the colour materials that were obtainable and commonly used during the period.[15] With non-invasive analytical methods, including infrared reflectography and macro-XRF scanning, conservators can identify the specific pigments (and, using other methods, binders) that went into a painting.[16] De Geest's piece has not been analysed, so such identification is not yet available. However, several contemporaneous pieces depicting flora and fauna that have been analysed offer useful insight into the common colour materials found in watercolours of natural history more widely.[17] By comparing the colours listed in historical manuals, the pigments identified by recent analyses in similar works, and the visual properties of the colours in De Geest' original folio, I selected the ten historically-oriented colours listed in the charts of the next pages to paint the reproduction. The flowers were painted with freshly made paints and the mixing of the selected colours reproduced the range of variations observed in the original folio.[18]

For watercolour paints to become a watercolour painting, the paints need to be applied onto a support. The manuscript folios by De Geest are painted on parchment, which is one of the two most common supports—another one being rag-based papers—for historical watercolours.[19] Parchment is a durable writing and painting material, made from the prepared skin of an animal (Fig. 76). Typical species for parchment-making include goats, sheep, and calves, with the last one offering the highest quality working surface and often reserved for important or luxurious items.[20] Most manuscript

15 A few largely quoted ones include Boltz, *Illuminierbuch*; Ter Brugge and Goeree, *Verlichteriekunde*; Norgate, *Miniatura*.

16 The ILLUMINATED project at The Fitzwilliam Museum offers a helpful introduction (https://www.fitzmuseum.cam.ac.uk/illuminated/lab/lab/analytical-methods) to different analytical methods suitable for examining illuminated manuscripts and similar types of paintings.

17 Including Montalbano et al., 'Painting on Parchment Besides Miniatures'; Ambers et al., 'John White's Watercolours'; Nany Turner and Karen Trentelman, 'Exquisite Views of Nature in a 16th-Century Book,' Getty Iris (blog), http://blogs.getty.edu/iris/to-hold-nature-in-the-hand-revealing-the-wonders-of-the-mira-monumenta-calligraphiae/.

18 Colour mixing has now been shown as a common practice for illumination, even though the extent of mixing is comparatively minimal to present-day practice, see Panayotova, 'Colour in Illuminated Manuscripts,' 17.

19 The Rome volume of De Geest's florilegia is painted on paper.

20 Cheese, 'From Pelt to Painted Page,' 75–78. The project Pergamenum21 (https://www.pergamenum21.eu/) has a helpful video showing a general parchment making process and what can be learned from this process from a transdisciplinary collaboration, see Parchment Making (https://youtu.be/_TmyEiVUTlg).

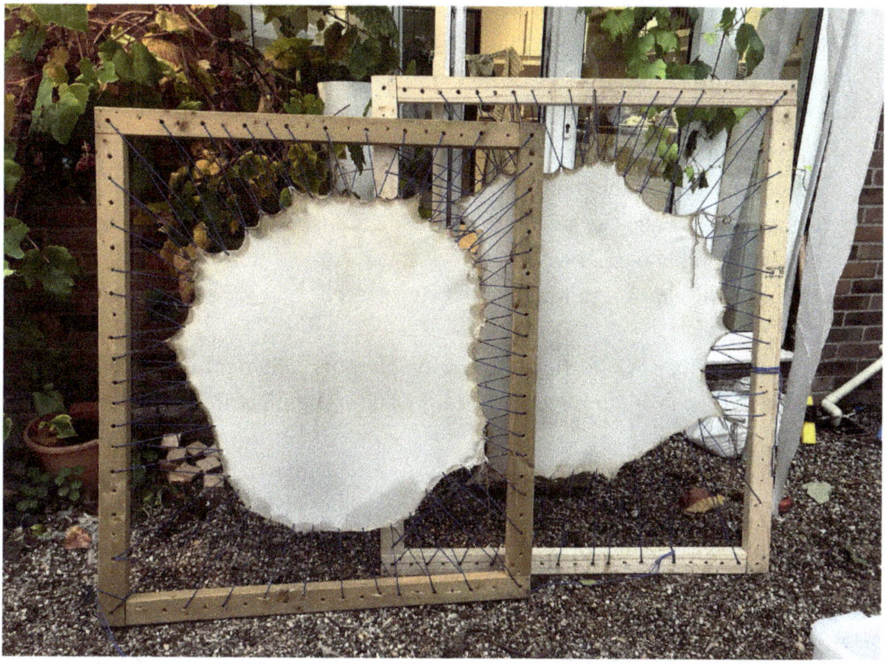

FIGURE 76 Parchments being stretched on frame
IMAGE CREDIT: SARA CHARLES, TEACHING MANUSCRIPTS

florilegia have calfskin parchment as the support for the painted folios, including the volume by De Geest.[21] With a piece of calfskin parchment and the selected colours, I made the fresh paints and reproduced De Geest's colourful flowers (Fig. 73) over the span of three eight-hour workdays.[22] It is true that the pigments and inks used in the reproduction most likely differ from the original, and many of the tools (such as brushes) are from present-day manufacturers, but that is not a problem because replicating the past to the very last detail is not the goal here. Rather, copying and re-painting De Geest' original is a way to closely examine and question the working processes of historical watercolours on a broad level through active bodily engagements.

21 Tongiorgi Tomasi, *An Oak Spring Flora*, 84, Borring et al., 'The Gottorfer Codex,' 171–172.

22 The piece of calf parchment on which I painted was made by Sara Charles, see the making process on her website Teaching Manuscripts (https://www.teachingmanuscripts.com/blank-page-6).

### 2.1 *How to Read the Collection*

H—Historical Colours

F—Floral Colours

E—Experimental Colours

- Each colour includes a photo of the pigment material/ dyestuff, pigment/clothlet sample, and watercolour paint swatch.
- Champagne chalk, the substrate for making most of the lake pigments for this collection, is shown below.

For each entry, the caption shows the following order:

> Figure number. Collection Inventory Number, Pigment Material/Dyestuff, *Botanical Name of Plant-based Dyestuff*, Maker/Manufacturer.

The order of the squared pictures is as follows:

> Left: picture of pigment material/dyestuff (not always the actual material used for making the samples); Middle: pigment/clothlet sample; Right: watercolour paint swatch.

FIGURE 77 Example for picture order, Champagne chalk, Kremer Pigmente (58000)

## 2.2 *Pigment Collections: Historical Colours*

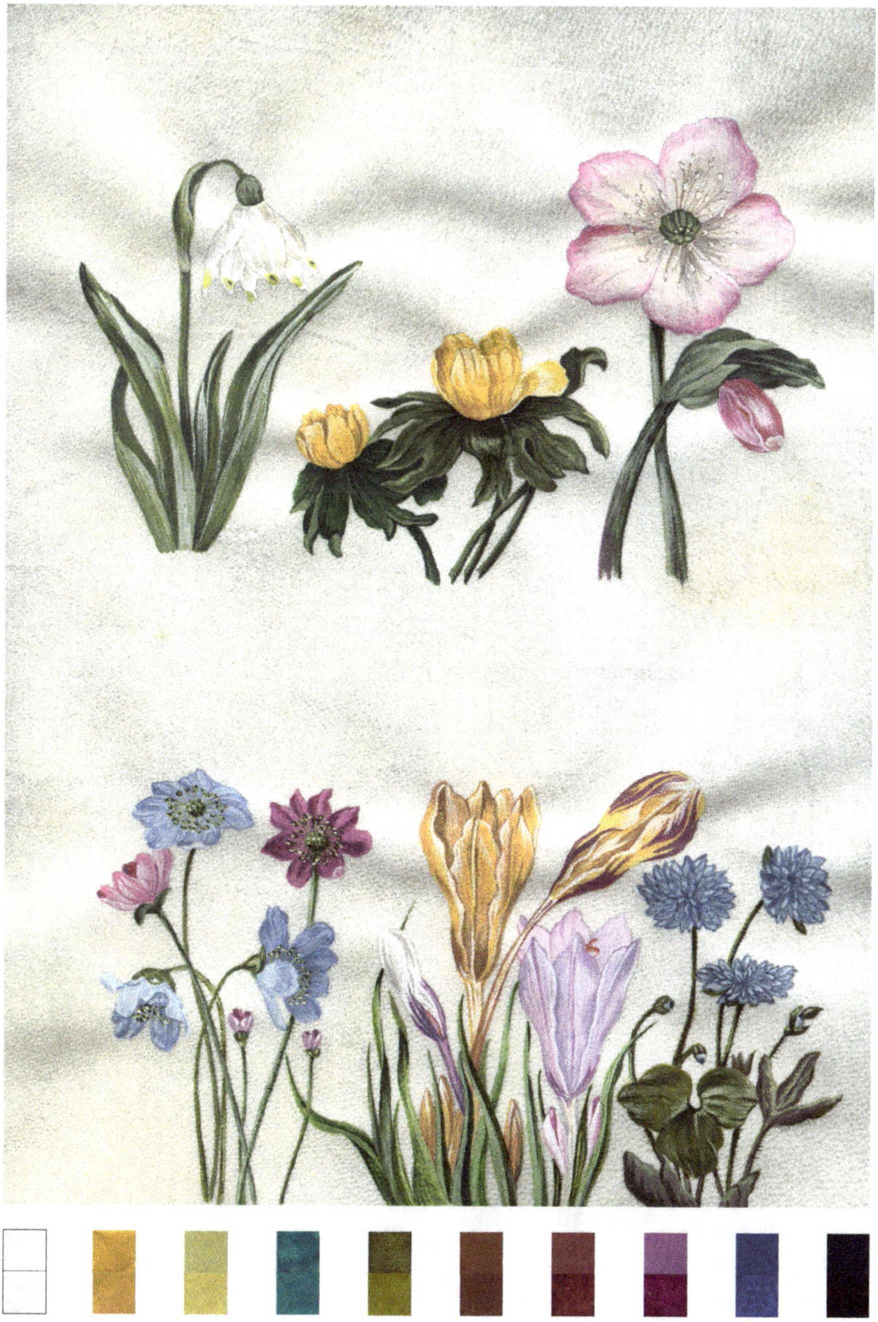

FIGURE 78 Reproduction of watercolour by Franciscus de Geest and the colours used

*Colours Used*

Snowdrop
- Lead White (H1)
- Sap Green (H5)
- Red Ochre (H6)
- Cochineal (H8)
- Indigo (H10)

Winter Aconites
- Lead White (H1)
- Orpiment (H2)
- Verdigris (H4)
- Sap Green (H5)
- Red Ochre (H6)
- Indigo (H10)

Hellebores
- Lead White (H1)
- Weld (H3)
- Sap Green (H5)
- Red Ochre (H6)
- Cochineal (H8)
- Smalt (H9)
- Indigo (H10)

Hepaticas
- Lead White (H1)
- Weld (H3)
- Verdigris (H4)
- Sap Green (HS)
- Red Ochre (H6)
- Madder Root (H7)
- Cochineal (H8)
- Smalt (H9)
- Indigo (H10)

Crocuses
- Lead White (H1)
- Orpiment (H2)
- Weld (H3)
- Verdigris (H4)
- Sap Green (HS)
- Red Ochre (H6)

- Madder Root (H7)
- Cochineal (H8)
- Smalt (H9)
- Indigo (H10)

Hepaticas (Double Flowers)

- Lead White (H1)
- Sap Green (HS)
- Madder Root (H7)
- Cochineal (H8)
- Smalt (H9)
- Indigo (H10)

FIGURE 79 H1, Lead White, Kremer Pigmente (46000)

FIGURE 80 H2, Orpiment, Kremer Pigmente (10700)

FIGURE 81 H3, Weld Yellow Lake, *Reseda luteola*, Jessie Wei-Hsuan Chen

FIGURE 82 H4, Verdigris, Verfmolen De Kat (6046-93-1)

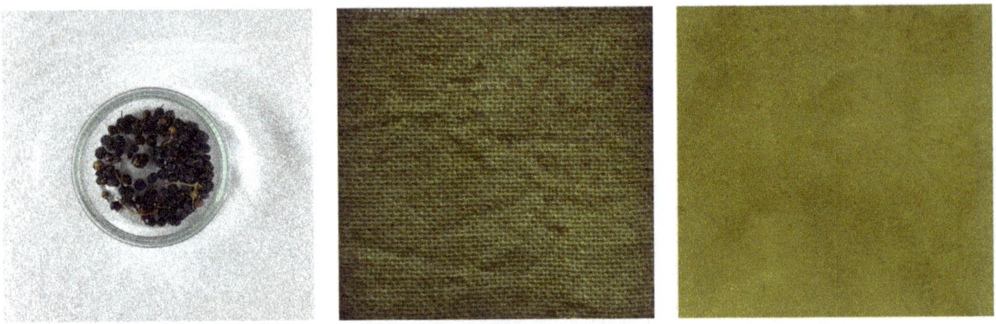

FIGURE 83 H5, Sap Green (made with ripe buckthorn berry, clothlet instead of pigment), *Rhammus sp.*, Jessie Wei-Hsuan Chen

FIGURE 84 H6, Red Ochre, Kremer Pigmente (11584)

FIGURE 85 H7, Madder Root Red Lake, *Rubia tinctoria*, Jessie Wei-Hsuan Chen

FIGURE 86 H8, Cochineal Red Lake, Jessie Wei-Hsuan Chen

FIGURE 87 H9, Smalt, Kremer Pigmente (10000)

FIGURE 88 H10, Indigo, *Indigofera tinctoria*, Kremer Pigmente, 36002

## 3 Keeping Plant Colours

The ten selected colours for painting the reproduction present a range of inorganic and organic colourants.[23] Overall, inorganic colours include earth, mineral, and metal-based pigments, and they tend to be stable and lightfast.[24] Red ochre is an earth pigment that is among the oldest colours that continue to be used today. Other common earth pigments include yellow ochre, green earth, (burnt) sienna, and (burnt) umber. Orpiment (yellow) is a mineral pigment, and azurite (blue), malachite (green),

23 A highly useful encyclopaedic source on pigments is Eastaugh et al., *Pigment Compendium*. The *Artists' Pigments* series has many in-depth scientific studies of individual pigments and their use throughout history, see Feller, ed., *Artists' Pigments*, vol. 1; Roy, ed., *Artists' Pigments*, vol. 2; FitzHugh, ed., *Artists' Pigments*, vol. 3.

24 The grouping of pigments can vary depending on the sources. For instance, some sources do not separate early and mineral pigments and place earth in the same category as mineral pigments.

and authentic ultramarine made of lapis lazuli all belong to this group as well. Lead white and verdigris are metal-based colours. Both are loose terms: lead white generally refers to pigments consisting of basic lead carbonate and verdigris to copper acetate. Metal-based colours, which also include massicot, minium, and artificial vermilion, for example, tend to give vivid colours. As their vibrancy is hard to match by natural pigments made of other materials, these colours were regularly used throughout many historical time periods, in spite of being (highly) toxic. Not belonging to any of the three groups, smalt is a blue artificial pigment obtained by grinding cobalt glass into powder. Indigo, sap green, and the lake pigments with weld, madder root, and cochineal are organic colours made by processing plants and insects through various methods. Organic colours are comparatively fugitive and light-sensitive.

Lake pigment, simply put, is a dyed substrate. The dyestuff provides the colour, and the substrate gives the soluble dye some substance so that the colourant can be used as a pigment to be ground and mulled with the desired binder, such as linseed oil for oil painting, egg yolk for tempera, and gum arabic for watercolour.[25] The yellow and red lakes for painting the reproduction were made with the dyes extracted from weld, madder root, and cochineal precipitated onto potash alum (potassium aluminium sulfate, $KAl(SO4)_2 \cdot 12\ H_2O$). Hydrated alumina is one of the most common substrates in early modern Europe, and traditionally lake pigment refers more specifically to pigments with dye precipitated onto this group of chemical compounds. As the substrate usually determines the property of the pigment, (red) lake pigments are mostly used as a glazing colour because alum is translucent and hence has a low covering power. The broader definition of lake pigment encompasses a wider variety of substrates that were or can be used for colour-making, both historically and at present. Dyed powdered chalk, ground eggshells, marble dust, and lead white can all fall into the lake pigment category.[26] Compared to alum, these substrates are more opaque and produce paler pigments (Fig. 89).

The sap green used for the reproduction is not a lake pigment, but simply a dye. This green came from adding a little potash alum to the dye extracted from dried ripe buckthorn berries. Alum functioned as a modifier here to change the brown buckthorn-berry dye into green, but the quantity was not enough to form pigment particles.[27] Because it is an organic solution, its shelf life can be short when simply left in a jar. Historically, sap green was usually made by reducing the juice of fresh ripe berries, mixed with alum, to a thick syrup consistency and stored in pig bladders, which

25 Kirby et al., *Natural Colorants for Dyeing and Lake Pigments*, 69.

26 Kirby et al., *Natural Colorants for Dyeing and Lake Pigments*, 76–77.

27 I followed the recipe from the Mayerne manuscript at the British Library (Sloane MS 2052, fol. 22r), translated in Fels, ed., *Lost secrets of Flemish Painting*, 159.

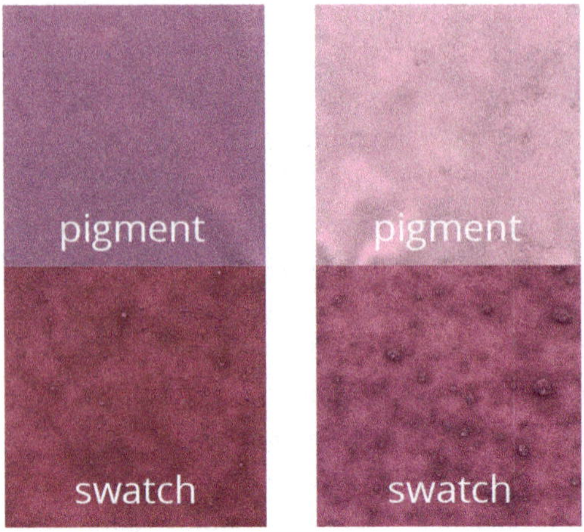


FIGURE 89
Left column: potash alum as substrate; right column: Champagne chalk as substrate

gave sap green its alternative name of 'bladder green.'[28] Pig bladders prolonged the time one could paint with sap green, especially when the berries were out of season for making fresh colours.[29] Because only dried berries were available to me, the sap green made for painting the reproduction is a thin liquid. Although it has different properties from historical sap green, it offered an opportunity to explore the clothlet as another historical method for storing the colour for future use. The basic concept of preparing a clothlet is to saturate a small piece of cloth with a dye. Because the cloth is not mordanted, the dye is not fixed onto the fibre. Clothlets can easily be stored in the dark once dried, and soaking clippings of the clothlet in a binder releases the colour (Fig. 90), turning the mixture into an ink for painting in thin washes.[30]

Lake pigments and clothlets are simple but effective methods for keeping dyes useful for a longer period, and the first step of making both involves extracting the dye from the dyestuff. Oftentimes, this step is essentially the same as the basic process of preparing a dyebath for textile dyeing. Generally speaking, a common way to extract the colourant is to immerse a dyestuff in water, sometimes with an assist added to turn the water acid or alkaline. Often the mixture is soaked before standing on heat

28 Eastaugh et al., *Pigment Compendium*338–339; Thompson, *The Materials of Medieval Painting*, 169–171.

29 On the relation between colour-making and seasonality, see Jenny Boulboullé, 'Seasonality and the (Re)creation of Early Modern Color Worlds,' The Recipes Project (blog), https://recipes.hypotheses.org/9359.

30 For examples of making and using clothlets, see recipes 28 and 38 in Neven, *The Strasbourg Manuscript*, 96–97; 108–109.

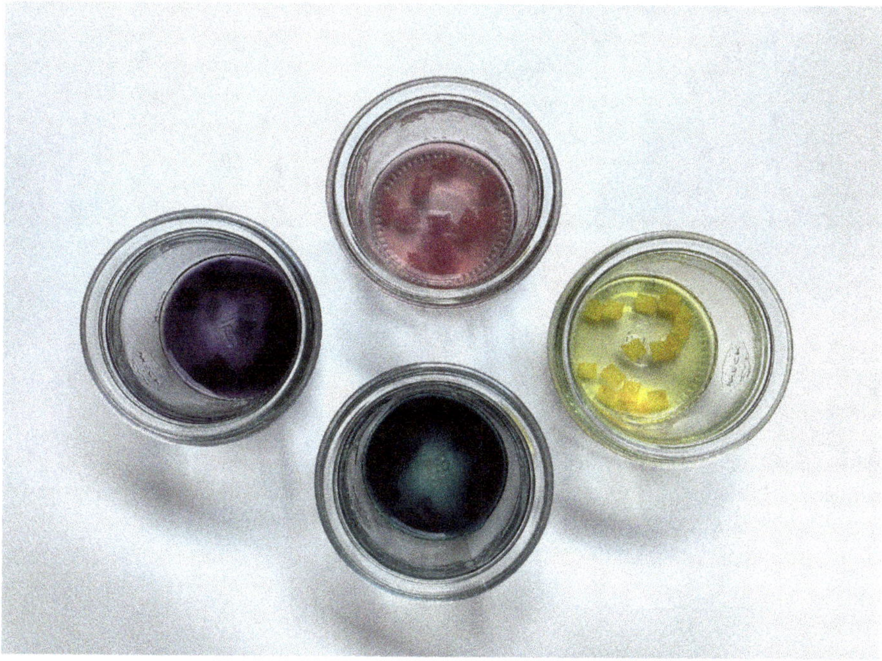

FIGURE 90 Inks made from soaking clippings of clothlets in gum arabic to release the dyes

for some time.[31] There are several factors that can affect the colour of the dye. The needed temperature and length for extracting colour vary for each dyestuff, but in general overheating tends to brown the dye. Different batches of the same dyestuff may contain various levels of colourant. It is common that using the same quantity of a dyestuff and following the same recipe may still result in different shades every time (Fig. 91). Moreover, adding a modifier, such as metal salts, can often change the colour of a dye by altering its pH level.[32] For example, adding alum turned the brown dyes extracted from the dried ripe buckthorn berries into sap green and yellow onion skin into a yellow that is similar to light yellow-ochre (Fig. 92).

Naturally, the quality of the resulting dye distinguished certain plants as serviceable and reasonably durable dyestuff. Without concern over quality in colour and durability, however, one can extract colour from several parts of most plants. What happens when, instead of using only the common dyeing plants, I extract the colours

31 Indigo, however, works in the opposite way that dyers need to reverse the indigotin, which can be used directly as a pigment, into a dye.

32 For example, see Medlej, *Wild Inks and Paints*, in which the artist and expert practitioner shows the range of colours one can get from a selection of plants.

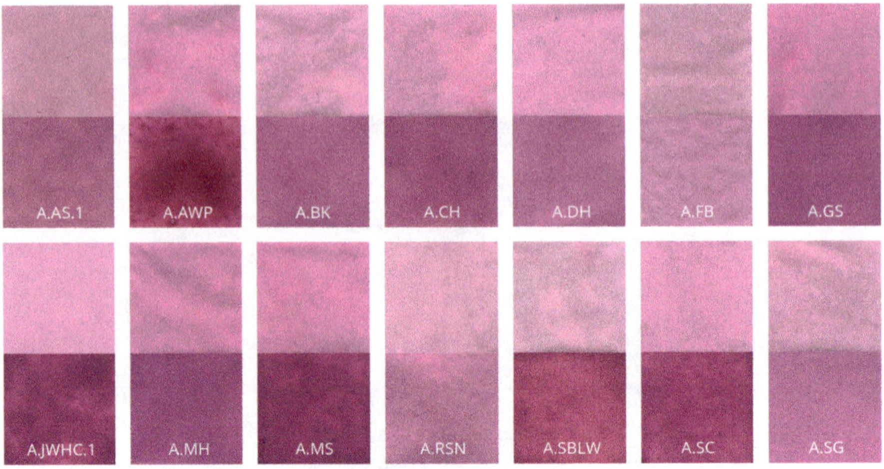


FIGURE 91 Different shades of red lake pigments made of redwood/Brazilwood following the same procedures and with the same amount of ingredients

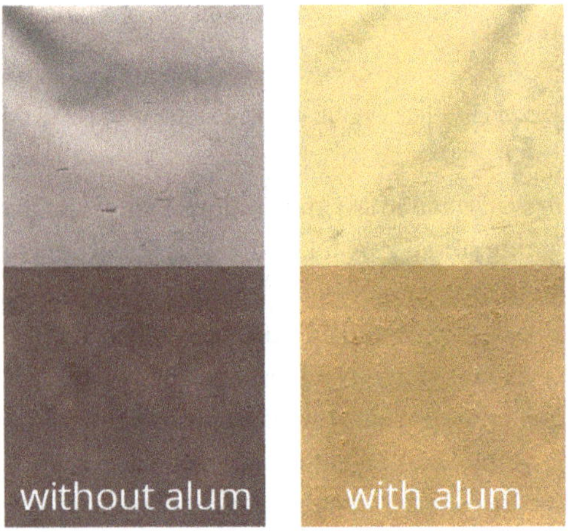


FIGURE 92 Yellow onion skin lake pigments without and with potash alum

from garden flowers that regularly appeared in early modern florilegia? This idea is not entirely divergent from historical practices. Many clothlets of the late medieval period were made of flower petals, only, rather than extracting the dyes through heating, the maker crushed the petals into a mush and filtered it before saturating a clothlet with the juice.[33] Combining the ideas of extracting colours from garden flower petals and playing with the two colour-keeping methods, the series of twelve small paintings (Fig. 93) came into being. The style of the watercolours was a result of spontaneously

33 For instance, recipe 28 in the Strasbourg Manuscript lists the juice of red poppies as the main ingredient, see Neven, *The Strasbourg Manuscript*, 98–99.

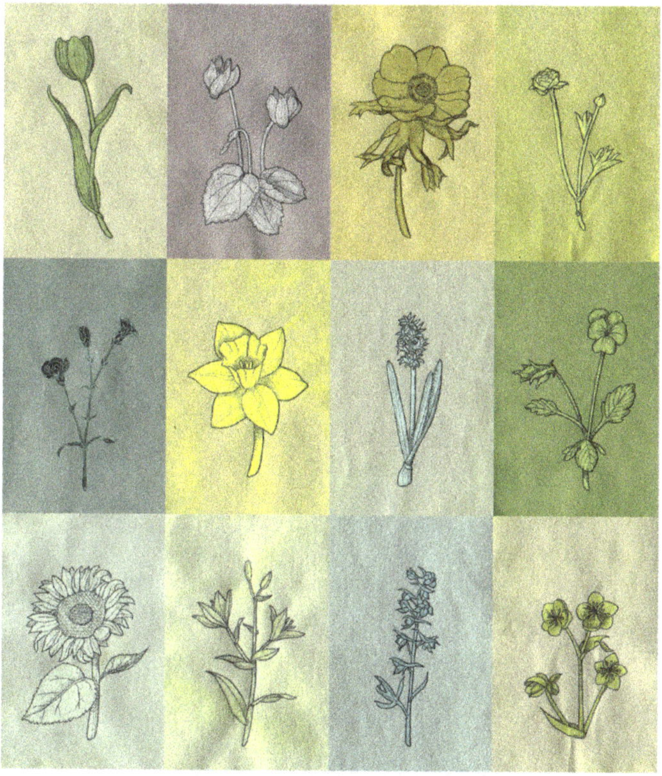

FIGURE 93 Twelve small watercolours painted with inks and paints made of the petals of the depicted flowers

reacting to the behaviours of these inks and paints to see where a playful process can lead.

I painted each watercolour with the colours made from the depicted flower, after each flower was made into both clothlets and lake pigments. The selection of flowers is solely based on what my local vendors could provide, and the order of the watercolours on the next pages are arranged chronologically according to when they were painted. All the colours were made with a standardised procedure. The clothlets were saturated with the juice of crushed flower petals, infused with a little water and alum. For making the lake pigments, the petals were simmered for 20 to 30 minutes in water. The dyebath was filtered and reheated with a little alum before being precipitated onto Champagne chalk. Both clothlets and lake pigments were left to dry and sit for a week or two. Afterwards, I reactivated the clothlets by soaking some clippings in gum arabic and used the inks to tint the backgrounds. The flat layer of paints on the flowers are made by tempering the lake pigments with gum arabic with a small mortar and pestle. To finish the watercolours, I added outlines and details to the flowers with graphite or white coloured pencil.

## 3.1 *Pigment Collections: Floral Colours*

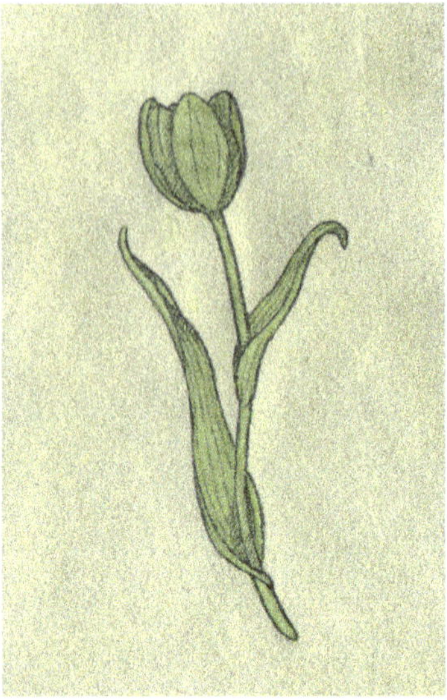

FIGURE 94
Tulip painted with colours extracted from tulip petals

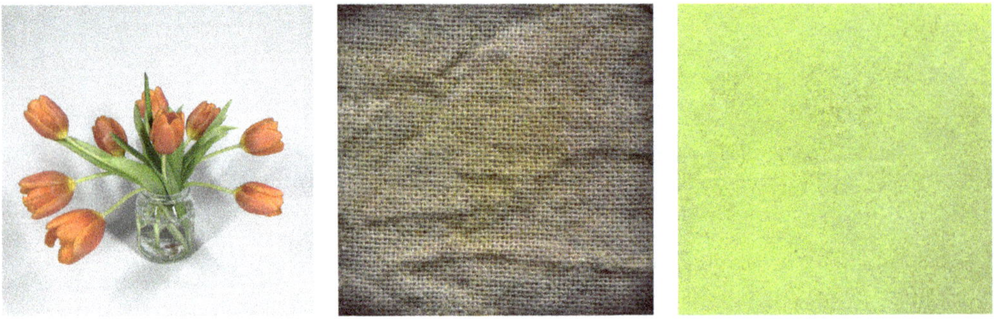

FIGURE 95 F1, Tulip (Clothlet), *Tulipa sp.*, Jessie Wei-Hsuan Chen

FIGURE 96 F2, Tulip (Pigment), *Tulipa sp.*, Jessie Wei-Hsuan Chen

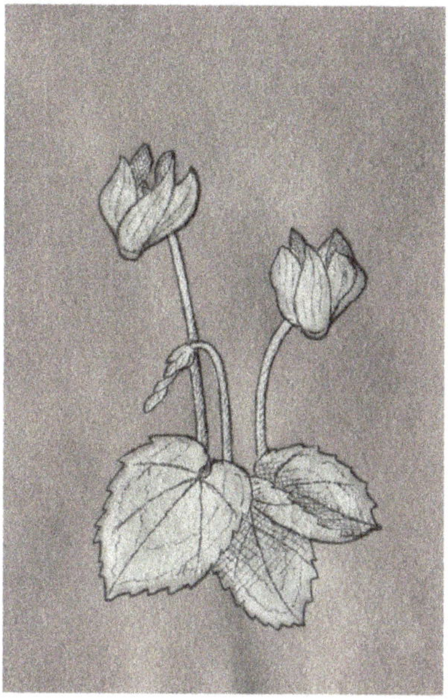

FIGURE 97
Cyclamen painted with colours extracted from cyclamen petals

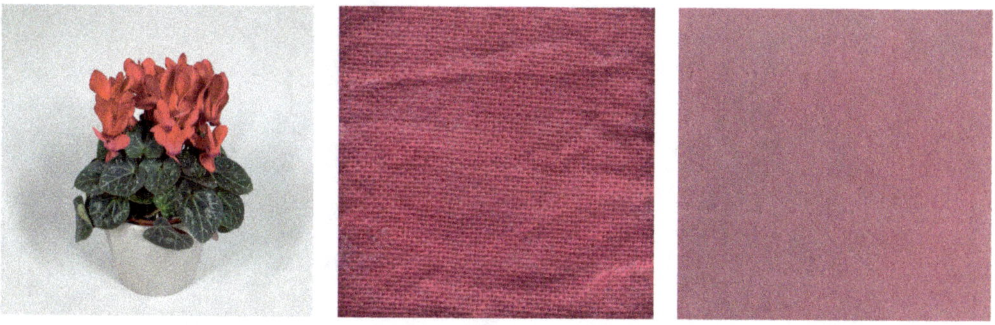

FIGURE 98 F3, Cyclamen (Clothlet), *Cyclamen sp.*, Jessie Wei-Hsuan Chen

FIGURE 99 F4, Cyclamen (Pigment), *Cyclamen sp.*, Jessie Wei-Hsuan Chen

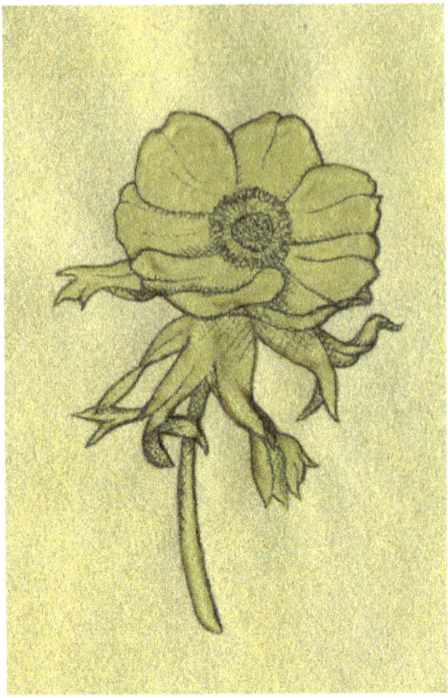

FIGURE 100
Poppy anemone painted with colours extracted from poppy anemone petals

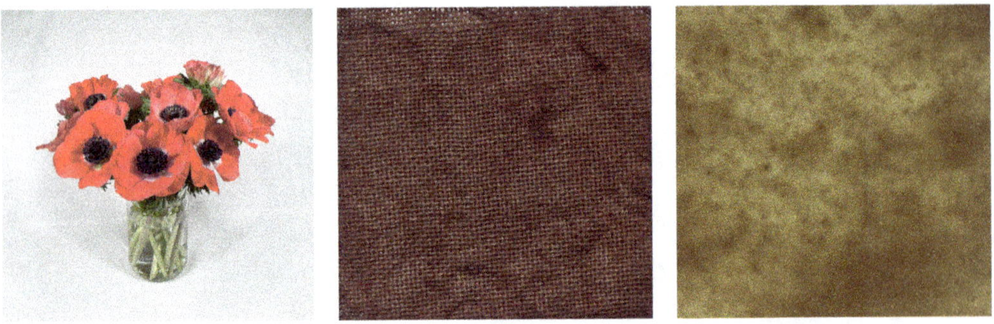

FIGURE 101 F5, Poppy Anemone (Clothlet), *Anemone coronaria*, Jessie Wei-Hsuan Chen

FIGURE 102 F6, Poppy Anemone (Pigment), *Anemone coronaria*, Jessie Wei-Hsuan Chen

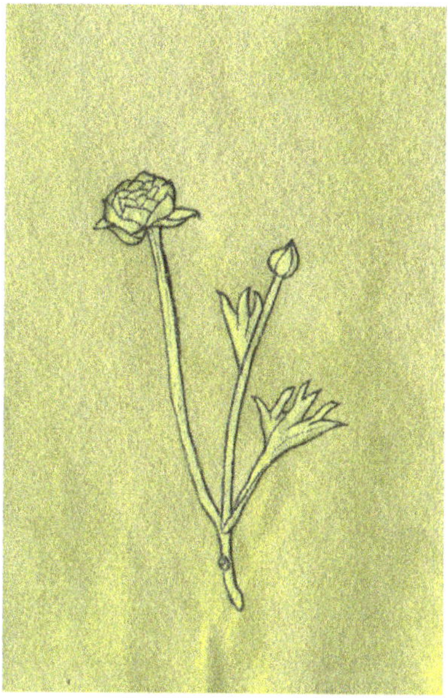

FIGURE 103
Persian buttercup painted with colours extracted from Persian buttercup petals

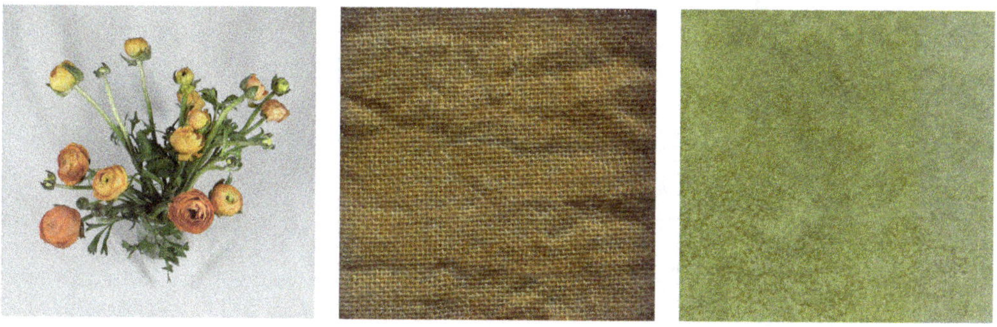

FIGURE 104 F7, Persian Buttercup (Clothlet), *Rhanunculus asiaticus*, Jessie Wei-Hsuan Chen

FIGURE 105 F8, Persian Buttercup (Pigment), *Rhanunculus asiaticus*, Jessie Wei-Hsuan Chen

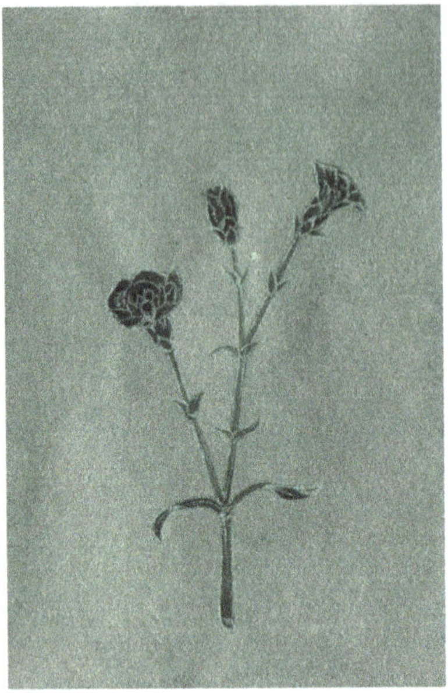

FIGURE 106
Carnation painted with colours extracted from carnation petals

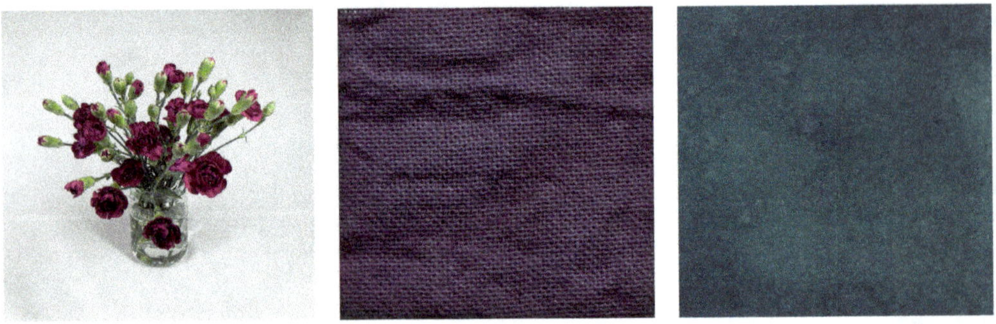

FIGURE 107 F9, Carnation (Clothlet), *Dianthus caryophyllus*, Jessie Wei-Hsuan Chen

FIGURE 108 F10, Carnation (Pigment), *Dianthus caryophyllus*, Jessie Wei-Hsuan Chen

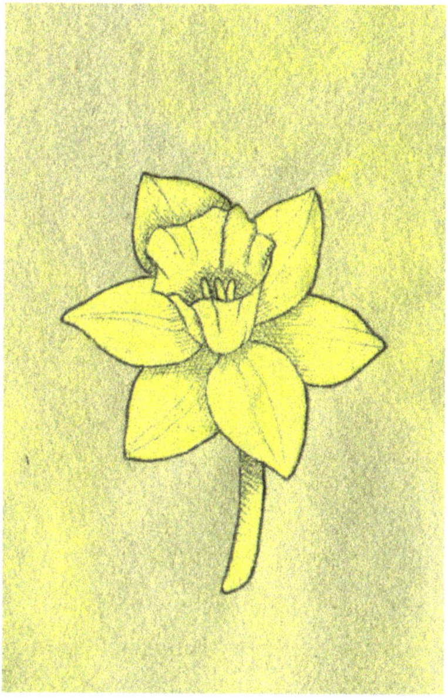

FIGURE 109
Narcissus painted with colours extracted from narcissus petals

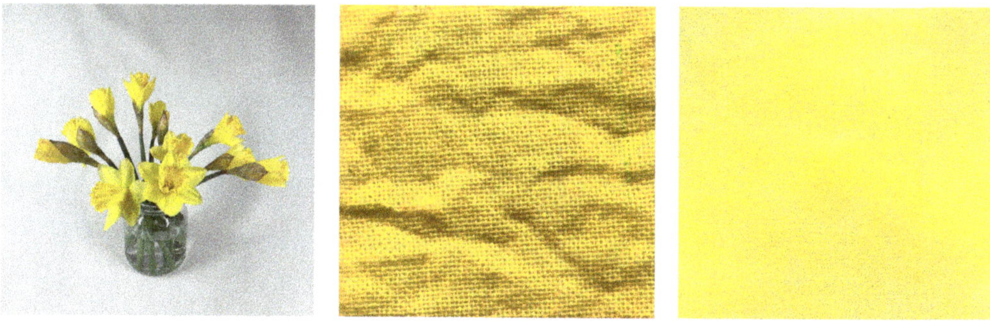

FIGURE 110 F11, Narcissus (Clothlet), *Narcissus sp.*, Jessie Wei-Hsuan Chen

FIGURE 111 F12, Narcissus (Pigment), *Narcissus sp.*, Jessie Wei-Hsuan Chen

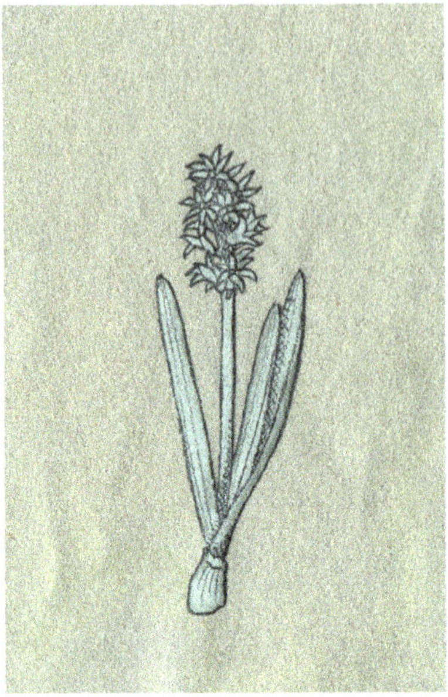

FIGURE 112
Hyacinth painted with colours extracted from hyacinth petals

FIGURE 113 F13, Hyacinth (Clothlet), *Hyacinthus sp.*, Jessie Wei-Hsuan Chen

FIGURE 114 F14, Hyacinth (Pigment), *Hyacinthus sp.*, Jessie Wei-Hsuan Chen

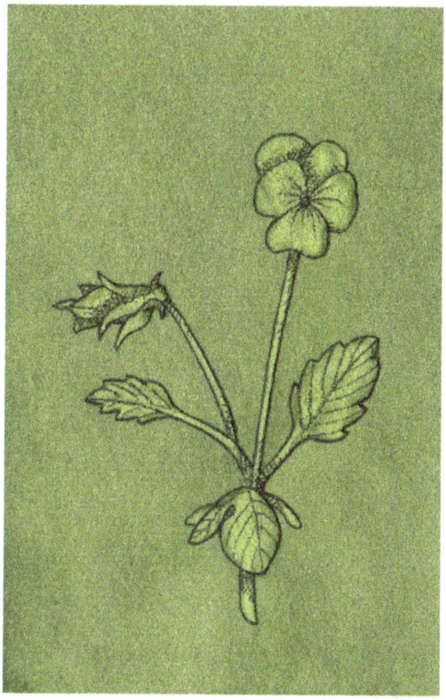

FIGURE 115
Pansy painted with colours extracted from pansy petals

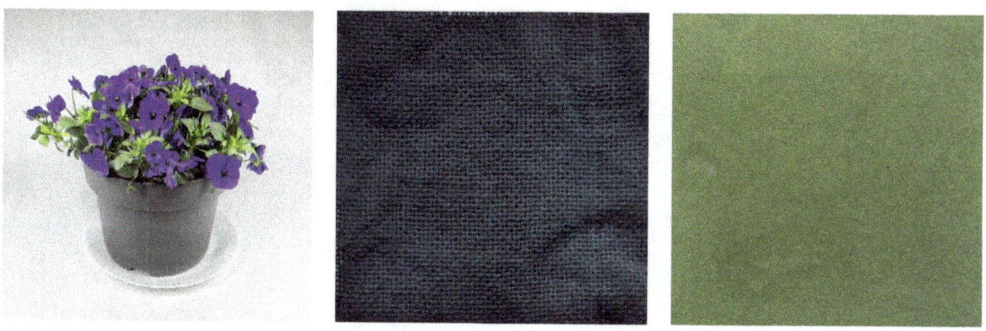

FIGURE 116 F15, Pansy (Clothlet), *Viola sp.*, Jessie Wei-Hsuan Chen

FIGURE 117 F16, Pansy (Pigment), *Viola sp.*, Jessie Wei-Hsuan Chen

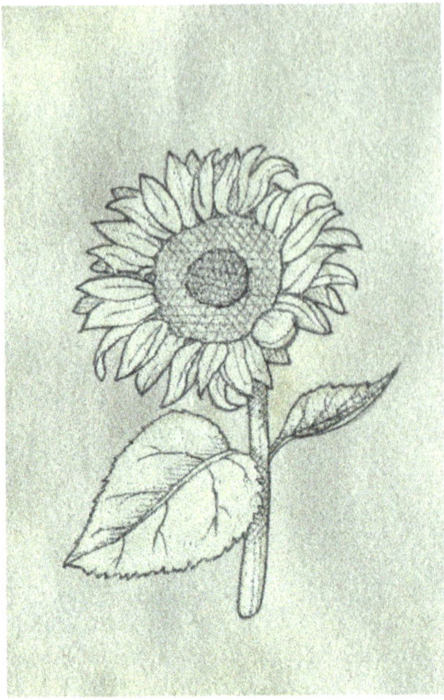

FIGURE 118
Sunflower painted with colours extracted from sunflower petals

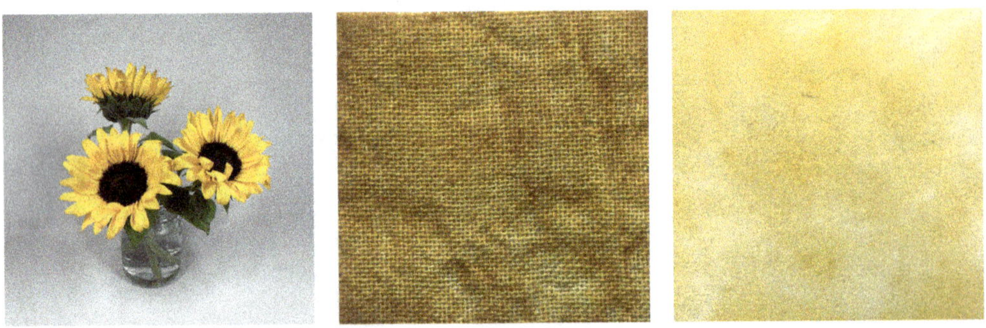

FIGURE 119 F17, Sunflower (Clothlet), *Helianthus annuus*, Jessie Wei-Hsuan Chen

FIGURE 120 F18, Sunflower (Pigment), *Helianthus annuus*, Jessie Wei-Hsuan Chen

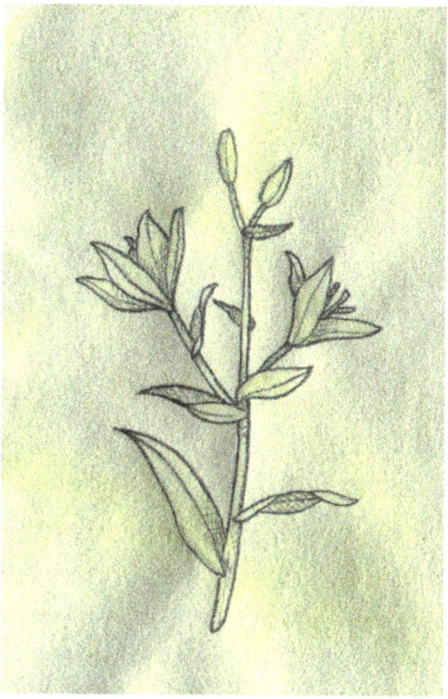

FIGURE 121
Lily painted with colours extracted from lily petals

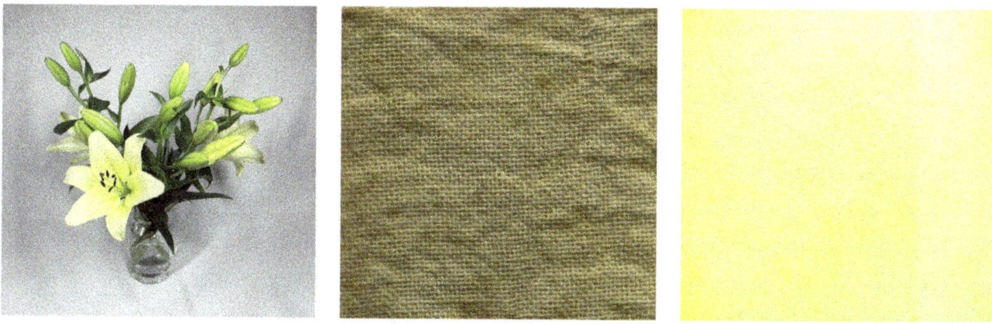

FIGURE 122 F19, Lily (Clothlet), *Lilium sp.*, Jessie Wei-Hsuan Chen

FIGURE 123 F20, Lily (Pigment), *Lilium sp.*, Jessie Wei-Hsuan Chen

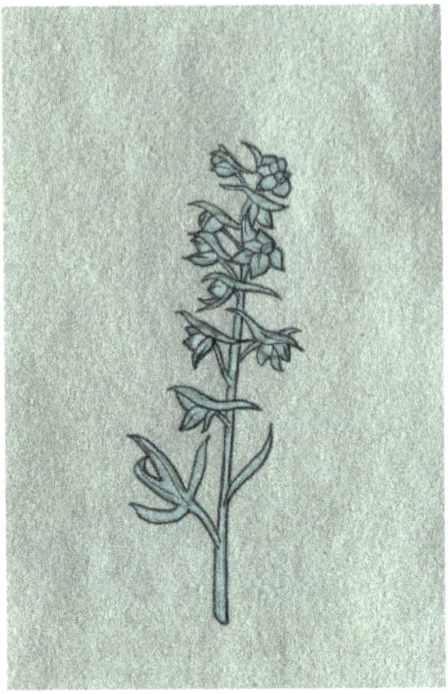

FIGURE 124
Larkspur painted with colours extracted from larkspur petals

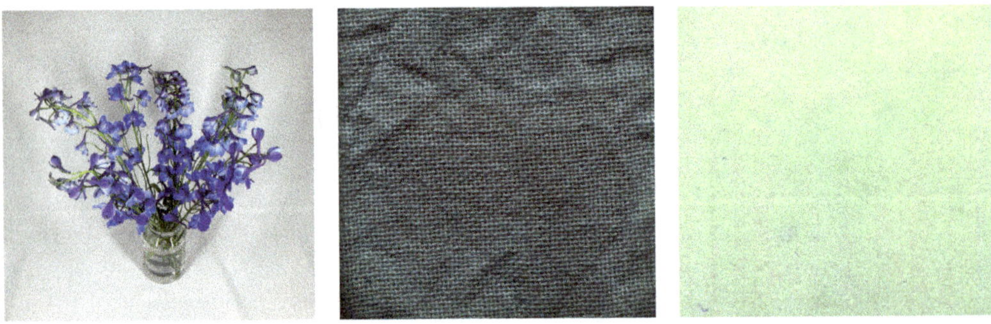

FIGURE 125 F21, Larkspur (Clothlet), *Delphinium sp.*, Jessie Wei-Hsuan Chen

FIGURE 126 F22, Larkspur (Pigment), *Delphinium sp.*, Jessie Wei-Hsuan Chen

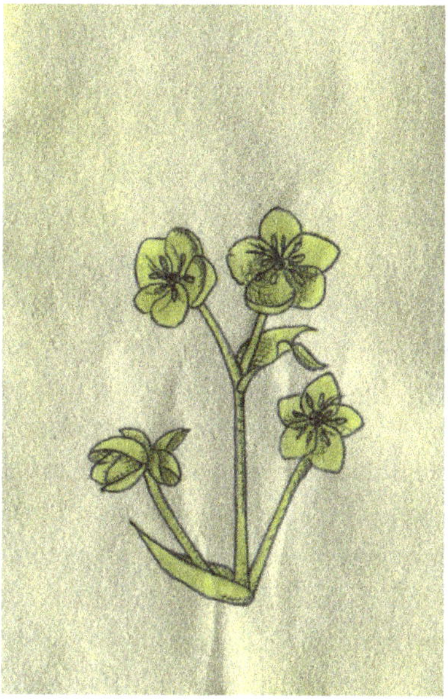

FIGURE 127
Hellebore painted with colours extracted from hellebore petals

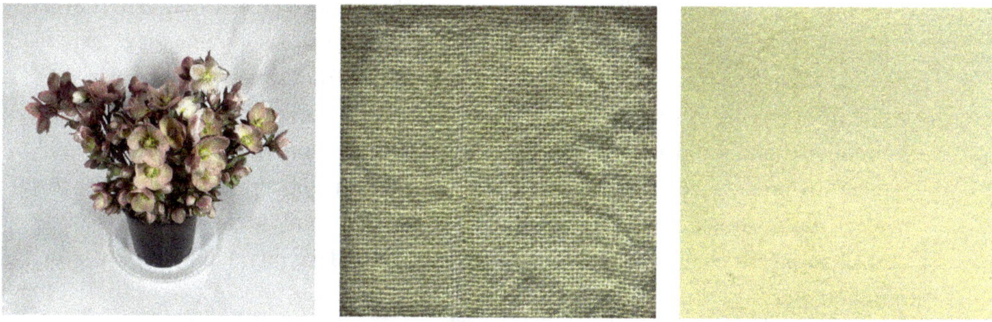

FIGURE 128 F23, Hellebore (Clothlet), *Helleborus sp.*, Jessie Wei-Hsuan Chen

FIGURE 129 F24, Hellebore (Pigment), *Helleborus sp.*, Jessie Wei-Hsuan Chen

## 4 Finding Colours, Fading Colours

Makers who are attentive to the materials they use might be uneasy when reading that I applied a relatively standardised process to produce the inks and pigments for the small watercolours. Every material has its own properties. Thus, each requires different handling to bring out its ideal qualities, and it usually entails many trials and errors to find how to best do so. If, as mentioned earlier, each plant needs different heating time (and temperature) to extract dye, the logical conclusion is that a standardised making process would inevitably produce bad colour from most plants. However, from my experimentations at the early stage of this project, although the colours obtained from this process are not optimal, they provide enough information of the colour a plant yields. For example, simmering an avocado pit for half an hour is too short to get a proper deep and rich pink dye, but it still produced a pink, albeit paler than the ideal. This mediocre colour might be a low-grade artist's material, but the standardised process offered feasibility in exploring more plant colours within a limited timeframe. It reduced the need for multiple tries and made planning workdays easier because it took more-or-less the same time to make a lake pigment out of each plant-based dyestuff.

Similarly, testing different modifiers, such as vinegar, baking soda, iron, and copper, to change the colours of a dyebath is not part of this project. Present-day expert dyers and other practitioners working with plant-based colours have done a lot of experimentation on this front, and a quick search on the Internet shows many examples.[34] Aside from serving as a substrate, potash alum can function as a modifier, as well as a mordant to better bind a colour to a surface and improve its permanency (to a degree), depending on how, how much, and when in the making process it is used.[35] For this project, alum worked as the only modifier (and possibly a mordant on a lower level), and Champagne chalk served as the substrate for the lake pigments with inventory numbers starting with F and E.[36]

Applying the standardised process and using the same ingredients for making every lake pigment created close conditions—a similar 'environment'—for the colours in this project. This not only shows the major colour one can extract from each of the tested plant, as alum is one of the most ubiquitous materials used in colour-making, it also exhibits how this environment affects the colours. Some dyes bind better with

34 Medlej, *Wild Inks and Paints*. See also the Resources tab on the website of the Slow Lane Studio (https://www.slowlanestudio.co.uk/resources).

35 Most sources that are written for specific groups of readers (dyers, conservators, etc.) usually mention the use of potash alum for only one or a couple purposes. For simple descriptions on the various functions of this metal salt, see Medlej, *Wild Inks and Paints*, 22.

36 Lake pigments with inventory numbers starting with H used alum as the substrate.

chalk and thrive better on this substrate than others. Having this overview of their conditions, one can always return to a specific plant and further experiment with it by adding different modifiers or precipitating onto other substrates.

For this project, the standardised process offered an additional opportunity to engage with people with various hands-on skill levels. Between February and March 2021, my collaborator, Anna Svensson, and I organised four online workshop sessions to virtually make lake pigments together in real time. The workshops aimed to explore conducting hands-on experimentations and engaging with performative methods in an online environment. The participants came from several fields, including history and its sub-fields, archaeobotany, contemporary art, (painting) conservation, and museum curation. While everyone is an expert in their field, many participants had never extracted colours and/or made lake pigments before. The written-down steps of the standardised process became a recipe, from which everyone made a magenta lake out of redwood/Brazilwood (Fig. 91) during the first half of the workshop.[37] The participants selected one or two plants they found interesting and extracted the dyes to make more lake pigments in the second half, and some of them contributed their results to the project, displayed on the following pages.

Collectively, the contributors and I tested around fifty plants and produced sixty colours. We found colours in plants that are both mundane and rare in Europe or North America.[38] We found colours in plants that are common (historical) dyestuffs and plants that may be rather creative for this purpose. There were also several pigments made with assorted plants. This collection of experimental colours is of a decent size, but they represent only the smallest fraction of the vast plant world, even though produced with a standardised process and a group of makers. 60 colours provided a rich palette, and I used nineteen of them to paint the colourful bouquet of tulips (Fig. 130). Similar to the early modern painting technique for reproducing Franciscus De Geest's watercolour, I started with a flat layer of mid-tone for local colours, and then brushed on shadows and highlights. Because there were plenty of hues and degrees of saturation from which to choose, I layered and blended with only pure colours and did not mix the paints.

More valuable perhaps than the produced colour samples and watercolours, however, were the observations, discussions, and questions from the workshops. In the

37 The redwood/Brazilwood for the workshop were shavings of *Haematoxylum brasiletto* (also known as Mexican logwood) according to the supplier Kremer Pigmente (under Product Information, https://www.kremer-pigmente.com/en/shop/dyes-vegetable-color-paints/natural-organic-dyes-vegetable-color-paints/36150-brazilwood-shavings.html).

38 While online workshop has many advantages, it creates other challenges. We could only accommodate participants from Europe and North America due to practicalities of shipping supplies for the experiments.

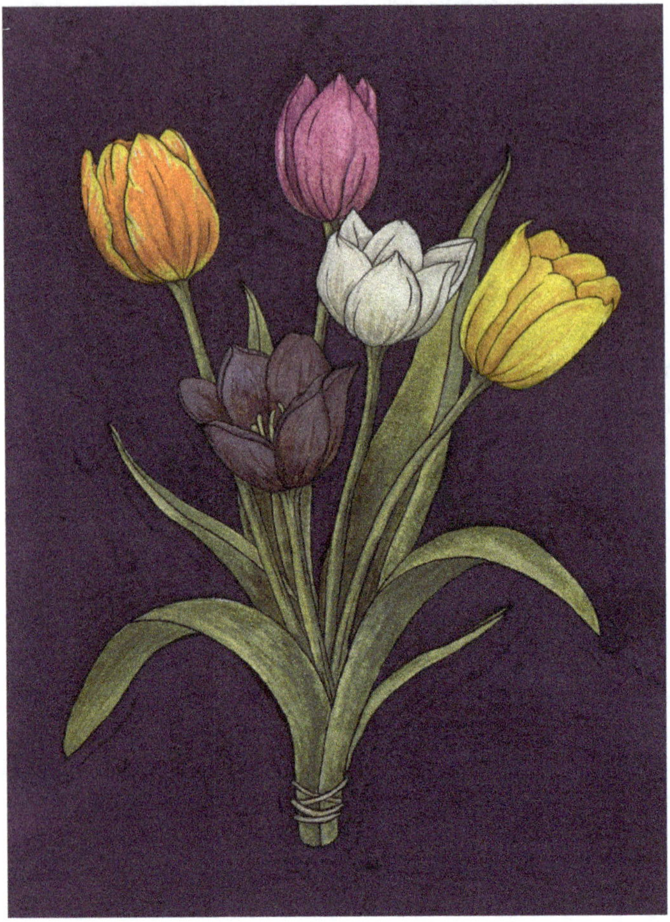

FIGURE 130 Watercolour of a tulip bouquet painted with experimental plant-based lake pigments

process of finding colours in these plants, we traversed the many layers of where and how plants have been used, both in our own lives and through the ages. For example, many colours from the collection are made of plants we consume through various means, as food, as spices, and/or as drinks. Most of these edible plants have a long local or global history that are often, but not always, connected to migration and trade.[39] The chemistry and technologies behind (historical) colour-making was another topic

39 See particularly current and new research and projects from workshop participants Sasha Gora (http://www.lsashagora.com/), Merit Honderlink, and Suzanne Bernhardt (http://suzannebernhardt.info/).

that was interesting for many. What kinds of issues did makers have to anticipate and work with when pure chemicals were not available? How was temperature regulated and measured before the common use of a thermometer? As making the lake pigments required the engagement with almost all senses, the taste and smell of plants also became part of the inquiries. With a perfumer in the virtual room, we learned about the very different processes and devices for extracting plant colours and scents.[40]

A recurring topic throughout the four workshop sessions was the fugitive nature of organic colours. Because many of the pigment samples and paint swatches easily fade when exposed to light, keeping them in closed pages usually helps with the preservation of the colours (Fig. 131), sometimes for many centuries, as it did for historical watercolours in manuscripts and albums. However, darkness does not prevent many colours in this project, especially those made with non-traditional dyestuffs, from deteriorating. Aside from light, oxidisation and humidity can also change the colour, often to a less desirable shade or hue. The small watercolour painted with the ink and pigment made of cyclamen petals has already become muted within a few months (Fig. 132). It is inevitable that the painting of the tulip bouquet will lose its vibrancy at some point.

The paintings and the colour collection in its physical form are living pieces. They will be ever-changing until all the colours turn into, one can speculate, browns and greys. How does one conserve something like this project that is, by nature, not conservable? Does one invest in a lot of resources to prevent it from changing? Or does one embrace its ephemerality, but document how it changes every so often?[41] Conversely, how about digital preservation?[42] The scanned and colour-corrected samples and swatches in this booklet will remain somewhat permanent to show how the colours looked when they were first made. However, the booklet can easily become obsolete when no digital program recognises the PDF format anymore. Unlike De Geest's florilegia, maybe with the exception of the reproduction, this project—both its material and digital form—presumably will not survive 400 years and will be gone without a trace.

---

40 See the work of Liza Witte (http://www.lizawitte.com/).

41 These are, of course, not new questions, but those that have been and continue to be largely discussed and debated by (object) conservators. One article that provided much food for thought for this project is Henderson, 'Beyond Lifetimes.'

42 Digital preservation aims to ensure that digital materials can continually be accessible in the future.

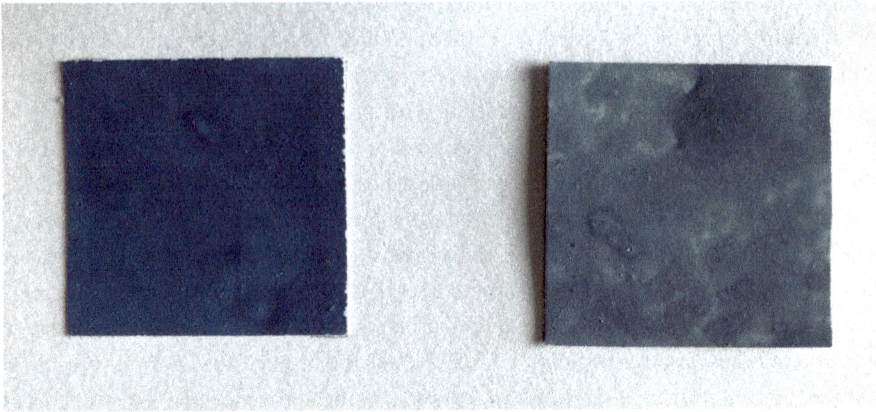

FIGURE 131 Swatches of red cabbage lake pigment painted on the same day and stored in different ways for five months. Left: kept in the dark; right: exposed in daylight (but not direct sunlight)

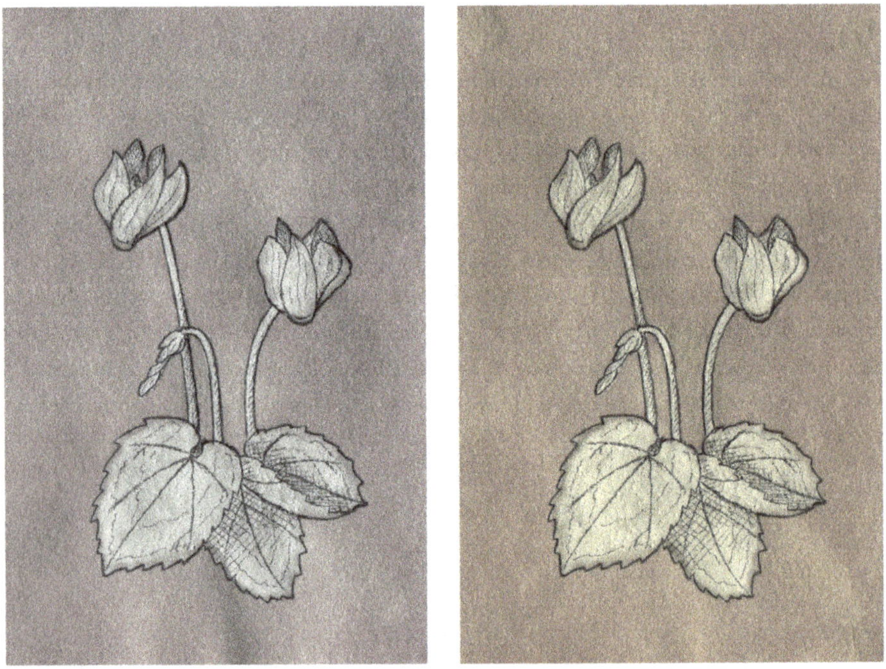

FIGURE 132 Left: watercolour of cyclamen scanned a week after it was painted; right: watercolour of cyclamen scanned about four-and-a-half months after it was painted

## 4.1 *Pigment Collections: Experimental Colours*

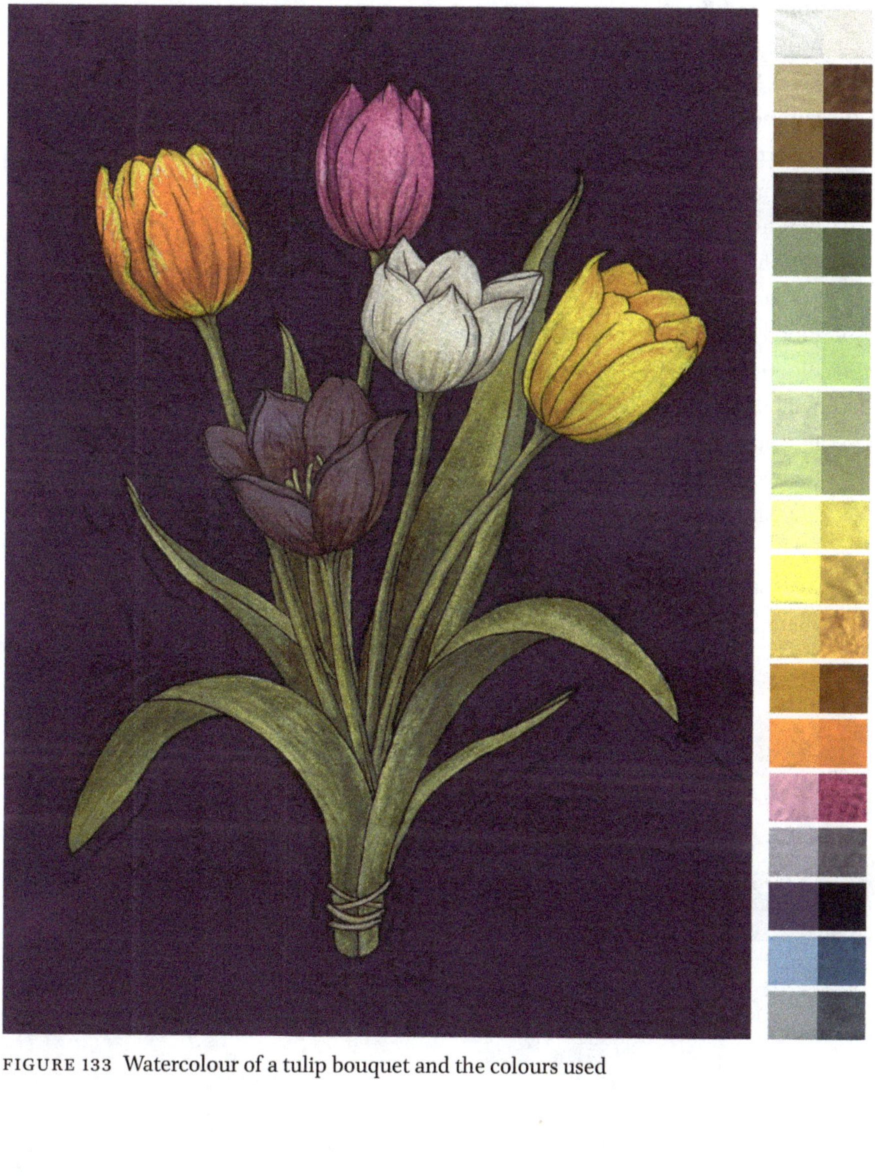

FIGURE 133 Watercolour of a tulip bouquet and the colours used

*Colours Used*

- Saffron (E1)
- Thai Tea (E8)
- Alder Cone (E9)
- Black Walnut Husk (E10)
- Lupine (E23)
- Assorted Petals (E24)
- Wisteria (E25)
- Bouquet (E26)
- Artichoke (E27)
- Spinach (E31)
- Blood Orange Peel (Fresh) (E36)
- Blood Orange Peel (Dried) (E39)
- Buckthorn Berry (Unripe) (E40)
- Bermuda Buttercup (E44)
- Redwood/Brazilwood (E48)
- Blackberry (E50)
- Logwood (ES1)
- Red Cabbage (E54)
- Black Turtle Bean (E56)

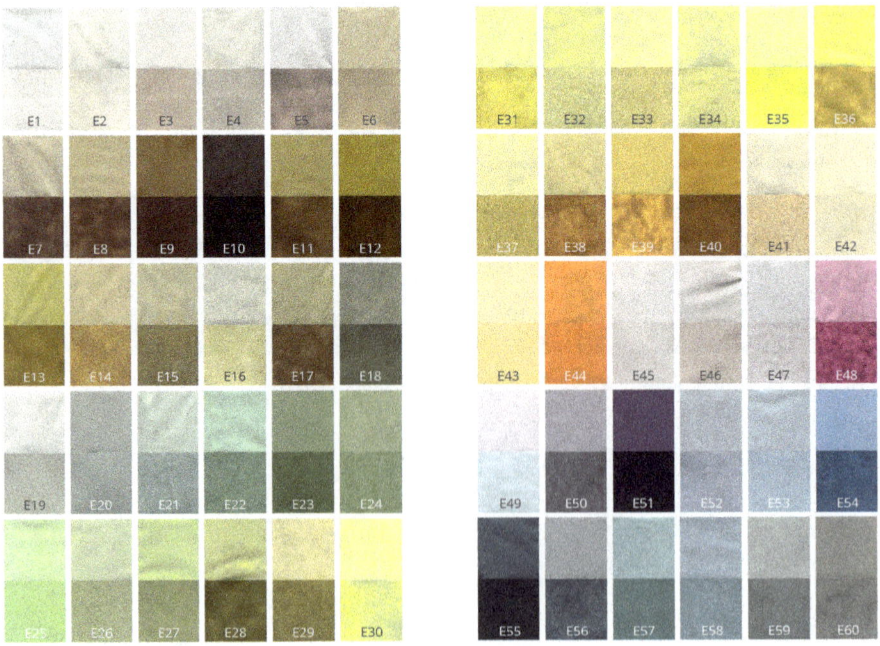


FIGURE 134 Overview of the collection of experimental colours

FIGURE 135 E1, Saffron (Dried), *Crocus sativus*, Jessie Wei-Hsuan Chen

FIGURE 136 E2, Cape Gooseberry (Fresh), *Physalis peruviana*, Ruth Sargent Noyes

FIGURE 137 E3, Red Sandalwood (Powdered), *Pterocarpus santanin*, Jessie Wei-Hsuan Chen

FIGURE 138 E4, Birch Bark (Fresh), *Betula sp.*, Brandy Kraft

FIGURE 139 E5, Avocado Pit (Dried), *Persea americana*, Jessie Wei-Hsuan Chen

FIGURE 140 E6, Beetroot (Fresh), *Beta vulgaris*, Dominik Hünniger

FIGURE 141 E7, Black Oak (Powdered), *Quercus velutina*, Jessie Wei-Hsuan Chen

FIGURE 142 E8, Thai Tea (Dried), *Camellia sinensis*, Jessie Wei-Hsuan Chen

FIGURE 143 E9, Alder Cone (Dried), *Alnus sp.*, Anna Svensson

FIGURE 144 E10, Black Walnut Husk (Dried), *Juglans nigra*, Sara Charles

FIGURE 145 E11, Mimosa (Dried), *Acacia dealbata*, Suzanne Bernhardt and Liza Witte

FIGURE 146 E12, Buckthorn Berry (Ripe, Dried), *Rhamnus sp.*, Jessie Wei-Hsuan Chen

FIGURE 147 E13, Pomegranate Peel (Fresh), *Punica granatum*, Jessie Wei-Hsuan Chen

FIGURE 148 E14, Turmeric (Powdered), *Curcuma longa*, Jessie Wei-Hsuan Chen

FIGURE 149 E15, Mahonia Bark (Fresh), *Mahonia sp.*, Anna Svensson

FIGURE 150 E16, Grape Tomato (Fresh), *Solanum lycopersicum*, Jessie Wei-Hsuan Chen

FIGURE 151 E17, Coffee (Powdered), *Cojfea sp.*, Jessie Wei-Hsuan Chen

FIGURE 152 E18, Velvet Roll-Rim (Fresh), *Tapinella atrotomentosa*, Anna Svensson

FIGURE 153 E19, Vanda Orchid (Fresh), *Vanda sp.*, Brandy Kraft

FIGURE 154 E20, Cranberry (Fresh), *Vaccinium macrocarpon*, Charlotte Hoitsma

FIGURE 155 E21, Cornflower (Dried), *Centaurea cyanus*, Jessie Wei-Hsuan Chen

FIGURE 156 E22, Assorted Carnations (Fresh), *Dianthus caryophyllus*, Jessie Wei-Hsuan Chen

FIGURE 157 E23, Lupine (Fresh, Precipitated onto Pure Calcium Carbonate), *Lupinus sp.*, Eva-Maria Spampinato

FIGURE 158 E24, Assorted Petals (Dried), Jessie Wei-Hsuan Chen

FIGURE 159 E25, Wisteria (Fresh, Precipitated onto Pure Calcium Carbonate), *Wisteria sp.*, Eva-Maria Spampinato

FIGURE 160 E26, Bouquet (Fresh, Tulip + Wild Blueberry), *Tulipa sp.* + *Vaccinium myrtillus*, Jessie Wei-Hsuan Chen

FIGURE 161 E27, Artichoke (Fresh), *Cynara cardunculus*, Sasha Gora

FIGURE 162 E28, Cape Gooseberry Husk (Fresh), *Physalis peruviana*, Ruth Sargent Noyes

FIGURE 163 E29, Mallow (Dried), *Malva sylvestris*, Anna Svensson

FIGURE 164 E30, Carrot Leaf (Fresh), *Daucus carota*, Merit Honderlink

FIGURE 165 E31, Spinach (Fresh), *Spinacia oleracea*, Jessie Wei-Hsuan Chen

FIGURE 166 E32, Lisianthus (Fresh), *Eustoma russellianum*, Jessie Wei-Hsuan Chen

FIGURE 167 E33, Ivy Leaf (Fresh), *Hedera sp.*, Suzanne Bernhardt and Liza Witte

FIGURE 168 E34, Safflower (Dried), *Carthamus tinctorius*, Jessie Wei-Hsuan Chen

FIGURE 169 E35, Parsley (Fresh), *Petroselinum crispum*, Dominik Hünniger

FIGURE 170 E36, Blood Orange Peel (Fresh), *Citrus sinensis*, Giulia Simonini

FIGURE 171 E37, Rosehip (Dried), *Rosa sp.*, Suzanne Bernhardt and Liza Witte

FIGURE 172 E38, Yellow Onion Skin (Dried), *Allium cepa*, Jessie Wei-Hsuan Chen

FIGURE 173 E39, Blood Orange Peel (Dried), *Citrus sinensis*, Giulia Simonini

FIGURE 174 E40, Buckthorn Berry (Unripe, Dried), *Rhamnus sp.*, Jessie Wei-Hsuan Chen

FIGURE 175 E41, Annatto (Dried), *Bixa orellana*, Suzanne Bernhardt and Liza Witte

FIGURE 176 E42, Papaya Peel (Fresh), *Carica papaya*, Francesca Bewer

FIGURE 177 E43, Bougainvillea (Fresh, Precipitated onto Pure Calcium Carbonate), *Bougainvillea sp.*, Eva-Maria Spampinato

FIGURE 178 E44, Bermuda Buttercup (Fresh, Precipitated onto Pure Calcium Carbonate), *Oxalis pes-caprae*, Eva-Maria Spampinato

FIGURE 179 E45, Chinese Money Plant (Fresh), *Pilea peperomioides*, Marleen Schans

FIGURE 180 E46, Avocado Pit (Fresh), *Persea americana*, Jessie Wei-Hsuan Chen

FIGURE 181 E47, Folium (Clothlet Dyed with Turnsole), *Chrozophora tinctoria*, Sara Charles

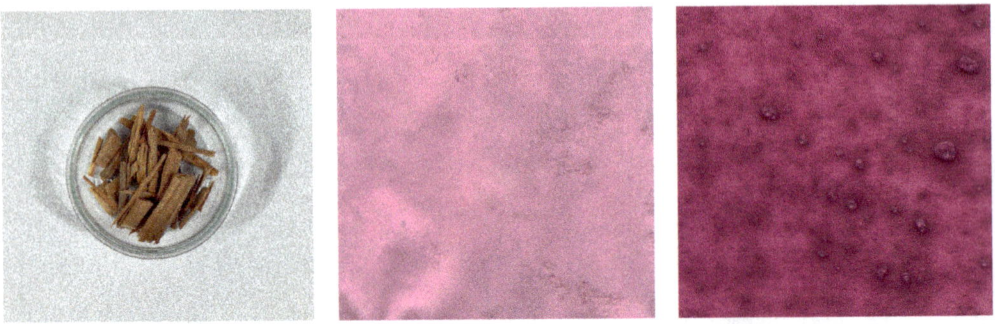

FIGURE 182 E48, Redwood/Brazilwood (Dried), *Haematoxylum brasiletto*, Jessie Wei-Hsuan Chen

FIGURE 183 E49, Assorted Plants 1 (Fresh Magnolia Petal + Raspberry), *Magnolia × soulangeana + Rubus idaeus*, Kristan Hanson

FIGURE 184 E50, Blackberry (Frozen), *Rubus futicosus*, Merit Honderlink

FIGURE 185 E51, Logwood (Dried), *Haematoxylum campechianum*, Jessie Wei-Hsuan Chen

FIGURE 186 E52, Raspberry (Fresh), *Rubus idaeus*, Jessie Wei-Hsuan Chen

FIGURE 187 E53, Pomegranate Seed (Fresh), *Punica granatum*, Jessie Wei-Hsuan Chen

FIGURE 188 E54, Red Cabbage (Fresh), *Brassica oleracea convar. capitata var. rubra*, Annemarie Kok

FIGURE 189 E55, Assorted Roses (Fresh, Red Naomi + Unknown Varieties), *Rosa sp.*, Annika Windahl-Pontén

FIGURE 190 E56, Black Turtle Bean (Dried), *Phaseolus vulgaris*, Jessie Wei-Hsuan Chen

FIGURE 191 E57, Peruvian Lily (Fresh), *Alstroemeria sp.*, Dominik Hünniger

FIGURE 192 E58, Assorted Plants 2 (Frozen, Red Cabbage + Beetroot), *Brassica oleracea convar. capitata var. rubra + Beta vulagris*, Jessie Wei-Hsuan Chen

FIGURE 193 E59, Red Grape Peel (Fresh), *Vitis vinifera*, Jessie Wei-Hsuan Chen

FIGURE 194 E60, Hibiscus (Dried), *Hibiscus sp.*, Francesca Bewer

## 5 Epilogue

Looking back at the making of this digital booklet and its material counterparts (Fig. 195), the process was unintentionally transhistorical. Being confronted with the fading and ever-changing colours of this project reminds one that most of what we see today from the past depends heavily on chance and other inconsistent aspects of survival, despite the great efforts archivists, curators, and keepers of collections have put in to preserve as much as possible. High material quality and suitable environment mark two factors that improve the chance for something to last centuries and longer, but they can also skew the types of evidence that were preserved. More often than not, records of all kinds are lost along the way, and reconstructing the past can never give a definitive picture and interpreting historical evidence needs to be done with care and nuance, which requires a lot of training. What is more important is to continue to ask big-and-small questions, carefully (re)assess our current understandings, and renew information where needed.

By delving into the technical making of the pictorial genre of florilegium, it has been shown that there is a whole world of materials behind the painted layers that transcends the specific time and place of the seventeenth-century Dutch Republic. The flowers Franciscus de Geest depicted are more widely available to the masses today, and there is still a sizeable group of artists and practitioners that use artist's materials, such as natural pigments, similar to the early modern painter. Furthermore, our connections and relationships with the plant world and nature continue to be a prominent theme for people from many professions dealing with the past, present, and future to explore, understand, and conserve.

The making process of this project also highlighted the importance of exchange and collaboration. The many conversations I had with conservators and historians working on adjacent topics shaped most of my decision-making when sourcing and selecting historical materials and techniques to reproduce De Geest's watercolour. Social media posts from present-day dyers and practitioners working with plants in various capacities offered a lot of insight and tips on the material properties and qualities of several plant-based dyestuffs. The questions raised during the workshop sessions, as shown in the previous section, and many more that remain on the back burner, are useful starting points for several potential projects in the future. Whether an expert or novice on a subject, everyone has something to contribute; revisiting basic information by explaining it to beginners can often lead to seeing something familiar under a new light. Similarly, while a concept or topic may have been thoroughly investigated or over-saturated in one field, it may be unexplored or inspirational in another. However, collaboration takes effort to communicate because oftentimes every group has its own 'language.' How historians view certain materials, concepts, and words can be drastically different from how artists and scientists (all in the broadest sense) do. With respect and open-mindedness, cross-fertilisation can collectively expand our knowledge on many fronts.

Even though this booklet is at its end, the cross-fertilisation does not have to be. For the workshop sessions, the #plantingpigments hashtag was created for the participants to follow the project and share their own process on social media. The hashtag will remain a virtual venue for sharing process, exchange ideas, and/or finding inspirations for anyone who is interested in engaging. When reflecting upon the main components of this project, it boiled down to three keywords: *plants*, *colours*, and *makings*. As stated in the prologue, through plant-based lake pigments, this booklet shows nothing more than one possibility to bring plants, colours, and makings together. There are easily hundreds of possibilities to combine two of the keywords, let alone all three. If this booklet has offered some inspiration or motivation for (re)visiting old and new ideas, it would be terrific to see your (process of) academic, scientific, professional, creative, or recreational endeavours with plants, colours, and/or makings through the #plantingpigments hashtag!

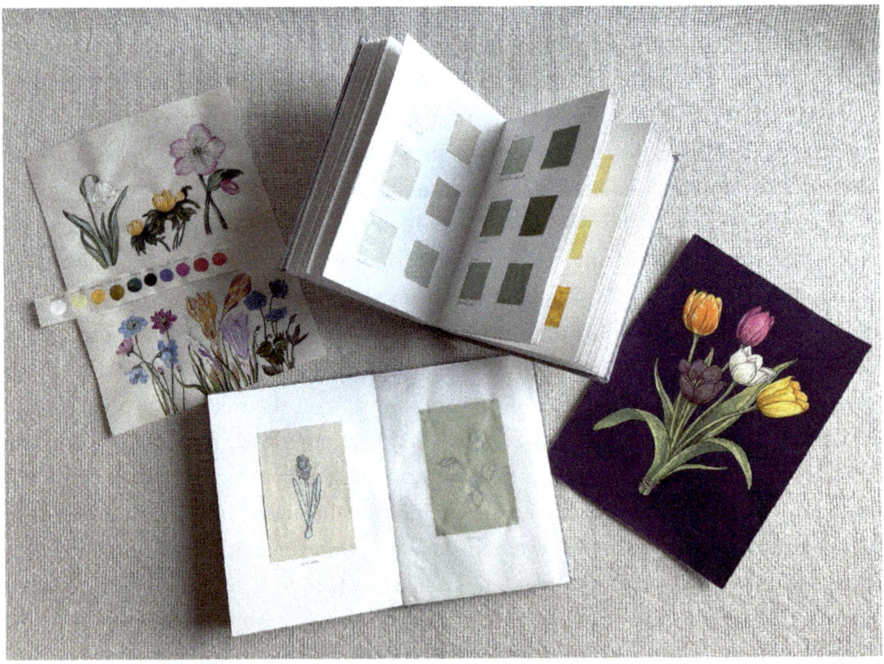

FIGURE 195 The physical paintings and colour samples of the Planting Pigments project

## 6 Acknowledgement

This was a project that carried me through some difficult times while dealing with several crises in the past year. I want to thank Anna Svensson for inviting and collaborating with me on 'A Green Turn?' panel series for STREAMS 2020/2021 (international conference for the Environmental Humanities). The invitation kickstarted the early version of this booklet and this current state debuted at the 2021 conference after the year-long postponement due to the Covid-19 pandemic. I am grateful to Sven Dupré and Florike Egmond for their (administrative) support in pursuing this project. Many thanks to the pigment contributors and workshop participants: their engagement added greatly to the development and enjoyment of making this booklet. Three of the workshop sessions were made possible by the Dutch Research Council (NWO) from the allotment of my PhD grant (Promoties in de Geesteswetenschappen, Project Number PGW.18.016). The Plant Humanities Initiative at the Dumbarton Oaks Research Library and Collection supported the last session. The Oak Spring Garden Foundation provided photos of Franciscus de Geest's watercolours for my research, including the one for me to reproduce and show in this booklet. Last but not the least, a warmest thank you to Sara Charles for supplying the parchment, Merit Hondelink for verifying the botanical names of the dyestuffs, Tamar Hestrin-Grader for copyediting the booklet, and Rhiannon Estwick for proofreading and helping with the text for the temporary project website.

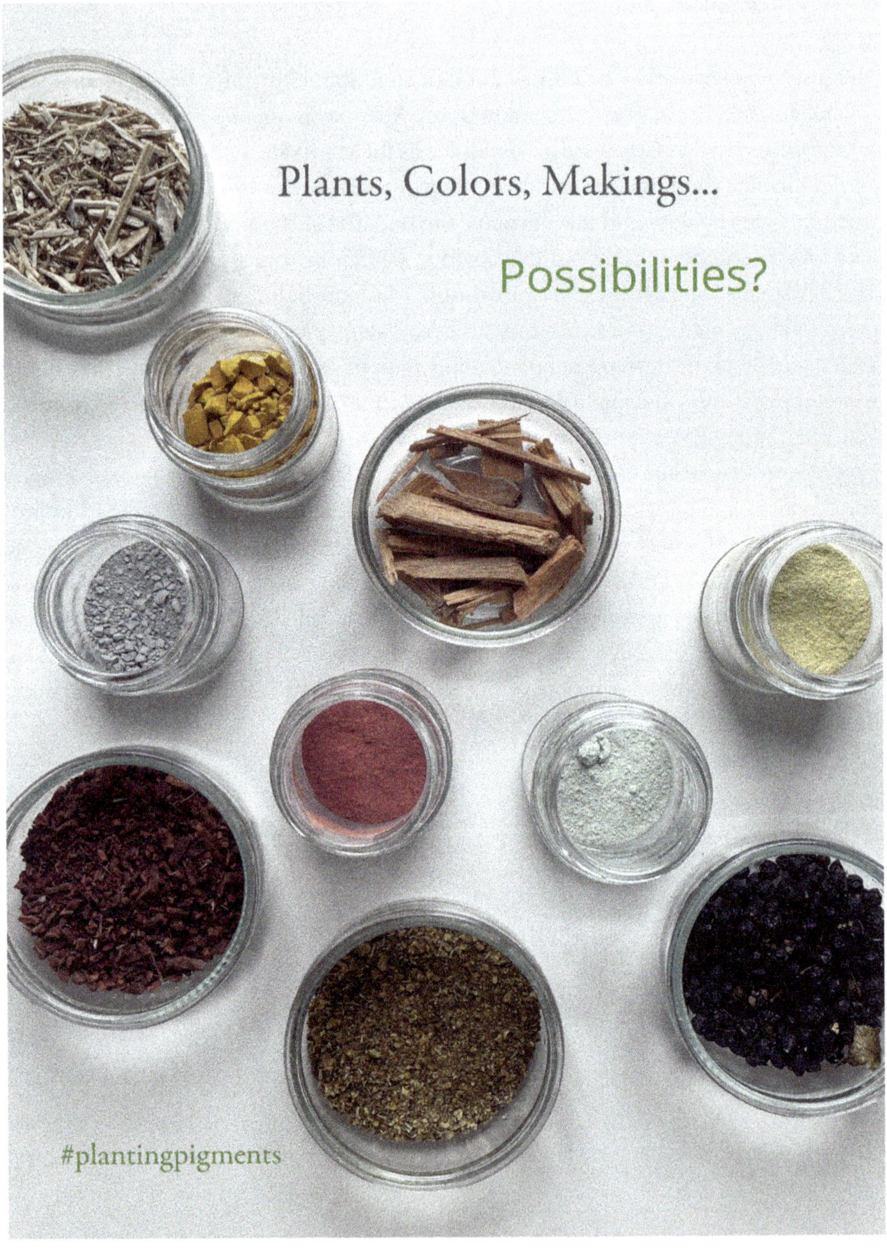


FIGURE 196 The original back cover of the booklet

APPENDIX 3

# *Hortus floridus Iesse, Or, The Imperfect Book of Flowers: A Twenty-First-Century Florilegium*

The *Hortus floridus Iesse* is a twenty-first-century reinterpretation of seventeenth-century European florilegia, often called flower books.

The florilegium, simply put, is an image collection of decorative flowers that is usually made in water- and body-colour or (hand-coloured) intaglio printing on paper or parchment. Seventeenth-century florilegia focus on flowers and decorative plants that were rare and curious, including plants that were then recently introduced to Europe from Asia Minor and the Americas, as well as garden cultivars from the thriving horticultural activities of the period. A flower book usually includes several varieties of the same flower, such as irises of different colours and narcissi with single and double flowers.

The *Hortus floridus Iesse* takes on this fascination with decorative flowers. However, instead of visually recording the rare and curious flowers of today, the collection is, for the most part, the result of the personal process of dealing with my many struggles during the Covid-19 pandemic. Decorative flowers were one of the few things that brought me joy on some of the hardest days. Drawing and painting these flowers provided momentary relief from the sorrow, sadness, and depression I was experiencing. It also (re)kindled the excitement to identify and learn more about the flowers that I did not know before.

Stylistically, the collection is much inspired by the imperfections in historical florilegia, exploring the idea that a flower picture can be unfinished, awkward, or even ugly at times, but still be charming. On an emotional level, the *Hortus floridus Iesse* is just as much about learning to let go of perfection and seeing value in being good *enough*.

## Inspirations

- *Hortus Eystettensis*, Basilius Besler, Nürnberg, 1613.
- *Hortus floridus*, Crispijn van de Passe II, Utrecht, 1614.
- Gottorfer Codex, Hans Simon Holtzbecker, 1649–1659, SMK—Statens Museum for Kunst, Copenhagen.
- Florilegium for Eberhard Anckelmann, Hans Simon Holtzbecker, 1660, The British Museum, London.

- Hortus amoenissimus, Franciscus de Geest, Leeuwarden, 1668, Biblioteca nazionale centrale di Roma, Rome.
- Jardin de rares et curieux fleurs, Franciscus de Geest, Leeuwarden, second half of the seventeenth century, Oak Spring Garden Foundation, Upperville, VA.
- Bloemenboek 'Plusieurs Especes de Fleurs dessinees d'apres le Naturel', multiple painters for Agneta Block, second half of the seventeenth century, Rijksmuseum, Amsterdam.

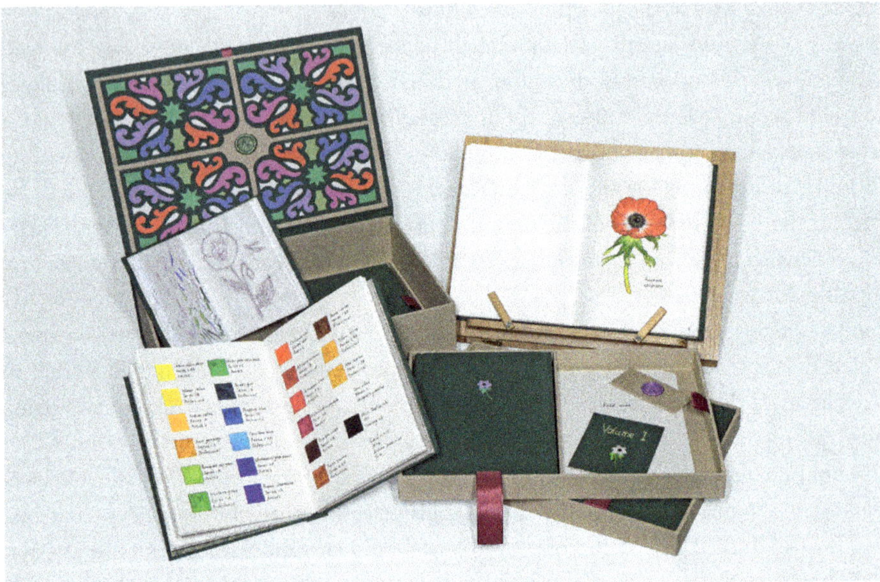


FIGURE 197 Overview of the *Hortus floridus Iesse*. The piece includes sixty watercolours of flowers bound in four volumes, an index of depicted flowers and material samples, process sketches and colour tests, and a box containing these objects

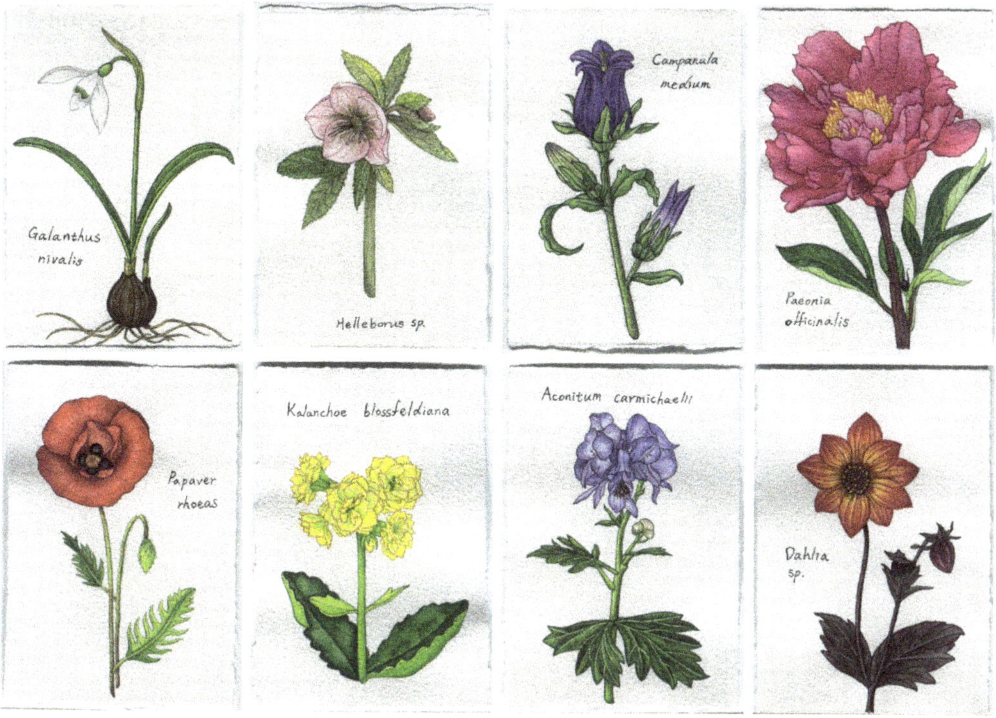


FIGURE 198 Selected pieces from the collection. The flowers in this florilegium are painted with tubes from the Winsor & Newton Artist's/Professional Watercolour Series (and zinc white from Designer's Gouache) on Arches cold pressed watercolour papers. The plant names are written in black Indian ink from the Winsor & Newton Drawing Ink Series

# Bibliography


Ackerman, James S., David E. Allen, William Ashworth Jr., Michael Baxandall, Gunnar Broberg, Allan Ellenius, Wolfgang Harms, Paul Hulton, Martin Kemp, David Knight, and Bengt-Olof Landin. *The Natural Sciences and the Arts: Aspects of Interaction from the Renaissance to the 20th Century: An International Symposium*. Almqvist & Wiksell, 1985.

Alex, Harald, ed. *Die schönsten Pflanzen-Aquarelle der Renaissance: Das Kräuterbuch des Johannes Kentmann von 1563*. Verlag Galerie Bilderwelt Reinhard Schultz, 2022.

Ambers, Janet, Duncan Hook, and Antony Simpson. 'John White's Watercolours: Analysis of the Pigments.' In *European Visions: American Voices*, edited by Kim Sloan, 67–71. British Museum, 2009.

Anderson, Christy, Anne Dunlop, and Pamela H. Smith, eds. *The Matter of Art: Materials, Practices, Cultural Logics, c.1250–1750*. Manchester University Press, 2015.

Arber, Agnes. *Herbals: Their Origin and Evolution: A Chapter in the History of Botany, 1470–1670*. Third Edition. Cambridge University Press, 1986.

Arens, Esther Helena. 'Flowerbeds and Hothouses: Botany, Gardens, and the Circulation of Knowledge in Things.' *Historical Social Research / Historische Sozialforschung* 40, no. 1 (2015): 265–283.

Atasoy, Nurhan. 'Links Between the Ottoman and the Western World on Floriculture and Gardening.' In *Botanical Progress, Horticultural Innovation and Cultural Changes*, edited by Michel Conan and W. John Kress, 61–79. Dumbarton Oaks Research Library and Collection, 2007.

Bakker, Boudewijn. 'Au Vif—Naar 't Leven—Ad Vivum: The Medieval Origin of a Humanist Concept.' In *Aemulatio: Imitation, Emulation and Invention in Netherlandish Art from 1500 to 1800: Essays in Honor of Eric Jan Sluijter*, edited by Anton W.A. Boschloo, Jacquelyn N. Coutré, Stephanie S. Dickey, and Nicolette C. Sluijter-Seijffert, 37–52. Waanders Publishers, 2011.

Balfe, Thomas, Joanna Woodall, and Claus Zittel, eds. *Ad Vivum?: Visual Materials and the Vocabulary of Life-Likeness in Europe before 1800*. Brill, 2019.

Barker, Nicolas. *Hortus Eystettensis: The Bishop's Garden and Besler's Magnificent Book*. The British Library, 1994.

Barrett, Timothy D. *European Hand Papermaking: Traditions, Tools, and Techniques*. Second Edition. The Legacy Press, 2019.

Bass, Marisa Anne, Anne Goldgar, Hanneke Grootenboer, and Claudia Swan. *Conchophilia: Shells, Art, and Curiosity in Early Modern Europe*. Princeton University Press, 2021.

Baumann, Kirsten, ed. *Der Gottorfer Codex: Blütenpracht und Weltanschauung*. Hirmer Verlag, 2014.

Bellavitis, Maddalena, ed. *Making Copies in European Art 1400–1600: Shifting Tastes, Modes of Transmission, and Changing Contexts*. Brill, 2018.

Bellingradt, Daniel. 'The Paper Trade in Early Modern Europe: An Introduction.' In *The Paper Trade in Early Modern Europe: Practices, Materials, Networks*, edited by Daniel Bellingradt and Anna Reynolds, 1–27. Brill, 2021.

Bendall, Sarah A. *Shaping Femininity: Foundation Garments, the Body and Women in Early Modern England*. Bloomsbury, 2022.

Bendall, Sarah A., and Serena Dyer, eds. *Embodied Experiences of Making in Early Modern Europe: Bodies, Gender, and Material Culture*. Amsterdam University Press, 2025.

Benkert, Davina. 'The 'Hortus Siccus' as a Focal Point: Knowledge, Environment, and Image in Felix Platter's and Caspar Bauhin's Herbaria.' In *Sites of Mediation: Connected Histories of Places, Processes, and Objects in Europe and Beyond, 1450–1650*, edited by Susanna Burghartz, Lucas Burkart, and Christine Göttler, 211–239. Brill, 2016.

Berkhout, Lenneke. '"De beste en kundigste enteniers van geheel Europa"' *Cascade bulletin voor tuinhistorie* 29 (2020): 7–32.

Berkhout, Lenneke. *Hoveniers van Oranje: Functie, werk en positie 1621–1732*. Verloren, 2020.

Berrens, Dominik. 'The Meaning of Flora.' *Humanistica Lovaniensia: Journal of Neo-Latin Studies* 68, no. 1 (2019): 237–249.

Bertram, Gitta, Nils Büttner, and Claus Zittel, eds. *Gateways to the Book: Frontispieces and Title Pages in Early Modern Europe*. Brill, 2021.

Besler, Basilius. *Hortus Eystettensis sive diligens et accurata omnium plantarum, florum, stirpium ex variis orbis terrae partibus singulari studio collectarum, quae in celeberrimis viridiariis arcem episcopalem ibidem cingentibus hoc tempore conspiciuntur, delineatio et ad vivum repraesentatio*. Nuremberg: s.n., 1613.

Besler, Basilius. *Hortus Eystettensis, sive, diligens et accurata omnium plantarum, florum, stirpium ex variis orbis terrae partibus singulari studio collectarum, quae in celeberrimis viridiariis arcem episcopalem ibidem cingentibus hoc tempore conspiciuntur, delineatio et ad vivum repraesentatio*. Nuremberg[?]: s.n., 1640.

Bilak, Donna, Jenny Boulboullé, Joel Klein, and Pamela H. Smith. 'The Making and Knowing Project: Reflections, Methods, and New Directions.' *West 86th: A Journal of Decorative Arts, Design History, and Material Culture* 23, no. 1 (2016): 35–55.

Blair, Ann M. *Too Much to Know: Managing Scholarly Information Before the Modern Age*. Yale University Press, 2011.

Bleichmar, Daniela. *Visible Empire: Botanical Expeditions and Visual Culture in the Hispanic Enlightenment*. University of Chicago Press, 2012.

Bleichmar, Daniela. *Visual Voyages: Images of Latin American Nature from Columbus to Darwin*. Yale University Press, 2017.

Bleiler, E.F., ed. *Early Floral Engravings: All 110 Plates from the 1612 'Florilegium' by Emanuel Sweerts*. Pictorial Archive Series. Dover Publications, 1976.

Blewett, Morwenna, and Lucy Wrapson. 'Introduction: Cabinet, Elaboratory, Gallery 1500–1800. The Preservation of Art and Material Culture in Europe.' *Notes and Records* 76, no. 2 (2021): 229–235.

Blok, Gualtherus. *Vijver-hof van Agneta Blok*. Amsterdam: Voor den aucteur, 1702.

Blunt, Wilfrid, and William T. Stearn. *The Art of Botanical Illustration*. Reprint Edition. ACC Art Books, 2015.

Boltz, Valentin. *Illuminierbuch: Darinn Begriffen Wie Man Alle Farben Machen Und Bereiten Soll* ... Augsburg: Johann Weh, 1675.

Borring, Niels, Christian Balleby Jensen, and Anja Scocozza. 'The Seventeenth-Century Florilegium Gottorfer Codex: Examination and Restoration of 1180 Gouache Flower Paintings on Parchment.' In *Care and Conservation of Manuscripts 14: Proceedings of the Fourteenth International Seminar Held at the University of Copenhagen 17th–19th October 2012*, edited by Matthew J. Driscoll, 137–48. Museum Tusculanum Press, 2014.

Bowen, Karen L., and Dirk Imhof. 'Reputation and Wage: The Case of Engravers Who Worked for the Plantin-Moretus Press.' *Simiolus: Netherlands Quarterly for the History of Art* 30, no. 3/4 (2003): 161–195.

Bredekamp, Horst. *The Lure of Antiquity and the Cult of the Machine: The Kunstkammer and the Evolution of Nature, Art and Technology*. Translated by Allison Brown. Markus Wiener Publishers, 1995.

Brenninkmeijer-De Rooij, Beatrijs. *Roots of 17th-Century Flower Painting: Miniatures, Plant Books, Paintings*. Primavera Press, 1996.

Bridson, Gavin D.R., and James J. White, eds. *Plant, Animal & Anatomical Illustration in Art & Science: A Bibliographical Guide from the 16th Century to the Present Day*. St. Paul's Bibliographies / Hunt Institute for Botanical Documentation, 1990.

Britz, Billie S. 'Environmental Provisions for Plants in Seventeenth-Century Northern Europe.' *Journal of the Society of Architectural Historians* 33, no. 2 (1974): 133–144.

Brunfels, Otto. *Herbarum vivae eicones: Ad natur[a]e imitationem, suma cum diligentia et artificio effigiat[a]e, una cum effectibvs earundem, in gratiam ueteris illius, & iamiam renascentis herbariae medicinae* ... Strasbourg: Johann Schott, 1530.

Bulckens, Koen. 'A Flower Book: Some Remarks on Nicolas Robert's "Variae Ac Multiformes Florum Species".' The Theater that was Rome: A Virtual Roman Library, 2017. https://library.brown.edu/projects/rome/essays/flowerbook/.

Bürger, Thomas, and Marina Heilmeyer. *Maria Sibylla Merian: Neues Blumenbuch / New Book of Flowers*. Facsimile Edition. Prestel, 1999.

Burke, Peter. *What Is the History of Knowledge?* Polity Press, 2016.

Bushart, Magdalena, Catalina Heroven, Michael Roth, and Martin Sonnabend, eds. *Maria Sibylla Merian und die Tradition des Blumenbildes von der Renaissance bis zur*

*Romantik*. Staatliche Museen zu Berlin—Preußische Kulturbesitz, Städel Museum / Hirmer Verlag GmbH., 2017.

Bushnell, Rebecca W. *Green Desire: Imagining Early Modern English Gardens*. Cornell University Press, 2018.

Camerarius, Joachim. *Hortus medicus et philosophicus: in quo plurimarum stirpium breues descriptiones, nouae icones non paucae, indicationes locorum natalium, obseruationes de cultura earum peculiares, atque insuper nonnulla remedia euporista, nec non philologica quaedam continentur*. Frankfurt: Iohannem Feyerabend, 1588.

Carrión, María M. 'Planted Knowledge: Art, Science, and Preservation in the Sixteenth-Century Herbarium from the Hurtado de Mendoza Collection in El Escorial.' *Journal of Early Modern Studies* 6, no. 1 (2017): 47–67.

Carroll, Angus. 'The 'Hortus Eystettensis': How Paper Type Can Help Decipher Printing History.' *The Papers of the Bibliographical Society of America* 103, no. 3 (2009): 391–400.

Cennini, Cennino. *The Craftsman's Handbook. The Italian 'Il Libro dell' Arte'*, translated by Daniel V. Thompson. Dover Publications, 1933.

Chapman, H. Perry, and Thijs Weststeijn. 'Connoisseurship as Knowledge. An Introduction.' *Netherlands Yearbook for History of Art* 69 (2019): 7–41.

Cheese, Edward. 'From Pelt to Painted Page: Western Medieval Manuscript Parchments.' In *The Art & Science of Illuminated Manuscripts: A Handbook*, edited by Stella Panayotova, 73–94. Harvey Miller Publishers, 2020.

Chen, Jessie Wei-Hsuan. 'The Blocks, the Tools, and the Printed Plants: The Making and Printing of Botanical Woodblocks at the Early Modern *Officina Plantiniana*.' Master's Thesis, Utrecht University, 2018.

Chen, Jessie Wei-Hsuan. 'A Woodblock's Career: Transferring Visual Botanical Knowledge in the Early Modern Low Countries.' *Nuncius* 35, no. 1 (2020): 20–63.

Chen, Jessie Wei-Hsuan. 'Before Black There Was Red: The Madder Illustration Woodblock for the Herbals of Rembert Dodoens (1517–1585).' In *Burgundian Black: Reworking Early Modern Colour Technologies*, edited by Jenny Boulboullé and Sven Dupré. EMC Imprint, 2022. https://burgundianblack.tome.press/chapter/before-black-there-was-red-the-black-woodblock-of-a-madder-illustration-for-the-herbals-of-the-mechelen-botanist-rembert-dodoens-1517-1585ar1/.

Chen, Jessie Wei-Hsuan. 'Remaking Sixteenth-Century Botanical Woodblocks: Embodied Artisanal Knowledge in Early Modern Woodcutting.' In *Embodied Experiences of Making in Early Modern Europe: Bodies, Gender, and Material Culture*, edited by Sarah A. Bendall and Serena Dyer, 163–184. Amsterdam University Press, 2025.

Clericuzio, Antonio. 'Plant and Soil Chemistry in Seventeenth-Century England: Worsley, Boyle and Coxe.' *Early Science and Medicine* 23, no. 5 (2018): 550–584.

Clifton, James. '"I Am Come into My Garden": The Canticle of Canticles and the Florilegium of Philips Galle and Adriaen Collaert.' In *The Art of Biblical Interpretation: Visual Portrayals of Scriptural Narratives*, edited by Heidi J. Hornik, Ian Boxall, and Bobbi Dykema, 37–66. The Society for Biblical Literature Press, 2021.

Clusius, Carolus. *Rariorum plantarum historia*. Antwerp: Officina Plantiniana, 1601.

Cohn, Marjorie B. *Wash and Gouache: A Study of the Development of the Materials of Watercolor*. The Center for Conservation and Technical Studies, Fogg Art Museum and The Foundation of the American Institute for Conservation, 1977.

Collaert, Adriaen. *Florilegium, ab Hadriano Collaert caelatum et a Philip. Galleo Editum. illustriss. eccelentissimoque dno D. Ioanni Medici. omnis generis elegantiarum admiratori et patrono, Philip. Gallaeus DD*. Antwerp: Philips Galle, c.1590.

Commelin, Caspar. *Horti Medici Amstelaedamensis Plantae Rariores et Exoticae*. Leiden: J. du Vivie, 1716.

Conan, Michel, and W. John Kress, eds. *Botanical Progress, Horticultural Innovation and Cultural Changes*. Dumbarton Oaks Research Library and Collection, 2007.

Cottesloe, Gloria, and Doris Hunt. *The Duchess of Beaufort's Flowers*. Webb & Bower, 1983.

Cribb, Phillip, and Michael Tibbs. *A Very Victorian Passion: The Orchid Paintings of John Day, 1863 to 1888*. Blacker Publishing / The Royal Botanic Gardens, Kew, 2004.

Dackerman, Susan, ed. *Painted Prints: The Revelation of Color in Northern Renaissance & Baroque Engravings, Etchings & Woodcuts*. The Baltimore Museum of Art, 2002.

Dackerman, Susan, ed. *Prints and the Pursuit of Knowledge in Early Modern Europe*. Harvard Art Museums / Yale University Press, 2011.

Daston, Lorraine. 'Type Specimens and Scientific Memory.' *Critical Inquiry* 31, no. 1 (2004): 153–182.

Daston, Lorraine. 'Epistemic Images.' In *Vision and Its Instruments: Art, Science, and Technology in Early Modern Europe*, edited by Alina Alexandra Payne, 13–35. The Pennsylvania State University Press, 2015.

Daston, Lorraine, and Katharine Park. *Wonders and the Order of Nature: 1150–1750*. Third Printing Edition. Zone Books, 2012.

Dauwalder, Lea. 'Felix Platter's Herbarium: The Preservation of a Historical "Bound Herbarium".' *Journal of Paper Conservation* 14, no. 4 (2013): 26–32.

Dauwalder, Lea, and Luc Lienhard. *Das Herbarium des Felix Platter: Die Älteste Wissenschaftliche Pflanzensammlung der Schweiz*. Haupt Verlag, 2016.

Davidson, Hilary. 'The Embodied Turn: Making and Remaking Dress as an Academic Practice.' *Fashion Theory: The Journal of Dress, Body and Culture* 23, no. 3 (2019): 329–362.

Davidson, Hilary, and Anna Hodson. 'Joining Forces: The Intersection of Two Replica Garments.' In *Textiles and Text: Re-Establishing the Links Between Archival and*

*Object-Based Research*, edited by Maria Hayward and Elizabeth Kramer, 204–210. Archetype Publications, 2007.

De Bry, Johann Theodor. *Florilegium novum hoc est: variorum maximeque rariorum florum ac plantarum singularium una cum suis radicibus & cepis, eicones diligenter in aere sculptae & ad vivum-ut plurimum expressae*. Oppenheim: Johann Theodor de Bry, 1612.

De Bry, Johann Theodor. *Amplificatio sive dilatatio florilegii nuper coepti auctique: iam vero variis atque elegantioribus floribus exornati*. Frankfurt: Johann Theodor de Bry, 1613.

De Bry, Johann Theodor, and Matthäus Merian. *Anthologia magna, sive florilegium novum et absolutum variorum maximeque rariorum germinum, florum ac plantarum cum ciconibus* ... Frankfurt: Officina Bryana, 1626.

De Bry, Johann Theodor, and Matthäus Merian. *Florilegium renovatum et auctum: variorum maximeque rariorum germinum, florum ac plantarum, quas pulchritudo, fragrantia, usus, varietas, differntia commendat, et non tantum noster hic, sed et adversus veteribusque ignotus orbis e foecundo suo procreat gremio, eicones elegantissimae, summa cum diligentia ad vivum expositae: additis eorum proprijs, veris ac genuinis nominibus*. Frankfurt: Matthäus Merian, 1641.

De Cuveland, Helga. *Der Gottorfer Codex von Hans Simon Holtzbecker*. Wernersche Verlagsgesellschaft, 1989.

De Jong, Erik. 'Nature and Art. The Leiden Hortus as "*Musaeum*".' In *The Authentic Garden: A Symposium on Gardens*, edited by Leslie Tjon Sie Fat and Erik De Jong, 37–60. Clusius Foundation, 1991.

De Koning, Jan, Gerda Van Uffelen, Alicja Zemanek, and Bogdan Zemanek, eds. *Drawn after Nature: The Complete Botanical Watercolours of the 16th-Century Libri Picturati*. KNNV Publishing, 2008.

De la Fuente Pedersen, Eva, and Hanne Kolind Poulsen, eds. *Flowers and World Views*. Statens Museum for Kunst, 2013.

De Nave, Francine, and Dirk Imhof, eds. *Botany in the Low Countries (End of the 15th Century–ca. 1650)*. The Plantin-Moretus Museum / The Stedelijk Prentenkabinet, 1993.

DiMeo, Michelle, and Sara Pennell, eds. *Reading and Writing Recipe Books, 1550–1800*. Manchester University Press, 2013.

Dodoens, Rembert. *Cruijdeboeck, in den welcken die geheele historie, dat es Tgheslacht, tfatsoen, naem, natuere, cracht ende werkinghe, van den Cruyden, niet alleen hier telande wassende, maer oock van den anderen vremden in der Medecijnen oorboorlijck, et grooter neersticheyt begrepen ende verclaert es, met der seluer cruyden natuerlijck naer dat leven conterfeytsel daer by ghestelt*. Antwerp: Jan van der Loe, 1554.

Dodoens, Rembert. *Florum, et coronariarum odoratarumque nonnullarum herbarum historia*. Antwerp: Christophe Plantin, 1568.

Dodoens, Rembert. *Stirpium historiae pemptades sex, sive libri XXX*. Antwerp: Christophe Plantin, 1583.

Dodoens, Rembert. *Cruydt-boeck, van Rembertus Dodonaeus, volgens sijne laetste verbeteringe: met biivoegsels achter elck capittel, uut verscheyden cruydtbeschrijvers: Item in't laetste een beschrijvinge vande Indiaensche gewassen, meest getrokken wt de schriften van Carolus Clusius*. Leiden: Officina Plantiniana, 1608.

Doherty, Tiarna. 'Painting Connoisseurship: *Liefhebbers* in the Studio.' *Netherlands Yearbook for History of Art* 69 (2019): 146–173.

Dupré, Sven, Anna Harris, Julia Kursell, Patricia Lulof, and Maartje Stols-Witlox, eds. *Reconstruction, Replication and Re-Enactment in the Humanities and Social Sciences*. Amsterdam University Press, 2020.

Dupré, Sven, and Christoph Lüthy. *Silent Messengers: The Circulation of Material Objects of Knowledge in the Early Modern Low Countries*. LIT Verlag, 2011.

Dupré, Sven, and Geert Somsen. 'The History of Knowledge and the Future of Knowledge Societies.' *Berichte zur Wissenschaftsgeschichte* 42, no. 2–3 (2019): 186–199.

Duthie, Ruth. 'English Florists' Societies and Feasts in the Seventeenth and First Half of the Eighteenth Centuries.' *Garden History* 10, no. 1 (1982): 17–35.

Duthie, Ruth. *Florists' Flowers and Societies*. Shire Publications, 1988.

Duthie, Ruth. 'The Planting Plans of Some Seventeenth-Century Flower Gardens.' *Garden History* 18, no. 2 (1990): 77–102.

Eastaugh, Nicholas, Valentine Walsh, Tracey Chaplin, and Ruth Siddall. *Pigment Compendium: A Dictionary and Optical Microscopy of Historical Pigments*. New Edition. Elsevier / Butterworth-Heinemann, 2008.

Egmond, Florike. 'Clusius, Cluyt, Saint Omer: The Origin of the Sixteenth-Century Botanical and Zoological Watercolours in Libri Pituratui A. 16–30.' *Nuncius* 20 (2005): 11–67.

Egmond, Florike. 'Curious Fish: Connections Between Some Sixteenth-Century Watercolours and Prints.' In *Early Modern Zoology: The Construction of Animals in Science, Literature and the Visual Arts*, edited by Karl A.E. Enenkel and Paul J. Smith, 245–272. Brill, 2007.

Egmond, Florike. 'Precious Nature: Rare Naturalia as Collector's Items and Gifts in Early Modern Europe.' In *Luxury in the Low Countries: Miscellaneous Reflections on Netherlandish Material Culture, 1500 to the Present*, edited by Rengenier C. Rittersma. Pharo, 2010.

Egmond, Florike. *The World of Carolus Clusius: Natural History in the Making, 1550–1610*. Pickering & Chatto, 2010.

Egmond, Florike. 'Names of Naturalia in the Early Modern Period: Between the Vernacular and Latin, Identification and Classification.' In *Translating Knowledge in the Early Modern Low Countries*, edited by Harold Cook and Sven Dupré, 131–162. LIT Verlag, 2012.

Egmond, Florike. 'Experimenting with Living Nature: Documented Practices of 16th-Century Naturalists and Naturalia Collectors.' *Journal of Early Modern Studies* 6, no. 1 (2017): 21–45.

Egmond, Florike. *Eye for Detail: Images of Plants and Animals in Art and Science, 1500–1630*. Reaktion Books, 2017.

Egmond, Florike. 'The Ad Vivum Conundrum: Eyewitnessing and the Artful Representation of Naturalia in Sixteenth-Century Natural Science.' In *Zeigen, Überzeugen, Beweisen: Methoden der Wissensproduktion in Kunstliteratur, Kennerschaft und Sammlungspraxis der Frühen Neuzeit*, edited by Elisabeth Oy-Marra and Irina Schmiedel, 33–62. Ad Picturam, 2020.

Egmond, Florike, and Sachiko Kusukawa. 'Circulation of Images and Graphic Practices in Renaissance Natural History: The Example of Conrad Gessner.' *Gesnerus* 73, no. 1 (2016): 29–72.

Eiermann, Wolf, ed. *Prachtvoll illuminirt: Das Handkolorit in der Druckgraphik (1493–1870)*. Hirmer Verlag, 2018.

Eklund, Hillary Caroline, ed. *Ground-Work: English Renaissance Literature and Soil Science*. Duquesne University Press, 2017.

Eriksen, Christoffer Basse, and Xinyi Wen. 'Colouring Flowers: Books, Art, and Experiment in the Household of Margery and Henry Power.' *The British Journal for the History of Science* 56, no. 1 (2023): 21–43.

Etheridge, Kay. *The Flowering of Ecology: Maria Sibylla Merian's Caterpillar Book*. Brill, 2020.

Evans, Mark. *Renaissance Watercolours: From Dürer to Van Dyck*. V&A Publishing, 2020.

Evans, Robert John Weston, and Alexander Marr, eds. *Curiosity and Wonder from the Renaissance to the Enlightenment*. Ashgate, 2006.

Febvre, Lucien, and Henri-Jean Martin. *The Coming of the Book: The Impact of Printing 1450–1800*. Verso Edition. Verso, 1997.

Feeser, Andrea, Maureen Daly Goggin, and Beth Fowkes Tobin, eds. *The Materiality of Color: The Production, Circulation, and Application of Dyes and Pigments, 1400–1800*. Routledge, 2017.

Fehrenbach, Frank. 'Cut Flowers.' *Nuncius* 32, no. 3 (2017): 583–614.

Felfe, Robert. '"Naer het leven": Eine sprachliche Formel zwischen bildgenerierenden Übertragungsvorgängen und ästhetischer Vermittlung.' In *Ad Fontes! Niederländische Knst des 17. Jahrhunderts in Quellen*, edited by Claudia Fritzsche, Karin Leonhard, and Gregor J.M. Weber, 165–195. Michael Imhof Verlag, 2013.

Feller, Robert L., ed. *Artists' Pigments: A Handbook of Their History and Characteristics*. Vol. 1. National Gallery of Art / Archetype Publications, 1986.

Fels, Donald C., ed. *Lost Secrets of Flemish Painting: Including the First Complete English Translation of the De Mayerne Manuscript, B.M. Sloane 2052*. Alchemist Publications, 2010.

Ferguson, Nicola. *Double Flowers: The Remarkable Story of Extra-Petalled Blooms*. Pimpernel Press, 2018.

Ferrari, Giovanni Battista. *De florum cultura libri IV*. Rome: Stephanus Paulinus, 1633.

Findlen, Paula. *Possessing Nature: Museums, Collecting, and Scientific Culture in Early Modern Italy*. University of California Press, 1994.

Findlen, Paula, and Anna Toledano. 'The Materials of Natural History.' In *Worlds of Natural History*, edited by H.A. Curry, N. Jardine, J.A. Secord, and E.C. Spary, 151–169. Cambridge University Press, 2018.

Fischer, Hubertus, Volker R. Remmert, and Joachim Wolschke-Bulmahn, eds. *Gardens, Knowledge and the Sciences in the Early Modern Period*. Springer, 2016.

FitzHugh, Elisabeth West, ed. *Artists' Pigments: A Handbook of Their History and Characteristics*. Vol. 3. National Gallery of Art / Oxford University Press, 1997.

Flannery, Maura C. *In the Herbarium: The Hidden World of Collecting and Preserving Plants*. Yale University Press, 2023.

Fleischer, Alette. 'Leaves on the Loose: The Changing Nature of Archiving Plants and Botanical Knowledge.' *Journal of Early Modern Studies* 6, no. 1 (2017): 117–135.

Fors, Hjalmar, Lawrence M. Principe, and H. Otto Sibum. 'From the Library to the Laboratory and Back Again: Experiment as a Tool for Historians of Science.' *Ambix* 63, no. 2 (2016): 85–97.

Fransen, Sietske, and Katherine M. Reinhart. 'The Practice of Copying in Making Knowledge in Early Modern Europe: An Introduction.' *Word & Image* 35, no. 3 (2019): 211–222.

Freedberg, David, and Enrico Baldini. *Citrus Fruit*. Vol. I. The Paper Museum of Cassiano Dal Pozzo, B. Harvey Miller, 1997.

Friedrich, Susanne, Arndt Brendecke, and Stefan Ehrenpreis, eds. *Transformations of Knowledge in Dutch Expansion*. De Gruyter, 2015.

Frumer, Yulia. 'What Is and Isn't in a Name.' *Berichte zur Wissenschaftsgeschichte* 42, no. 2–3 (2019): 150–166.

Fuchs, Leonhart. *De historia stirpium commentarii insignes*. Basel: Michael Isengrin, 1542.

Gaskell, Philip. *A New Introduction to Bibliography*. Oak Knoll Press, 2012.

Gebhardt, Andrew. *Holland Flowering: How the Dutch Flower Industry Conquered the World*. Amsterdam University Press, 2014.

Gibson, James J. *The Ecological Approach to Visual Perception*. Lawrence Erlbaum Associates, 1986.

Gilias, Guy, Cornelis Van Tilburg, and Vincent Van Roy, eds. *Rembert Dodoens: Een zestiende-eeuwse kruidenwetenschapper, zijn tijd- en vakgenoten en zijn betekenis*. Garant, 2016.

Glardon, Philippe. 'Review of Florilegium. Das Ist: Ein Blumen Buch, Frankfurt 1616, Staatsbibliothek Bamberg. Mit Einer Einleitung von Werner Dressendörfer. Erlangen, Harald Fischer Verlag, [2003]. 1 CD-ROM. I 86.21. ISBN 3-89131-423-X.' *Gesnerus* 62, no. 1–2 (2005): 135–136.

Goedings, Truusje. 'Ingekleurde Prenten En Boeken.' In *Voor Nederland Bewaard: De Verzamelingen van Het Koninklijk Oudheidkundig Genootschap in het Rijksmuseum*, edited by Jan Frederik Heijbroek, 163–177. De Prom, 1995.

Goedings, Truusje. *'Afsetters en meester-afsetters': De kunst van het kleuren 1480–1720*. Uitgeverij Vantilt, 2015.

Goldgar, Anne. *Tulipmania: Money, Honor, and Knowledge in the Dutch Golden Age*. The University of Chicago Press, 2007.

Goldman, Paul. *Looking at Prints, Drawings and Watercolours: A Guide to Technical Terms*. British Museum, 1988.

Goody, Jack. *The Culture of Flowers*. Cambridge University Press, 1993.

Graser, Jenny, and Martin Sonnabend. 'Johann Walter der Altere und das Florilegium des Grafen Johannes von Nassau-ldstein.' In *Maria Sibylla Merian und die Tradition des Blumenbildes von der Renaissance bis zur Romantik*, edited by Magdalena Bushart, Catalina Heroven, Michael Roth, and Martin Sonnabend, 101–115. Staatliche Museen zu Berlin—Preußische Kulturbesitz, Städel Museum / Hirmer Verlag, 2017.

Greene, Wesley, and Barbara Temple Lombardi. *Vegetable Gardening the Colonial Williamsburg Way: 18th-Century Methods for Today's Organic Gardeners*. Rodale, 2012.

Grenda-Kurmanow, Magdalena. 'Adhesives Used in Herbaria: Current Practice with Regard to What We Know from Written Sources on Mounting Herbarium Specimens and Conservation.' *Taxon* 70, no. 1 (2021): 1–15.

Griffiths, Antony. *The Print Before Photography: An Introduction to European Printmaking, 1550–1820*. British Museum Press, 2016.

Grootenboer, Hanneke. *The Rhetoric of Perspective: Realism and Illusionism in Seventeenth-Century Dutch Still-Life Painting*. University of Chicago Press, 2005.

Hagendijk, Thijs. 'Learning a Craft from Books: Historical Re-Enactment of Functional Reading in Gold- and Silversmithing.' *Nuncius* 33, no. 2 (2018): 198–235.

Harald Fischer Verlag. *Camerarius-Florilegium*. Meisterwerke der Buchillustration. CD-ROM. Harald Fischer Verlag, 2004.

Harley, Rosamond Drusilla. *Artists' Pigments c. 1600–1835: A Study in English Documentary Sources*. Butterworths, 1970.

Harney, Marion, ed. *Gardens and Landscapes in Historic Building Conservation*. Wiley, 2014.

Heiser, Charles B. 'The Sunflower among the North American Indians.' *Proceedings of the American Philosophical Society* 95, no. 4 (1951): 432–448.

Heiser, Charles B. 'The Origin and Development of the Cultivated Sunflower.' *The American Biology Teacher* 17, no. 5 (1955): 161–167.

Heiser, Charles B. 'The Domesticated Sunflower in Old Mexico?' *Genetic Resources and Crop Evolution* 45 (1998): 447–449.

Henderson, Jane. 'Beyond Lifetimes: Who Do We Exclude When We Keep Things for the Future?' *Journal of the Institute of Conservation* 43, no. 3 (2020): 195–212.

Hendriksen, Marieke M.A. 'Rethinking Performative Methods in the History of Science.' *Berichte zur Wissenschaftsgeschichte* 43 (2020): 313–322.

Hill, John. *Of the Origin and Production of Proliferous Flowers. With the Culture at Large for Raising Double from Single, and Proliferous From the Double.* London: s.n., 1759.

Hofmann, Mara, and Caroline Zöhl. *Hortus Eystettensis: Studien zur Entstehung des Kupferstichwerks und zum Exemplar des Andrea Vendramin.* Universitätsbibliothek Heidelberg, 2003.

Huisman, Gerda C. 'Franciscus de Geest and His "Delightful Gardens".' In *Hortus Amoenissimus Omnigenis Floribus, Plantis, Stirpibus ... by Franciscus de Geest, Biblioteca Nazionale Centrale Di Roma MS Varia 291*, edited by Alessandro Menghini, Facsimile Edition, 23–54. Aboca, 2012.

Hunt, John Dixon. 'Curiosities to Adorn Cabinets and Gardens.' In *The Origins of Museums: The Cabinet of Curiosities in Sixteenth- and Seventeenth-Century Europe*, edited by Oliver Impey and Arthur MacGregor, 193–203. Clarendon Press, 1985.

Hyde, Elizabeth. *Cultivated Power: Flowers, Culture, and Politics in the Reign of Louis XIV.* University of Pennsylvania Press, 2005.

Impey, Oliver, and Arthur MacGregor, eds. *The Origins of Museums: The Cabinet of Curiosities in Sixteenth- and Seventeenth-Century Europe.* Clarendon Press, 1985.

Ingold, Tim. *Making: Anthropology, Archaeology, Art and Architecture.* Routledge, 2013.

Jalobeanu, Dana. 'The Toolbox of the Early Modern Natural Historian: Note-Books, Commonplace-Books and the Emergence of Laboratory Records.' *Journal of Early Modern Studies* 4, no. 1 (2015): 107–123.

Jarvis, Charles E. '"The Most Common Grass, Rush, Moss, Fern, Thistles, Thorns or Vilest Weeds You Can Find": James Petiver's Plants.' *Notes and Records* 74 (2019): 303–328.

Jenner, Thomas. *A Book of Drawing, Limning, Washing, or Colouring of Maps and Prints: And the Art of Painting, with the Names and Mixtures of Colours Used by the Picture-Drawers. Or, the Young-Mans Time Well Spent ...* London: M. Simmons, 1652.

Johnson, Lisa. 'Pieter de la Court van der Voort and Innovations in Pineapple Cultivation in Early Eighteenth-Century Gardens.' *Garden History* 47, no. 1 (2019): 23–41.

Johnstone, Katharine. 'Some Cultivated Anemones and Their Histories.' *Scientific Horticulture* 24 (1972/73): 14–28.

Jorink, Eric, and Bart Ramakers, eds. 'Art and Science in the Early Modern Netherlands.' *Netherlands Yearbook for the History of Art* 61 (2011).

Kaufmann, Thomas Da Costa. 'The Perspective of Shadows: The History of the Theory of Shadow Projection.' *Journal of the Warburg and Courtauld Institutes* 38 (1975): 258–287.

Kaufmann, Thomas DaCosta, and Virginia Roehrig Kaufmann. 'The Sanctification of Nature: Observations on the Origins of *Trompe l'oeil* in Netherlandish Book Painting of the Fifteenth and Sixteenth Centuries.' *The J. Paul Getty Museum Journal* 19 (1991): 43–64.

Kenny, Neil. *Curiosity in Early Modern Europe: Word Histories*. Harrassowitz Verlag, 1998.

Keunecke, Hans-Otto, ed. *Hortus Eystettensis: Zur Geschichte Eines Gartens und Eines Buches*. Schirmer/Mosel, 1989.

Kingsbury, Noël. *Hybrid: The History and Science of Plant Breeding*. The University of Chicago Press, 2009.

Kinross, Robin. *A4 and Before: Towards a Long History of Paper Sizes*. NIAS, 2009.

Kirby, Jo, Maarten Van Bommel, and André Verhecken. *Natural Colorants for Dyeing and Lake Pigments: Practical Recipes and Their Historical Sources*. Archetype Publications, 2014.

Kistemaker, Renée, Natalja Kopaneva, and Annemiek Overbeek, eds. *Peter de Grote en Holland: Culturele en wetenschappelijke betrekkingen tussen Rusland en Nederland ten tijde van tsaar Peter de Grote*. Thoth / Amsterdam Historisch Museum, 1996.

Klein, Ursula. 'Experimental History and Herman Boerhaave's Chemistry of Plants.' *Studies in History and Philosophy of Science* 34, no. 4 (2003): 533–567.

Knight, Leah. *Of Books and Botany in Early Modern England: Sixteenth-Century Plants and Print Culture*. Ashgate, 2009.

Kolind Poulsen, Hanne. *The Green Florilegium*. Prestel, 2013.

Kusukawa, Sachiko. 'Image, Text and *Observatio*: The *Codex Kentmanus*.' *Early Science and Medicine* 14, no. 4 (2009): 445–475.

Kusukawa, Sachiko. *Picturing the Book of Nature: Images, Text, and Argument in Sixteenth-Century Human Anatomy and Medical Botany*. The University of Chicago Press, 2012.

Kusukawa, Sachiko, and Ian Maclean, eds. *Transmitting Knowledge: Words, Images, and Instruments in Early Modern Europe*. Oxford University Press, 2006.

Laird, Mark. 'Greenhouse Technologies and Horticulture: The 1st Duchess of Beaufort's Badminton Florilegium (1703–5) and J.J. Dillenius's Hortus Elthamensis (1732).' In *Technology and the Garden*, edited by Michael G. Lee and Kenneth I. Helphand, 55–77. Dumbarton Oaks Research Library and Collection, 2014.

Laird, Mark. *A Natural History of English Gardening, 1650–1800*. Yale University Press, 2015.

Landau, David, and Peter Parshall. *The Renaissance Print, 1470–1550*. Yale University Press, 1994.

Lauterbach, Iris. 'Commerce and Erudition: Civic Self-Representation Through Botany and Horticulture in Germany, Sixteenth to Eighteenth Centuries.' In *Gardens, Knowledge and the Sciences in the Early Modern Period*, edited by Hubertus Fischer, Volker R. Remmert, and Joachim Wolschke-Bulmahn, 319–341. Springer, 2016.

Lavezzi, Elisabeth. 'Catherine Perrot, Peintre Savant En Miniature, Les Leçons Royales de 1686 et de 1693.' In *Femmes Savantes, Savoirs de Femmes. Du Crépuscule de la Renaissance à l'aube Des Lumières*, edited by Colette Nativel, 229–245. Droz, 1999.

Learner, Mary. 'Embroidering the New Science: Seventeenth-Century Florilegia and Botanical Study.' *Nuncius* 35, no. 3 (2020): 685–717.

Lee, Michael G., and Kenneth I. Helphand, eds. *Technology and the Garden*. Dumbarton Oaks Research Library and Collection, 2014.

Lefèvre, Wolfgang, Jürgen Renn, and Urs Schoepflin, eds. *The Power of Images in Early Modern Science*. Birkhäuser, 2003.

Lehmann, Ann-Sophie. 'How Materials Make Meaning.' *Netherlands Yearbook for History of Art* 62 (2012): 6–27.

Lehmann, Ann-Sophie. 'An Alphabet of Colors: Valcooch's Rules and the Emergence of Sense-Based Learning Around 1600.' *Netherlands Yearbook for History of Art* 68 (2019): 168–203.

Leong, Elaine Yuen Tien. *Recipes and Everyday Knowledge: Medicine, Science, and the Household in Early Modern England*. The University of Chicago Press, 2018.

Levitt, Ruth. '"A Noble Present of Fruit": A Transatlantic History of Pineapple Cultivation.' *Garden History* 42, no. 1 (2014): 106–119.

Littger, Klaus Walter, and Werner Dressendörfer. *Florilegium: The Book of Plants—The Complete Plates*. Taschen, 2016.

Lobelius, Matthias. *Plantarum, seu, Stirpium historia*. Antwerp: Christophe Plantin, 1576.

Lobelius, Matthias. *Kruydtboeck, oft, Beschrÿuinghe van Allerleye Ghewassen, Kruyderen, Hesteren, ende Gheboomten*. Antwerp: Christophe Plantin, 1581.

Lobelius, Matthias, and Pierre Pena. *Stirpium Adversaria Nova, Perfacilis Vestigatio Luculentaque Accesio Ad Priscorum, Praesertim Dioscoridis & Recentiorum, Materiam Medicam* ... London: Thomas Purfoot, 1571.

Lüthy, Christoph, and Alexis Smets. 'Words, Lines, Diagrams, Images: Towards a History of Scientific Imagery.' *Early Science and Medicine* 14 (2009): 398–439.

Mabry, Makenzie E., Nuala Caomhanach, R. Shawn Abrahams, Michelle L. Gaynor, Kasey Khanh Pham, Tanisha M. Williams, Kathleen S. Murphy, Vassiliki Betty Smocovitis, Douglas E. Soltis, and Pamela S. Soltis. 'Building an Inclusive Botany: The "Radicle" Dream.' *Plants, People, Planet* 6, no. 3 (2024): 544–557.

MacGregor, Arthur, ed. *Naturalists in the Field: Collecting, Recording and Preserving the Natural World from the Fifteenth to the Twenty-First Century*. Brill, 2018.

Mantel, T.T. '"The Leyden-gardener". Het tuintraktaat van Henrick van Oosten.' In *Bloeiende kennis: Groene ontdekkingen in de Gouden Eeuw*, edited by Esther van Gelder. Verloren, 2012.

Marchand, Suzanne. 'How Much Knowledge Is Worth Knowing? An American Intellectual Historian's Thoughts on the Geschichte Des Wissens.' *Berichte zur Wissenschaftsgeschichte* 42, no. 2–3 (2019): 126–149.

Margócsy, Dániel. *Commercial Visions: Science, Trade, and Visual Culture in the Dutch Golden Age*. University of Chicago Press, 2014.

Marr, Alexander. 'Knowing Images.' *Renaissance Quarterly* 69, no. 3 (2016): 1000–1013.

Marr, Alexander, and Christopher P. Heuer. 'Introduction. The Uncertainty of Epistemic Images.' *21: Inquiries into Art, History, and the Visual* 1, no. 2 (2020): 251–255.

Martin, Laura J. *Wild by Design: The Rise of Ecological Restoration*. Harvard University Press, 2022.

Masson, Georgina. 'Italian Flower Collectors' Gardens in Seventeenth-Century Italy.' In *The Italian Garden*, edited by David R. Coffin, 63–80. Dumbarton Oaks, 1972.

McKeown, Joanne M. 'Self-Portrait of an Artist: Translation and the Creative Process of Catherine Perrot.' *Translation Review* 114, no. 1 (2022): 38–46.

Meder, Joseph. *The Mastery of Drawing*. Abaris Books, 1978.

Medlej, Joumana. *Wild Inks and Paints: A Seasonal Palette*. Joumana Medlej, 2021.

Melograni, Anna. 'Manuscript Materials: Cost and the Market for Parchment in Renaissance Italy.' In *Trade in Artists' Materials. Markets and Commerce in Europe to 1700*, edited by Jo Kirby, Susie Nash, and Joanna Cannon, 199–221. Archetype Publications, 2010.

Menghini, Alessandro, ed. *Hortus Amoenissimus, Omnigenis Floribus, Plantis, Stirpibus … by Franciscus de Geest, Biblioteca Nazionale Centrale Di Roma MS Varia 291*. Facsimile Edition. Aboca, 2012.

Merian, Maria Sibylla. *Neues Blumenbuch: Allen Kunstverstandigen Liebhabern zu Lust, Nutz und dienst, mit Fleiss verfertiget*. Nuremberg: s.n., 1680.

Merian, Maria Sibylla. *Metamorphosis insectorum Surinamensium, ofte verandering der Surinaamsche insecten, waar in de Surinaamsche rupsen en wormen met alle des zelfs veranderingen na het leven afgebeeld en beschreven worden …* Amsterdam: Gerard Valck, 1705.

Merian, Maria Sibylla. *Neues Blumenbuch*. 4th Edition. Insel Verlag, 2017.

Montalbano, Letizia, Michela Piccolo, and Maria Grazia Vaccari. 'Painting on Parchment Besides Miniatures: Scientific Analyses and a Study of the Artistic Techniques of Giovannino de' Grassi's Model Book.' In *Painting Techniques: History, Materials and Studio Practice*, edited by Ashok Roy and Perry Smith, 55–58. The International Institute for Conservation of Historic and Artistic Works, 1998.

Moran, Bruce T. 'Preserving the Cutting Edge: Traveling Woodblocks, Material Networks, and Visualizing Plants in Early Modern Europe.' In *The Structures of Practical Knowledge*, edited by Matteo Valleriani, 393–419. Springer, 2017.

Munroe, Jennifer. '"My Innocent Diversion of Gardening": Mary Somerset's Plants.' *Renaissance Studies* 25, no. 1 (2011): 111–123.

Nadri, Ghulam A. *The Political Economy of Indigo in India, 1580–1930: A Global Perspective*. Brill, 2016.

Nickelsen, Kärin. 'The Challenge of Colour: Eighteenth-Century Botanists and the Hand-Colouring of Illustrations.' *Annals of Science* 63, no. 1 (2006): 3–23.

Nissen, Claus. *Die botanische Buchillustration: Ihre Geschichte und Bibliographie*. 1994 Reprint Edition. Maurizio Martino Publisher, 1994.

Norgate, Edward. *Miniatura: or, The Art of Limning*. Edited by Martin Hardie. Clarendon Press, 1919.

Ogilvie, Brian W. *The Science of Describing: Natural History in Renaissance Europe*. The University of Chicago Press, 2006.

Olariu, Dominic. 'Herbs under Pressure: Plant Illustrations and Nature Printing in the First Half of the Fifteenth Century.' In *Naturalismen: Kunst, Wissenschaft und Ästhetik*, edited by Robert Felfe and Maurice Saß, 9–32. De Gruyter, 2019.

Oldenburger-Ebbers, C.S., and J. Heniger. *Ornamental Plants in 16th and 17th Century Gardens: An Annotated List of Plants to Assist with the Proper Reconstruction and Maintaince of Historic Gardens from These Centirues, to Which Introductory Remarks Have Been Added.* Biohistorical Institute Miquel Huis, 1975.

Oltrogge, Doris. '"Pro Lazurio Auricalco et Alii Correquisitis pro Illuminacione": The Werden Accounts and Other Sources on the Trade in Manuscript Materials in the Lower Rhineland and Westphalia around 1500.' In *Trade in Artists' Materials. Markets and Commerce in Europe to 1700*, edited by Jo Kirby, Susie Nash, and Joanna Cannon, 189–198. Archetype Publications, 2010.

Oltrogge, Doris. 'Writing on Pigments in Natural History and Art Technology in Sixteenth-Century Germany and Switzerland.' In *Early Modern Color Worlds*, edited by Tawrin Baker, Sven Dupré, Sachiko Kusukawa, and Karin Leonhard, 47–69. Brill, 2015.

Östling, Johan, David Larsson Heidenblad, and Anna Nilsson Hammar, eds. *Forms of Knowledge: Developing the History of Knowledge*. Nordic Academic Press, 2020.

Panayotova, Stella. 'Colour in Illuminated Manuscripts.' In *Colour: The Art & Science of Illuminated Manuscripts*, edited by Stella Panayotova, Deirdre Jackson, and Paola Ricciardi, 15–23. Harvey Miller Publishers, 2016.

Panayotova, Stella, ed. *The Art & Science of Illuminated Manuscripts: A Handbook*. Harvey Miller Publishers, 2020.

Panayotova, Stella, Deirdre Jackson, and Paola Ricciardi, eds. *Colour: The Art & Science of Illuminated Manuscripts*. Harvey Miller Publishers, 2016.

Pavord, Anna. *The Tulip*. Bloomsbury Publishing, 1999.

Pavord, Anna. *The Naming of Names: The Search for Order in the World of Plants*. Bloomsbury Publishing, 2005.

Peacock, John. *The Look of Van Dyck: The Self-Portrait with a Sunflower and the Vision of the Painter*. Ashgate, 2006.

Perrot, Catherine. *Les lecons royales, ou la manière de peindre en mignature les fleurs & les oyseaux: par l'explication des livres de fleurs & d'oyseaux de feu Nicolas Robert, fleuriste*. Paris: Jean B. Nego, 1686.

Pey, Ineke, and Ernst Homburg. *Een kabinet vol kleur: De collectie schildersmaterialen van de Amsterdamse verfhandelaar Michiel Hafkenscheid (1772–1846)*. Uitgeverij Vantilt, 2018.

Phipps, Elena. 'Cochineal Red: The Art History of a Color.' *The Metropolitan Museum of Art Bulletin* 67, no. 3 (2010): 4–48.

Pickering, Victoria. 'Mary Somerset, Duchess of Beaufort.' In *The Collectors: Creating Hans Sloane's Extraordinary Herbarium*, edited by Mark Carine, 60–65. Natural History Museum, 2020.

Poelhekke, J.J., and A.J.C. Oomen. 'Elf brieven van Agnes Block in de universiteitsbibliotheek te Bologna.' *Mededelingen van het Nederlands Instituut te Rome* 32 (1965): 2–28.

Pon, Lisa. *Raphael, Dürer, and Marcantonio Raimondi: Copying and the Italian Renaissance Print*. Yale University Press, 2004.

Pouryan, Ashkan Avali. 'Iranian World Plant Species in the European Network of Botanical Information Exchange in the Sixteenth Century.' *Early Science and Medicine* 27, no. 4 (2022): 351–396.

Powell-Warren, Catherine. 'Locating Early Modern Women's Participation in the Public Sphere of Botany: Agnes Block (1629–1704) and Networks in Print.' *Early Modern Low Countries* 4, no. 2 (2020): 234–258.

Powell-Warren, Catherine. 'Pineapple Lady: Expertise and Exoticism in Agnes Block's Self-Representation as Flora Batava.' In *Women, Collecting, and Cultures Beyond Europe*, 95–99. Routledge, 2022.

Powell-Warren, Catherine. *Gender and Self-Fashioning at the Intersection of Art and Science: Agnes Block, Botany, and Networks in the Dutch 17th Century*. Amsterdam University Press, 2023.

Preciado, Kathleen, ed. *Retaining the Original: Multiple Originals, Copies and Reproductions*. National Gallery of Art, 1989.

Prinz, Wolfram, and Andreas Beyer, eds. *Die Kunst und das Studium der Natur vom 14. zum 16. Jahrhundert*. VCH, 1987.

Pugliano, Valentina. 'Ulisse Aldrovandi's Color Sensibility: Natural History, Language and the Lay Color Practices of Renaissance Virtuosi.' In *Early Modern Color Worlds*, edited by Tawrin Baker, Sven Dupré, Sachiko Kusukawa, and Karin Leonhard, 70–108. Brill, 2015.

Raj, Kapil. 'Thinking Without the Scientific Revolution: Global Interactions and the Construction of Knowledge.' *Journal of Early Modern History* 21 (2017): 445–458.

Remond, Jaya. 'Artful Instruction: Pictorializing and Printing Artistic Knowledge in Early Modern Germany.' *Word & Image* 36, no. 2 (2020): 101–134.

Remond, Jaya. 'The Pictorial Idioms of Nature: Image Making as Phytographic Translation in Early Modern Northern Europe.' In *Premodern Experience of the Natural World in Translation*, edited by Katja Krause, Maria Auxent, and Dror Weil, 273–296. Routledge, 2022.

Reynolds, Anna. '"Worthy to Be Reserved": Bookbindings and the Waste Paper Trade in Early Modern England and Scotland.' In *The Paper Trade in Early Modern Europe: Practices, Materials, Networks*, edited by Daniel Bellingradt and Anna Reynolds, 342–368. Brill, 2021.

Rijks, Marlise. 'Scales, Skins, and Carapaces in Antwerp Collections.' In *The Matter of Mimesis: Studies of Mimesis and Materials in Nature, Art and Science*, edited by Marjolijn Bol and E.C. Spary, 295–320. Brill, 2023.

Robin, Johannes. *Catalogus stirpium tam indigenarum quam exoticarum quae Lutetiae coluntur ...* Paris: Philippi a Prato, 1601.

Roth, Dietrich, ed. *Die Blumenbücher des Hans Simon Holtzbecker und Hamburgs Lustgärten: Hans Simon Holtzbecker, Hamburger Blumenmaler des 17. Jahrhunderts: Botanische, garten- und kunsthistorische Aspekte*. Goecke & Evers, 2003.

Roth, Dietrich, ed. *Das Moller-Florilegium. Hans Simon Holtzbeckers Blumenalbum für den Bürgermeister Barthold Möller*. Hirmer Verlag, 2007.

Rowley, Gordon D. 'The Duchess of Beaufort's Succulent Plants.' *Bradleya* 5 (1987): 1–16.

Roy, Ashok, ed. *Artists' Pigments: A Handbook of Their History and Characteristics*. Vol. 2. National Gallery of Art / Oxford University Press, 1993.

Roy, Ashok, and Perry Smith, eds. *Painting Techniques: History, Materials and Studio Practice*. The International Institute for Conservation of Historic and Artistic Works, 1998.

Russell, Andrew L., and Lee Vinsel. 'After Innovation, Turn to Maintenance.' *Technology and Culture* 59, no. 1 (2018): 1–25.

Saunders, Gill. *Picturing Plants: An Analytical History of Botanical Illustration*. Zwemmer / Victoria and Albert Museum, 1995.

Savage, Spencer. 'The Hortus Floridus of Crispijn Vande Pas the Younger.' *The Library* 4, no. 3 (1923): 181–206.

Savage, Spencer. *Garden of Flowers: All 104 Engravings from the Hortus Floridus of 1614*. Pictorial Archive Series. Dover Publications, 2002.

Sayer, Robert. *The Florist : Containing sixty plates of the most beautiful flowers regularly disposed in their succession of blowing. To which is added an accurate description of their colours with instructions for drawing and painting according to nature: Being a new work intended for the use & amusement of gentlemen and ladies delighting in that art*. London: Printed for Robt. Sayer, T. Bowles, and John Bowles & Son, 1760.

Schattmann, Anne. 'Von der Leidenschaft für Blumen—das Florilegium des Caspar Anckelmann.' In *Maria Sibylla Merian und die Tradition des Blumenbildes von der Renaissance bis zur Romantik*, edited by Magdalena Bushart, Catalina Heroven, Michael Roth, and Martin Sonnabend, 117–125. Staatliche Museen zu Berlin—Preußische Kulturbesitz, Städel Museum / Hirmer Verlag, 2017.

Schiebinger, Londa. *Plants and Empire: Colonial Bioprospecting in the Atlantic World.* Harvard University Press, 2004.

Schiebinger, Londa, and Claudia Swan, eds. *Colonial Botany: Science, Commerce, and Politics in the Early Modern World.* University of Pennsylvania Press, 2005.

Schmidt, Benjamin. *Inventing Exoticism: Geography, Globalism, and Europe's Early Modern World.* University of Pennsylvania Press, 2015.

Schmidt, Benjamin, and Thijs Weststeijn, eds. *The Globalization of Netherlandish Art.* Brill, 2024.

Segal, Sam. *Flowers and Nature: Netherlandish Flower Painting of Four Centuries.* SDU Publishers, 1990.

Segal, Sam, and Klara Alen. *Dutch and Flemish Flower Pieces: Paintings, Drawings and Prints up to the Nineteenth Century*. 2 Vols. Brill, 2020.

Segre, Ada V., ed. *The Gardens at San Lorenzo in Piacenza, 1656–1665: A Manuscript Planting Notebook: With a Study, Transcription, and Translation.* Dumbarton Oaks Research Library and Collection, 2006.

Selosse, Philippe. 'The Role of Carolus Clusius (1526–1609) as Translator in the Emergence of a Taxonomic Terminology in Botany.' In *Translating Knowledge in the Early Modern Low Countries*, edited by Harold Cook and Sven Dupré, 349–368. LIT Verlag, 2012.

Sikkens-De Zwaan, Marisca. 'Magdalena Poulle (1632–99): A Dutch Lady in a Circle of Botanical Collectors.' *Garden History* 30, no. 2 (2002): 206–220.

Silver, Larry, and Elizabeth Wyckoff, eds. *Grand Scale: Monumental Prints in the Age of Dürer and Titian.* Davis Museum and Cultural Center, 2008.

Skogh, Lisa. 'The Varied Role of the Amateur.' *Nuncius* 31 (2016): 489–498.

Smith, Pamela H. *The Body of the Artisan: Art and Experience in the Scientific Revolution.* The University of Chicago Press, 2004.

Smith, Pamela H. *From Lived Experience to the Written Word: Reconstructing Practical Knowledge in the Early Modern World.* The University of Chicago Press, 2022.

Smith, Pamela H., Amy R.W. Meyers, and Harold J. Cook, eds. *Ways of Making and Knowing: The Material Culture of Empirical Knowledge.* The University of Michigan Press, 2014.

Smith, Pamela H., and Benjamin Schmidt, eds. *Making Knowledge in Early Modern Europe: Practices, Objects, and Texts, 1400–1800.* The University of Chicago Press, 2007.

Smith, Pamela H., and Paula Findlen, eds. *Merchants & Marvels: Commerce, Science, and Art in Early Modern Europe*. Routledge, 2002.

Spielmann, Heinz, and Jan Drees. *Gottorf im Glanz des Barock: Kunst und Kultur am Schleswiger Hof 1544–1713*. 4 Vols. Schleswig-Holsteinisches Landesmuseum, 1997.

Staubermann, Klaus, ed. *Reconstructions: Recreating Science and Technology of the Past*. National Museums Scotland, 2011.

Stefanaki, Anastasia, Henk Porck, Ilaria Maria Grimaldi, Nikolaus Thurn, Valentina Pugliano, Adriaan Kardinaal, Jochem Salemink, Gerard Thijsse, Claudine Chavannes-Mazel, Erik Kwakkel, and Tinde van Andel. 'Breaking the Silence of the 500-Year-Old Smiling Garden of Everlasting Flowers: The En Tibi Book Herbarium.' *PLoS ONE* 14, no. 6 (2019): e0217779. https://doi.org/10.1371/journal.pone.0217779.

Stijnman, Ad. *Engraving and Etching, 1400–2000: A History of the Development of Manual Intaglio Printmaking Processes*. Archetype Publications, 2012.

Stijnman, Ad, and Elizabeth Savage, eds. *Printing Colour 1400–1700: History, Techniques, Functions and Receptions*. Brill, 2015.

Strocchia, Sharon T. *Forgotten Healers Women and the Pursuit of Health in Late Renaissance Italy*. Harvard University Press, 2019.

Swan, Claudia. 'Ad Vivum, Naer Het Leven, From the Life: Defining a Mode of Representation.' *Word & Image* 11, no. 4 (1995): 353–372.

Swan, Claudia. 'The Uses of Realism in Early Modern Illustrated Botany.' In *Visualizing Medieval Medicine and Natural History, 1200–1550*, edited by Jean Ann Givens, Alan Reeds, and Alain Touwaide, 239–249. Ashgate Publishing, 2016.

Sweerts, Emanuel. *Florilegium Emanuelis Sweerti Septimonti Batavi Amsteledami commorantis, tractans de variis floribus, et aliis indicis plantis ad viuum delineatum in duabus partibus, et quatuor linguis concinnatum*. Frankfurt: Anthonium Kempner, 1612.

Sweerts, Emanuel. *Florilegium amplissimum et selectissimum, quo non tantum varia diversorum florum praestantissimorum et nunquam antea exhibitorum genera, sed et rarae quamplurimae indicarum plantarum, et radicum formae, ad vivum partibus duabus, quatuor etiam linguis offeruntur et delineantur*. Amsterdam: Joannem Janssonium, 1647.

Tallian, Timea. 'John White's Materials and Techniques.' In *European Visions: American Voices*, edited by Kim Sloan, 72–76. British Museum, 2009.

Taylor, Paul. *Dutch Flower Painting: 1600–1720*. Yale University Press, 1995.

Taylor, Paul. 'The Birth of the Amateur.' *Nuncius* 31 (2016): 499–522.

Ter Brugge, Geerard, and Willem Goeree. *Verlichterie-Kunde, of recht gebruyk der Water-Verwen. In welke des selfs kennis, en volkomen gebruyk tot de schilder-kunde, ende de illuminatie ofte verlichterie noodigh zijnde, kortelijk werden geleert* … Middelburgh: Willem Goeree, 1670.

*The Paper Museum of Cassiano dal Pozzo*. Olivetti, 1993.

Thiers, Barbara M. *Herbarium: The Quest to Preserve & Classify the World's Plants*. Timber Press, 2020.

Thijsse, Gerard. '"Everlasting Gardens": Origin, Spread and Purpose of the First Herbaria.' In *The Green Middle Ages: The Depiction and Use of Plants in the Western World 600–1600*, edited by Claudine A. Chavannes-Mazel and Linda IJpelaar, 72–107. Amsterdam University Press, 2023.

Thijsse, Gerard, and Margreet Wesseling. 'Herman Boerhaave: A Search for His Herbarium Collections.' *Taxon* 70, no. 1 (2021): 170–181.

Thompson, Daniel V. *The Materials of Medieval Painting*. Reprint Edition. Martino Fine Books, 2019.

Thurner, Mark, and Juan Pimentel, eds. *New World Objects of Knowledge: A Cabinet of Curiosities*. University of London Press, 2021.

Tice, Lisa Neal. 'Collecting in the Garden: Inventories of Casini in Late Sixteenth- and Early Seventeenth-Century Rome.' *Journal of the History of Collections* 23, no. 2 (2011): 315–331.

Tongiorgi Tomasi, Lucia. 'Image, Symbol and Word on the Title Pages and Frontispieces of Scientific Books from the Sixteenth and Seventeenth Centuries.' *Word & Image* 4, no. 1 (1988): 372–382.

Tongiorgi Tomasi, Lucia. 'Gherardo Cibo: Visions of Landscape and the Botanical Sciences in a Sixteenth-Century Artist.' *The Journal of Garden History* 9, no. 4 (1989): 199–216.

Tongiorgi Tomasi, Lucia. *An Oak Spring Flora: Flower Illustration from the Fifteenth Century to the Present Time—A Selection of the Rare Books, Manuscripts, and Works of Art in the Collection of Rachel Lambert Mellon*. Oak Spring Garden Library, 1997.

Tongiorgi Tomasi, Lucia. '"La Femminil Pazienza": Woman Painters and Natural History in the Seventeenth and Early Eighteenth Centuries.' In *The Art of Natural History: Illustrated Treatises and Botanical Paintings, 1400–1850*, edited by Therese O'Malley and Amy R.W. Meyers, 159–185. National Gallery of Art, 2010.

Tongiorgi Tomasi, Lucia. *The Royal Lessons by Catherine Perrot: A Pioneering Woman Flower Painter at the Court of Louis XIV*. Oak Spring Garden Foundation, 2022.

Tongiorgi Tomasi, Lucia, and Gretchen A. Hirschauer. *The Flowering of Florence: Botanical Art for the Medici*. National Gallery of Art, 2002.

Valleriani, Matteo, ed. *The Structures of Practical Knowledge*. Springer, 2017.

Vallet, Pierre. *Le jardin du roy tres chrestien Henry IV, roy de France et de Navarre dedie a la royne*. Paris: Pierre Vallet, 1608.

Vallet, Pierre. *Le jardin du roy tres chrestien, Loys XIII, roy de France et de Navare dedie a la royne*. Paris: Au logis de l'auteur, 1623.

Van Andel, Tinde, Rutger A. Vos, Ewout Michels, and Anastasia Stefanaki. 'Sixteenth-Century Tomatoes in Europe: Who Saw Them, What They Looked Like, and Where They Came From.' *PeerJ* 10 (2022): e12790. https://doi.org/10.7717/peerj.12790.

Van de Graft, C. Catharina. *Agnes Block: Vondels nicht en vriendin.* A.W. Bruna & Zoon's Uitgevers-Mij, 1943.

Van de Passe II, Crispijn. *Den Blom-hof, inhoudende de rare oft ongemeene blommen die op ten tegenwoordighen tijdt by lief-hebbers in estimatie ghehouden werden: ghedeelt naer de vier deelen des jaers.* Utrecht: voor Crispijn vande Pas, 1614.

Van de Passe II, Crispijn. *Hortus floridus, In quo rariorum & minus vulgarium florum Icones ad vivam veramq[ue] formam accuratissime delineatae Et secundum quatuor anni tempora divisae exhibentur Incredibili labore ac diligentia / Crisp. Passaei junioris Delineatae ac suum in ordinem redactae.* Arnhem: Ioannem Ianssonium Bibliopolam, 1614.

Van de Passe II, Crispijn. *Jardin de fleurs, contenant en soy les plus rares et plus excellentes fleurs que pour le present les amateurs dicelles tiennent en grande estime et dignite: divisees selon les quatre saisons de l'an.* Utrecht: Crispian de Pas, 1614.

Van de Passe II, Crispijn. *A Garden of Flowers, wherein very lively is contained a true and perfect discription of al the flowers contained in these foure followinge bookes. As also the perfect true manner of colouringe the same with theire naturall coloures, beinge all in theire seasons the most rarest and excellentest flowers, that the world affordeth: ministringe both pleasure and delight to the spectator, and most espetially to the wel affected practisioner ...* Utrecht: Salomon de Roy, for Crispian de Passe, 1615.

Van de Spiegel, Adriaan. *Isagoges in rem herbariam libri duo.* Leiden: Officina Elzeviriana, 1633.

Van Delft, Marieke, and Hans Mulder. *Maria Sybilla Merian. Metamorphosis Insectorum Surinamensium.* Lannoo, 2016.

Van Eikema Hommes, Margriet, Jolanda De Bruijn, Erma Hermens, and Arie Wallert. 'Still-Life Sources.' In *Still Lifes: Techniques and Style: An Examination of Paintings from the Rijksmuseum*, edited by Arie Wallert, 25–37. Rijksmuseum / Waanders Publishers, 1999.

Van Gelder, Esther. 'Tussen hof en keizerskroon: Carolus Clusius en de ontwikkeling van de botanie aan Midden-Europese hoven (1573–1593).' PhD Dissertation, Leiden University, 2011.

Van Groesen, Michiel. 'The De Bry Collection of Voyages (1590–1634): Editorial Strategy and the Representations of the Overseas World.' PhD Dissertation, University of Amsterdam, 2007.

Van Oosten, Hendrik. *De Nieuwe Nederlandse Bloem-Hof, ofte de Sorgvuldige Hovenier, onderwijsende, hoe men allerhande soorten van de curieuste bloemgewassen; als tulpen, angelieren, hyacinthen, narcissen &c. gelukkiglijk sal voortsetten, en veilig tegen 't verderf bewaaren ...* Leiden: J. du Vivie & I. Severinus, 1700.

Van Oosten, Hendrik. *The Dutch Gardener: Or the Compleat Florist. Containing, the Most Successful Method of Cultivating All Sorts of Flowers; the Planting, Dressing, and Pruning of All Manner of Fruit Trees. Together with a Particular Account of the Nursing of Lemon and Orange Trees in Northern Climates*. London: Printed for D. Midwinter and T. Leigh, 1703.

Van Ruyven-Zeman, Zsuzsanna. 'Pieter Kouwenhorn, 'uytnemend teykenaar ende gelase-sgrijver' en het carton van het universiteitsglas in de Pieterskerk te Leiden.' *Oud Holland* 117 (2004): 162–179.

Van Zanen, Sylvia. *Planten op Papier: Het Pionierswerk van Carolus Clusius (1526–1609)*. Walburg Pers, 2019.

Vander Ploeg Fallon, Melinda K. 'Petronella de La Court and Agneta Block: Experiencing Collections in Late Seventeenth-Century Amsterdam.' *Aurora* 4 (2003): 95–108.

Vandervellen, Pascale. *The Golden Age of Flemish Harpsichord Making: A Study of the MIM's Ruckers Instruments*. Exhibitions International, 2018.

Veldman, Ilja M. *Crispijn de Passe and His Progeny (1564–1670): A Century of Print Production*. Translated by Michael Hoyle. Sound & Vision Publishers, 2001.

Vervliet, Hendrik D.L. *Vine Leaf Ornaments in Renaissance Typography: A Survey*. Oak Knoll Press, 2012.

Vinsel, Lee, and Andrew L. Russell. *The Innovation Delusion: How Our Obsession with the New Has Disrupted the Work That Matters Most*. Currency, 2020.

Voet, Leon. *The Golden Compasses: The History of the House of Plantin-Moretus*. 2 Vols. Vangendt & Co., 1969.

Voet, Leon. *The Plantin Press, 1555–1589: A Bibliography of the Works Printed and Published by Christopher Plantin at Antwerp and Leiden*. Vol. 3. Van Hoeve, 1980.

Voet, Leon. 'Christopher Plantin as a Promoter of the Science of Botany.' In *Botany in the Low Countries (End of the 15th Century–ca. 1650)*, edited by Francine De Nave and Dirk Imhof, 39–45. The Plantin-Moretus Museum / The Stedelijk Prentenkabinet, 1993.

Völler von Gellhausen, Ulrich. *Florilegium, das ist: Ein Blumen Buch, darinnen allerhand Blümlein gantz artig …* Frankfurt: M. Weixner, 1616.

Von Schlosser, Julius. *Art and Curiosity Cabinets of the Late Renaissance: A Contribution to the History of Collecting*. Edited by Thomas DaCosta Kaufmann. Translated by Jonathan Blower. Getty Research Institute, 2021.

Warner, Marjorie F. 'The Augmentatio, 1614, of De Bry's Florilegium Novum.' *Libri* 6 (2009): 29–32.

Weiss, Daniel H., and Max Hollein. 'Report from the President and the Director.' *Annual Report of the Trustees of the Metropolitan Museum of Art*, no. 149 (2018): 6–10.

Wheelock Jr., Arthur K. *From Botany to Bouquets: Flowers in Northern Art*. National Gallery of Art, 1999.

Wickert, Konrad. *Das Camerarius-Florilegium*. Universitätsbibliothek Erlangen-Nürnberg, 1993.

Wieck, Roger S. *Miracles in Miniature: The Art of the Master of Claude de France*. The Morgan Library & Museum, 2014.

Wijnands, D.O. *The Botany of the Commelins: A Taxonomical, Nomenclatural and Historical Account of the Plants Depicted in the Moninckx Atlas and in the Four Books by Jan and Caspar Commelin on the Plants in the Hortus Medicus Amstelodamensis 1682–1710*. A.A. Balkema, 1983.

Wille, Helena. 'The Discovery of the Scientific Heritage of Karel van Sint Omaars (1533–1569). The Libri Picturati A 16–30 in The Jagiellon Library in Krakow.' *Scientiarum Historia* 22, no. 2 (1996): 67–80.

Wille, Helena. 'Botanical Collectors and Collections in the Low Countries.' In *Dodonaeus in Japan: Translation and the Scientific Mind in the Tokugawa Period*, edited by W.F. Vande Walle and Kazuhiko Kasaya, 77–95. Leuven University Press, 2001.

Wille, Helena. 'De albums van Karel van Sint Omaars: Een aristocratisch natuurobservator uit de Renaissance.' PhD Dissertation, KU Leuven, 2008.

Winiwarter, Verena. 'The Art of Making the Earth Fruitful: Medieval and Early Modern Improvements of Soil Fertility.' In *Ecologies and Economies in Medieval and Early Modern Europe*, edited by Scott Bruce, 93–114. Brill, 2010.

Wragge-Morley, Alexander. *Aesthetic Science: Representing Nature in the Royal Society of London, 1650–1720*. University of Chicago Press, 2020.

# Index

www.ingramcontent.com/pod-product-compliance
Lightning Source LLC
Chambersburg PA
CBHW070252260926
51849CB00002BA/394
* 9 7 8 9 0 0 4 7 3 5 1 3 2 *